D0560802

Bacterial Membranes and Walls

MICROBIOLOGY SERIES

Series Editor

ALLEN I. LASKIN

ESSO Research and Engineering Company
Linden, New Jersey

Volume 1 Bacterial Membranes and Walls, *edited by Loretta Leive*

Other Volumes in Preparation

Bacterial Membranes and Walls

Edited by Loretta Leive

National Institute of Arthritis, Metabolism, and Digestive Diseases
National Institutes of Health
Bethesda, Maryland

MARCEL DEKKER, INC. New York

MARCEL DEKKER, INC.

270 Madison Avenue, New York, New York 10016

LIBRARY OF CONGRESS CATALOG CARD NUMBER: 73-82622

ISBN: 0-8247-6085-9

Current printing (last digit):
10 9 8 7 6 5 4 3 2

PRINTED IN THE UNITED STATES OF AMERICA

CONTENTS

PART I

BIOSYNTHESIS AND ASSEMBLY

PART III

MORPHOGENESIS AND REPRODUCTION

LIST OF CONTRIBUTORS

JEAN-MARIE GHUYSEN, Service de Microbiologie, Faculté de Médecine, Institut de Botanique, Université de Liège, Liège, Belgium

ULF HENNING, Max-Planck-Institut für Biologie, Tübingen, Germany; and Friedrich-Miescher-Laboratorium der Max-Planck-Gesellschaft, Tübingen, Germany

LEON A. HEPPEL, Department of Biological Chemistry, School of Medicine, University of Maryland, Baltimore, Maryland; and Section of Biochemistry and Molecular Biology, Cornell University, Ithaca, New York

H. R. KABACK, The Roche Institute of Molecular Biology, Nutley, New Jersey

S. E. LURIA, Department of Biology, Massachusetts Institute of Technology, Cambridge, Massachusetts

LEONARD MINDICH, The Public Health Research Institute of the City of New York, Inc., New York, New York

HIROSHI NIKAIDO, Department of Bacteriology and Immunology, University of California, Berkeley, California

ARTHUR B. PARDEE, Departments of Biochemical Sciences and Biology, Princeton University, Princeton, New Jersey

BARRY P. ROSEN, Department of Biological Chemistry, School of Medicine, University of Maryland, Baltimore, Maryland; and Section of Biochemistry and Molecular Biology, Cornell University, Ithaca, New York

ULI SCHWARZ, Max-Planck-Institut für Biologie, Tübingen, Germany; and Friedrich-Miescher-Laboratorium der Max-Planck-Gesellschaft, Tübingen, Germany

GERALD D. SHOCKMAN,* "Chair de Actualité Scientifique 1971-1972," Service de Microbiologie, Faculté de Médecin, Institut de Botanique, Université de Liège, Liège, Belge

ALEXANDER TOMASZ, The Rockefeller University, New York, New York

PO CHI WU, Departments of Biochemical Sciences and Biology, Princeton University, Princeton, New Jersey

DAVID R. ZUSMAN, Departments of Biochemical Sciences and Biology, Princeton University, Princeton, New Jersey

*On leave from Department of Microbiology and Immunology, Temple University, School of Medicine, Philadelphia, Pennsylvania.

INTRODUCTION

TOWARD A STUDY OF SUPRAMOLECULAR STRUCTURE AND FUNCTION

Progress toward an understanding of the biochemistry and biology of the bacterial envelope is at a crucial point: chemical and morphological structures that comprise the envelope can no longer be dealt with as unrelated entities. The investigator must now ask how these structures interact dynamically — both temporally during their assembly and functionally during their operation. He therefore must integrate information from fields once considered separate and distinct. This book attempts to provide such information in a form that will facilitate its integration and incorporation into the design of future experiments.

In the following paragraphs, a general outline of the development of this field is given, with especial focus on the topics covered in this volume. Original works or previous reviews are referred to only when this volume contains little of the relevant material. This approach is intended to provide both background material and a framework for the use of the book. Finally, examples are given to show how past and present developments suggest questions for future research and how the organization of this book is related to these questions.

More than 30 years ago, researchers in many fields began to study surface functions. These descriptions were informative and often extremely elegant; for instance, the early work of Monod and his collaborators on transport [1] ; the work of Schlessinger, Delbruck, and others on the kinetics of phage adsorption [2] ; and the studies of Lederberg and others on the mating of bacteria [3].

Studies of structure began just slightly later and proceeded in parallel, as chemical isolation and analysis of surface components defined the composition of many of the macromolecules of bacterial membranes and walls. One may cite the pioneering work of Salton and others on peptidoglycan (Chapter 2), Luderitz and Westphal on lipopolysaccharide (Chapter 3), Baddiley on teichoic acids [4] , and many laboratories on lipids ([5] , Chapter 1).

Overlapping these early efforts were experiments defining the biosynthetic reactions that construct these macromolecules. Most of the pathways involved in the syntheses of peptidoglycan, lipopolysaccharides, lipids, and teichoic acids have now been outlined; but this period of research is not finished, as portions of these biosynthetic pathways are still unknown and under active investigation in many laboratories. An additional benefit of experiments to elucidate biosynthesis is further refinement of our knowledge of the sequence of many of these molecules.

Parenthetically it should be noted that the molecules most resistant to analysis in membranes and walls have been the proteins. In part because of technical difficulties in solubilizing proteins while retaining their function, only a fraction of membrane-bound enzymes have been defined as to function and very few have been purified. The question of whether there exist truly "structural" proteins (as opposed to proteins whose enzymatic function we have not yet discovered) is still unanswered (see Chapter 2).

As knowledge of structure and biosynthesis increased, it became partially possible to identify structures participating in various processes. For instance, adsorption sites for some phages can now be located and their chemical composition defined ([6, 7]; Chapter 3); receptor sites that bind colicins (Chapter 6) and those that bind DNA during transformation (Chapter 7) are being defined, isolated, and purified; and the interaction of cells during mating, by means of their pili, is under investigation [8]. Some mechanisms operative in transport can now be defined, although the processes involved are not yet completely understood (Chapters 4 and 5), and the biochemical events in chemotaxis are being explored [9].

Several major areas of research are ready for exploration. Four such areas are outlined below, two of which are concerned with control, or the coordination of cellular processes to permit orderly growth, and two with translocation, or the movement of molecules and information through and within envelope structures.

1. Translocation of the Products of Biosynthesis. The biosynthesis of surface macromolecules often takes place in sites far removed from their ultimate location. Thus, protein made on ribosomes ends up, not only in cytoplasmic membrane (to which it has access via the cytoplasm), but also outside this membrane — for instance, in the periplasmic space outside the cytoplasmic membrane of gram-negative cells; in the outer membrane of gram-negative cells; or as the extracellular enzymes of many gram-positive cells. To reach these last two locations, the proteins must, in addition to passing the cytoplasmic membrane, also pass the peptidoglycan barrier. Similarly, lipids and lipopolysaccharides of the outer membrane, which are made in the cytoplasmic membrane, must also be translocated

across space and structure. Finally, the peptidoglycan itself is assembled from components that are initially cytoplasmic, are polymerized on membrane-bound carrier lipid, and must then be integrated into a covalently-linked girdle on the opposite side of this membrane. The mechanism of translocation of all these products from their sites of synthesis to their final location is as yet unknown.

Important properties of the milieu through which these products travel have been deducted from experiments, on model systems, and in vivo, indicating that lipids in bilayers and in membranes are very fluid and that molecules can diffuse with extreme rapidity within them, although they are restricted in reversing orientation from one face of the membrane to the other [10, 11, 12]. These findings provide both predictions and restrictions for postulates to explain how products reach their final location, and also raise new problems concerning how these products are ultimately anchored.

2. Control of Biosynthetic Pathways Relative to Each Other. The pathways of biosynthesis of membrane components are becoming well defined, but we have scanty knowledge of whether and how feedback inhibition and repression mechanisms operate within these pathways, and less knowledge of how pathways are coupled so that membranes retain their normal composition during growth. A provocative finding bearing on this problem is the discovery that in the assembly of both peptidoglycan and lipopolysaccharide, two very different molecules, the same or virtually the same lipid carrier, polyisoprenol phosphate (Chapters 2 and 3), is employed. Such a phenomenon provides an immediate mechanism for simultaneous regulation of production of these two surface components, but the mechanism of this regulation is still not understood. It is probable that other common features in the biosynthesis of membrane macromolecules will emerge during further study.

3. Translocation of Material and Information from the Outside to the Inside of the Cell. This topic includes areas as diverse as active transport, chemotaxis, DNA transmission in mating, DNA adsorption in phage infection and transformation, and the mechanism of colicin action. In all cases, molecules normally extrinsic to the cell combine with receptors and initiate processes that result in the translocation of the initiating molecule to a site within the cell. (In the case of colicin action and chemotaxis it is not completely clear in all cases whether the molecule itself or the information it conveys is transmitted into the cell.) Headway has been made in defining at least some of the membrane components receiving the impinging molecule and/or providing the energy to effect the translocation, but the intermediate steps and the methods of coupling produced energy to the process remain unknown.

4. Control of Cellular Division and Morphology. Above and beyond the coupling of biosynthetic pathways so that membranes and walls retain their needed composition, synthesis of these structures must be intimately linked with DNA synthesis, cell division, and shape determination. Our understanding of this field is truly in its infancy, with the recent identification of some mutants blocked in various aspects of division and of some enzymes involved in surface growth and possibly shape determination (see Chapters 2, 8, and 9).

The sections of this volume have been arranged so that the knowledge of the past may provoke questions for the future, including those outlined above. Thus, the first section of the book, Biosynthesis and Assembly, contains three chapters that summarize the structure and biosynthesis of many important membrane and wall structures and focus on the questions of translocation during assembly and spatial arrangement, as well as on the questions of control during biosynthesis. The second section of the book, Interaction with the Environment, contains four chapters concerning the translocation of material from the outside to the inside of the cell. The third section, Morphogenesis and Reproduction, contains material applicable to the control of cell division and morphology.

Needless to say, the varied aspects of wall and membrane structure are so intimately related that information relevant to investigating each of these research areas appears in various parts of the book. Furthermore, because of necessary limitations of space, this book omits or slights many important aspects of envelope structure: for instance, flagellar structure and function, chemotaxis, the integration of DNA with envelope replication, and teichoic acid biosynthesis.

Despite these necessary omissions, it is hoped that both the information in this book, and also the arrangement of this information, will lead readers not only to a larger and more comprehensive grasp of the field but also to insights and correlations that can advance future knowledge.

Loretta Leive

REFERENCES

1. G. N. Cohen and J. Monod, Bact. Rev., 21, 169 (1957).

2. G. S. Stent (ed.), Papers on Bacterial Viruses, Little, Brown, Boston, 1960.

3. J. Lederberg and E. L. Tatum, Science, 118, 169 (1953).

4. J. Baddiley, Accounts Chem. Res., 3, 98 (1970).

5. J. E. Cronan and P. R. Vagelos, Biochim. Biophys. Acta, 265, 25 (1972).

6. W. Weidel, Ann. Rev. Microbiol., 12, 27 (1958).

7. W. Weidel, H. Frank, and H. H. Martin, J. Gen. Microbiol., 22, 158 (1960).

8. C. C. Brinton, Crit. Rev. Microbiol., 1, 105 (1971).

9. J. Adler, Science, 166, 1588 (1969).

10. L. D. Frye and M. Edidin, J. Cell Sci., 7, 319 (1970).

11. R. Kornberg and H. M. McConnell, Proc. Nat. Acad. Sci. U. S., 68, 2569 (1971).

12. C. J. Scandella, P. Devaux, and H. M. McConnell, Proc. Nat. Acad. Sci. U. S., 69, 2056 (1972).

PART I

BIOSYNTHESIS AND ASSEMBLY

Chapter 1

SYNTHESIS AND ASSEMBLY OF BACTERIAL MEMBRANES

Leonard Mindich

The Public Health Research Institute of the City of New York, Inc.
New York, New York

I. INTRODUCTION

The bacterial membrane is the surface element responsible for maintaining the osmotic stability of the cell. When cell wall structures are removed by drug or enzyme treatment, the cell loses its characteristic shape but maintains its ability to grow, to transport various substances, and under certain conditions, to divide. Such a cellular form is called a protoplast and its membrane is called the protoplast (or plasma) membrane. If this membrane is damaged, the cell lyses [1]. Membranes are composed primarily of lipid and protein in the proportion of 1:5. They constitute from 10 to 26% of the cell dry weight [2].

Gram-negative bacteria such as E. coli have an additional membrane exterior to the plasma membrane and the peptidoglycan cell wall; this structure is called the outer membrane. Both the protoplast membrane and outer membrane contain proteins and lipids; however, they differ in that the former carries the enzymes for electron transport and many biosynthetic and transport functions, whereas the latter contains the lipopolysaccharide somatic antigen and lacks biosynthetic, transport, and electron transport functions [3,4]. This chapter is concerned primarily with the protoplast membrane, its composition, and the manner in which its assembly occurs and is regulated. This is not to imply that the reader will, in fact, know how membrane assembly takes place upon completing this review. Our knowledge of this area is still very primitive. However, the general approaches to the relevant questions are outlined and the knowledge that we do have is presented. Several excellent reviews have appeared in the past few years that examine membrane models [5, 5a], bacterial membrane structure [6, 6a], membrane structure and function [7], the biogenesis of membranes [8], reconstitution of membranes [8a], and phospholipid metabolism and function [8b].

II. COMPOSITION OF BACTERIAL MEMBRANES

Most of the studies on membrane composition have been done on material isolated from gram-positive bacteria, since until recently the difficulty in separating the outer membrane from the protoplast membrane in preparations from gram-negative bacteria precluded reliable analysis of either one. Membranes contain from 20 to 30% lipid and the remainder is mostly protein [2, 9]. Membranes of stable L-forms usually have higher lipid concentrations than those from the parent cultures [10,11]. Polysaccharide components such as teichoic acids [12] and the lipopolysaccharide of the outer membrane [3] are sometimes found in the plasma membrane. The teichoic acids in the membrane differ from those in the cell wall and may constitute a specific membrane component; the polysaccharides can be partially accounted for as precursor material for the outer membrane, but contamination of membrane fractions with wall material is difficult to rule out.

A. Proteins

Isolated membranes contain many enzymatic activities. Some of these activities are found only in membrane and others are distributed between the cytoplasm, other cell structures, and the membrane. The various enzymatic activities form a continuous distribution with respect to their association with the cell membrane, some enzymes being found completely in the soluble fraction and others, such as succinic dehydrogenase, found completely in the membrane fraction. Some proteins, such as cytochrome c, which we know are associated with membranes, are easily removed during membrane isolation [13]. It is likely that the membrane protein content usually measured is a minimal one. Among the proteins or activities found to be highly associated with membrane are flavoprotein dehydrogenases, cytochromes, oxidases, ATPase (see Salton [6] and Gellman et al. [14] for extensive listings), transport activities for sugars, amino acids, vitamins, and inorganic ions [15], enzymes of phospholipid [16], peptidoglycan [17], teichoic acid [12], and polysaccharide synthesis. The bacterial membrane contains approximately 10 to 20% of the cellular protein.

Analysis of membrane proteins by acrylamide gel electrophoresis shows an expectedly complex pattern with many different proteins. The outer membrane of *E. coli* shows a simpler gel pattern than the inner membrane [18]. Although many of the membrane proteins have known functions, the question of the possibility of proteins the only role of which is in membrane structure, has come up from time to time. Claims of membrane structural proteins have been made for retinal membranes, for mitochondria of higher organisms [19], and for *Neurospora* [20]; however, these studies generally have not been confirmed or have turned out to be the result of experimental artifacts [5]. Structural membrane proteins have not been identified in bacteria, and in the case of *Halobacterium halobium* purple membrane, where only one protein is found, this protein is complexed with retinal and may serve as a photoreceptor for the cell [21]. The various functional proteins may also play a structural role, but to prove that a membrane protein functions critically as a structural protein requires the use of genetic techniques to demonstrate that conditional defects in a particular protein lead to conditional defects in membrane structure.

B. Lipids

Lipids are the most characteristic component of biological membranes. In many of the gram-positive bacteria, there is no lipid in the cell except in the protoplast membrane. The lipid composition of membranes has been reviewed by Kates [22] and MacFarlane [23]. Although some differences are found, the fatty acid composition of isolated membranes or "envelope" fractions is very similar to that for whole cells [24]. The major lipid components of bacterial membranes are phospholipids. In particular, the predominant components are phosphatidylglycerol and its amino acyl derivatives and phosphatidylethanolamine. Phosphatidylserine is found in smaller amounts; cardiolipin is usually found in small amounts in normal cells, but its level increases under conditions of stress; phosphatidylcholine is rare in bacteria as are sphingolipids [25]. Glycolipids and diglycerides are often found in small amounts in bacterial membranes. Polyisoprenoids are usually found in membranes and their role in peptidoglycan and polysaccharide synthesis has been demonstrated [26]. Carotenoids are also found in the bacterial membranes [27].

The reason for the multiplicity of types of phospholipid is not understood. It seems likely that phospholipids with different charge distributions might play different roles in the cell physiology, in particular with respect to transport and membrane structure. Phosphatidylglycerol is found in all bacteria, but phosphatidylethanolamine is not [28]. A mutant has been isolated in B. subtilis that has a very low phosphatidylethanolamine content [29]; in this case the growth of the organism is very poor in synthetic media but is almost normal in complex media, in spite of its low phosphatidylethanolamine content. This indicates that the compound plays an important role in the cell physiology, but it may not be indispensable.

The degree to which amino acyl derivatives of phosphatidylglycerol are found in gram-positive cells appears to be correlated with the pH of the medium. It has been postulated that the amino acyl derivates (usually basic amino acids such as lysine and ornithine) serve the cell by altering the charge distribution in the membrane lipids [30]. But the role of the different lipids is not only structural; various biosynthetic reactions [31, 32] and transport functions [33] seem to require particular phospholipids. The diversity of phospholipids might be necessary for the activation of many different membrane-bound activities.

The fatty acid composition of the various lipid classes is rather similar within the same culture; however, the fatty acid composition differs from species to species and even in the same strain of bacteria under different cultural conditions [34]. There are basic patterns of fatty acid composition among the families of bacteria, e.g., the Enterobacteriaceae have unsaturated fatty acids such as palmitoleic and vaccenic along with straight-chain fatty acids such as palmitic acid (reviewed by Kates [22], and Cronan and Vagelos [8b]). The members of Bacillus have methyl branched-chain fatty acids such as iso and anteiso molecules of predominantly 15- and 17-carbon atoms [35] along with smaller amounts of branched even-numbered and straight-chain fatty acids. Bacteria also have fatty acids containing a cyclopropane ring. Mutants have been obtained that require unsaturated fatty acids in organisms that usually synthesize them, indicating that these fatty acids play a particular role in the lipid composition of the membrane. Mutants have also been found in B. subtilis that require precursors of branched-chain fatty acids [36]. One of the major functions ascribed to the unsaturated, branched, or cyclopropane-ring fatty acids is to increase the fluidity of the hydrocarbon portion of the membrane by decreasing the critical temperature for the liquid-gel phase transition so that the hydrocarbon portion of the bilayers is in a liquid form at growth temperatures [37]. Mutants of E. coli that are auxotrophic for unsaturated fatty acids can be grown at 37°C in the presence of trans-octadecenoic acid or cis-Δ^6-octadecenoic acid. When these cells are shifted to 27° or 18°C, respectively, they die rapidly. This is interpreted as being due to the fact that lipids in membranes containing these fatty acids lose their fluidity at the lower temperatures [38]. Psycrophilic strains of Bacillus have cis-Δ^5-unsaturated fatty acids constituting from 18 to 32% of the total fatty acid content [39].

III. ORGANIZATION OF THE MEMBRANE

The basic structure of the membrane is considered to be a bilayer of phospholipid molecules, with the apolar fatty acid chains oriented perpendicularly to the plane of the membrane, and the polar groups at the outside. The membrane proteins are of a globular form with a considerable amount of α-helix content and they are exterior to the polar groups of the lipid molecules. Some of the membrane proteins extend into the hydrophobic center of the lipid bilayer and can be visualized by electron microscopy of freeze-etched preparations (see review of Singer and Nicolson [5a]). The model is essentially that of Danielli and Davson [40]. Electron micrographs of bacterial membranes resemble those found in eukaryotic cells; a railroad-track double line with a thickness of approximately 75 Å is observed. The membrane can be destroyed by detergents and the components can reassemble to form membranelike structures [8a, 41]. Isolated membranes can reseal their edges to form vesicles with the permeability properties of the cells from which they were derived [42]. In the past several years alternative models have been proposed for membranes. The relative merits of these models have been critically evaluated in an excellent review by Hendler [5]. These models propose that the lipids are arranged as micelles [43, 44], or that the membrane is composed of lipoprotein subunits [45, 46]. The bilayer model has received considerable support from several independent lines of evidence, including X-ray diffraction studies on dispersions of Mycoplasma laidlawii membranes [47] and erythrocyte membranes [48], which show a central region of low electron density expected for the apolar portion of the bilayer; and calorimetric studies that indicate a reversible thermotropic gel-liquid crystal phase transition in membranes at temperatures that correspond to those at which the same type of transition occurs in bilayers formed from the extracted lipids [49]. These studies have been further strengthened by the use of membranes in which the normal fatty acids in phospholipids have been replaced by other fatty acids resulting in different critical temperatures for the phase transitions [50]. Electron-spin resonance studies [51] have shown that the lipid molecules are oriented with their long axes perpendicular to the plane of the membrane.

If the membrane has the structure of a lipid bilayer with protein coating both surfaces, we can ask whether the composition of both surfaces is the same. Some differences are easily apparent between the inner and outer surfaces of biological membranes. Mitochondrial membranes have the F_1 component of the ATPase on only one surface [52]. Bacterial membranes seem to have similar particles attached to one surface of the membrane [53]. Other proteins of eukaryotic membranes seem to have an unequal distribution with respect to inner and outer surfaces of the membrane as evidenced by differential sensitivity to reagents that do not penetrate the membrane. It is also possible that the lipid distribution is not symmetrical in biological membranes, but no studies on this problem have been published.

IV. CONTROL OF SYNTHESIS OF MEMBRANE PROTEINS

Bacterial membranes have many different proteins, the function of only a small fraction having been identified. The synthesis of some of the known proteins is subject to control mechanisms that are fairly well understood. In the case of transport proteins (reviewed by Lin [54]; see also Chapter 5), the M protein of E. coli, which is necessary for lactose permeation, is subject to negative control, as are the other proteins of the lac operon [55], whereas the arabinose transport system is under positive control in the same manner as the other genes of the arabinose regulon. The synthesis of some amino acid transport proteins is inducible; others are constitutive. The synthesis of membrane-localized transport proteins of the phosphotransferase system is controlled in a manner similar to other enzymes needed for the catabolism of the various sugars [56]. Many of the elements of the electron transport system are present only under special circumstances. The synthesis of specific cytochromes and oxidases occurs in S. aureus and H. influenza only when the cells are growing aerobically [57, 58, 59]. The induction of bacteriochlorophyll synthesis occurs in Rhodopseudomonas when cells are shifted to semianaerobic conditions; concomitant with this induction is an increase in the rate of phospholipid synthesis [60]. The presence or absence of these activities does not influence the bulk composition of the membranes to any great extent, probably because of the large number of constitutive enzymes in membranes.

The extent to which the amount of membrane protein per cell is controlled is not known at the present time. Few studies on this important problem have been made (2), and although there does not appear to be great variation, it is necessary that studies be carried out on membranes from exponentially growing cells under various growth rates. Cultures that are growing slowly do not seem to have extra membrane when observed with the electron microscope. Mutants have been found that show an unbalanced production of membrane under certain conditions [61]. Infection of E. coli with amber mutants of phage fd leads to the production of excessive membranous material [62]. But the nature of this imbalance is presently obscure. The relative amounts of lipid and protein do not seem to differ greatly in fast and slowly growing cells, but reliable data as to the constancy of the membrane composition have not been published. If the composition is rather constant, we can expect that there is some mechanism for keeping it so. We know that the concentrations of individual proteins in the membrane can change drastically, but these proteins constitute only a small portion of the total membrane protein complement.

The question has been posed as to whether there is a general control mechanism for the synthesis of the bulk of membrane proteins, or a control of the incorporation of these proteins into cell membrane. Mutants in which lipid synthesis can be controlled experimentally are useful for such experiments. Mutants that cannot synthesize glycerol are particularly useful in that lipid synthesis can be rapidly stopped by removing glycerol from the culture medium. When glycerol auxotrophs of B. subtilis or S. aureus are deprived of glycerol, the synthesis of membrane proteins continues coordinately with that of cytoplasmic proteins. These proteins are also incorporated into the membrane. Although protein synthesis in these deprived cells stops after a doubling in mass, the membrane proteins have also doubled without a concomitant increase in membrane lipids. In this way membranes are formed that have a lipid content of 12% compared to the normal content of 20% [63]. Identifiable proteins or activities such as the M protein of E. coli [64], succinic dehydrogenase [63], citrate transport of B. subtilis [65], and enzyme II of the S. aureus lactose transport system [66] are incorporated into membranes in the absence of any increase in net content of membrane lipids. These results suggest the absence of a mechanism controlling the synthesis and incorporation of membrane proteins that senses the composition of the membrane. They also indicate that membranes of the same strain of bacteria can function with grossly different compositions. However, since these cells usually do not increase their mass beyond one doubling, there must be something defective in these lipid-poor membranes.

It appears, then, that proteins can be incorporated into the membrane in the absence of lipid synthesis. But are there any special requirements for incorporation of proteins into the membrane? Smith and White [13] found that free cytochrome c accumulates under conditions causing hyperproduction in _Hemophilus_. Kung and Henning [66a] showed that the soluble fractions of D^- and L-lactate dehydrogenases and L-α-glycerophosphate dehydrogenase vary with the carbon source utilized and the genetic backgrounds of the strains. These results indicate that these proteins bind to some factors in the membrane whose quantities can become limiting. The nature of these factors is not known. Several interesting experiments have been performed with _Saccharomyces cerevisiae_ that may be quite relevant. The mitochondria of these organisms contain an ATPase when growth occurs aerobically in a medium of 0.8% glucose. When the organism is grown in media of 5% glucose, the ATPase is present at low levels. If an organism growing at the higher sugar concentration is shifted to the low concentration, the ATPase increases in activity in the mitochondria. If the cells are derepressed in the presence of chloramphenicol, which inhibits mitochondrial protein synthesis, then several of the components of the ATPase appear in the cytoplasm of the cells but not in the mitochondria [67]. When the mitochondrial membranes are assayed for the ability to bind these components (F_1 and oligomycin sensitivity coupling protein), it is found that binding activity increases after derepression and that this increase is inhibited by either chloramphenicol, which inhibits mitochondrial protein synthesis, or cyclohexamide, which inhibits cytoplasmic protein synthesis; however, if the cells are derepressed sequentially, first in chloramphenicol and then in cyclohexamide, the binding activity increases [68]. Tzagoloff interprets these results as an indication that the binding protein synthesis is controlled by cytoplasmic protein synthesis, but that the protein(s) itself is made in the mitochondria; the ATPase components made in the presence of chloramphenicol do not attach to the mitochondrial membrane because the binding factor is not present. We have here a case in which a specific protein is necessary for attaching another protein to the membrane. Another interesting aspect of these experiments is the finding that subsequent incubation in the presence of cyclohexamide does not lead to the binding of the soluble ATPase components to the membrane [68]. It appears that incorporation occurs only concurrently with the synthesis. It is tempting to speculate that the mode of introducing the cytoplasmically synthesized ATPase components into the mitochondrion is via a mechanism similar to that proposed by Redman and Sabatini [69] for the synthesis and translocation of pancreatic enzymes through the membranes of the endoplasmic reticulum. They demonstrated that the proteins are synthesized by ribosomes sitting on the membrane and that the nascent peptide chain passes directly through the membrane. If the mitochondria produce a protein that binds those ribosomes engaged in the synthesis of the ATPase, then the nascent peptides would pass directly into the mitochondrial inner membrane.

There is no information currently available as to whether or not membrane proteins are synthesized by ribosomes attached to membranes. However, we do know that some membrane proteins, such as the E. coli M protein, are made on polysomes that are also involved in the synthesis of cytoplasmic proteins, which suggests that the same ribosomes can function for the synthesis of both types of protein. It is certainly possible that the nascent peptide chains are in contact with the cell membrane.

Membrane proteins may attain their final conformations only after being incorporated into the membrane. The experiments of Schlesinger [70] demonstrated that alkaline phosphatase molecules traverse the membrane as single peptide units and form their final complex structure in the periplasmic space. This may be the mechanism for placing proteins on the outside surface of the membrane as well as into the periplasmic space.

Although special mechanisms might be necessary for the correct placement of membrane proteins and lipids into the membrane structure, it is not unlikely that the components may find their way without any facilitation simply on the basis of their mutual affinities. Membranes that are disaggregated by various means can reaggregate to form membranelike structures [8a, 41, 71, 72, 73]. It is possible to form membranelike vesicles from lipids and purified mitochondrial proteins such as cytochrome oxidase [74], indicating that unique structural proteins are not necessarily required to form these structures.

V. CONTROL OF LIPID SYNTHESIS

Whereas proteins are not unique to the bacterial membranes, the lipids and especially the phospholipids, are found only in membrane structures. The pathways for the formation of fatty acids have been worked out in detail and this material has been reviewed by Stumpf [75] and Majerus and Vagelos [76]. The synthesis of bacterial fatty acids is briefly as follows. Malonyl-CoA is formed by the carboxylation of acetyl-CoA through the action of acetyl-CoA carboxylase, a biotin containing enzyme complex that has been extensively studied by Vagelos and his colleagues [77]. Malonyl-CoA and acetyl-CoA each react with acyl carrier protein (ACP) [78] to form thioesters of ACP. These compounds condense to form acetoacetyl-S-ACP with the release of carbon dioxide and acyl carrier protein. Acetoacetyl-S-ACP is then transformed sequentially to D(-)-β-hydroxybutyryl-S-ACP by reduction with NADPH, dehydrated to form crotonyl-S-ACP, and then reduced again with NADPH to form butyryl-S-ACP, whereupon another two carbons are added from malonyl-S-ACP and the cycle repeats. In some gram-positive bacteria the first reaction of malonyl-S-ACP is not with acetyl-S-ACP but with ACP derivatives of branched-chain amino acids such as isobutyrate, isovalerate, and 2-methylbutyrate [79, 80]. In these bacteria the fatty acids are predominantly iso- and anteiso-C_{15} and -C_{17}. In those bacteria that have unsaturated fatty acids, there is a specific enzyme, β-hydroxy-decanoyl thioester dehydrase that yields cis-decanoyl thioesters instead of trans compounds that are the normal intermediates in chain elongation. The cis compound remains unsaturated during subsequent elongation of the chain.

Bacteria also have cyclopropane fatty acids that may serve the same purpose as the branched-chain or unsaturated fatty acids. This purpose seems to be the increase in fluidity of the lipid bilayer at normal growth temperatures [81]. The cyclopropane fatty acids are derived from olefinic fatty acids [82]. Mutants that cannot make unsaturated fatty acids [83] or branched-chain fatty acids [36] cannot form colonies in their absence, thus attesting to the importance of the fluidity of the lipid bilayer. The transfer of fatty acids to sn-glycero-3-phosphate can occur from CoA derivatives or from ACP derivatives [84, 85, 86], but it is assumed that in vivo the fatty acids are transferred directly from ACP to the glycerolphosphate to form first monoacyl glycero-3-phosphate and then phosphatidic acid. The two reactions are known to be mediated by different enzymes because it is possible to isolate mutants that are conditionally defective in each step [87, 88]. The enzymes of fatty acid biosynthesis are in the soluble portion of the cell, but the transacylases that form glycerol esters seem to be membrane bound [87]. Acyl carrier protein seems to be localized near the surface of cells on the basis of autoradiographic experiments [89]. Phosphatidic acid seems to be the precursor for the various phosphatides found in bacteria (Fig. 1). Phosphatidic acid reacts with cytidine triphosphate to form CDP-diglyceride, which then reacts with sn-glycero-3-phosphate to form phosphatidylglycerolphosphate, which then loses the terminal phosphate to form phosphatidylglycerol. CDP-diglyceride can also react with serine to form phosphatidylserine. Phosphatidylserine is decarboxylated to form phosphatidylethanolamine. This pathway has been demonstrated in E. coli [90, 91, 92, 93] and in B. megaterium [94]. The enzymes involved in the transformations and synthesis of the phospholipids appear to be membrane bound and the products remain in the membrane. In E. coli the enzymes are localized in the plasma membrane and activity is low in the outer membrane, with the exception of CDP-diglyceride:L-serine phosphatidyltransferase (phosphatidylserine synthetase), which is rather loosely bound to membrane [95], and is found, inexplicably, primarily associated with ribosomes [95a].

Control of Fatty Acid Synthesis

The most economical way to control the extent of lipid synthesis would be to regulate fatty acid synthesis. Two types of control are necessary: one is required for coordination of lipid synthesis with protein synthesis, the other is for limiting the production of fatty acids because of exogenous supply or low rates of complex lipid synthesis. The first system would probably involve the sensing of amino acid supply or protein synthesis, perhaps in a manner analogous to the control of RNA synthesis in stringent bacteria. The second system would involve some type of feedback control by fatty acids themselves or their derivatives.

Several steps in fatty acid synthesis are amenable to control and several control mechanisms have been suggested. The activity of acetyl-CoA-carboxylase is the rate-limiting reaction for rat adipose tissue fatty acid synthetase *in vitro* [96]. Citrate and isocitrate act as allosteric effectors on acetyl-CoA-carboxylase, causing a reversible aggregation of the enzyme into an active filamentous form [97]. However, it is not clear that these compounds control the activity of the enzyme *in vivo*, since their effects are observed at concentrations higher than those presumably available to the enzyme in the cell [76]. Acetyl-CoA-carboxylase isolated from *E. coli* is not activated by citrate or similar compounds [77]. Majerus et al. [98] found that alterations in the activity of acetyl-CoA-carboxylase in rat liver and hepatoma cells was due to changes in the amount of enzyme per cell and not to differential activation. Measurements of acetyl-CoA-carboxylase activity in bacteria synthesizing fatty acids at different rates have not been reported. A stimulation of *E. coli* fatty acid synthetase by phosphate and phosphate sugars has been reported, but the level at which this occurs or the relevance for *in vivo* control has not been clarified [99]. The involvement of feedback inhibition in the control of fatty acid synthesis in higher organisms has been suggested, but support is rather meager [76]. Fatty acid synthesis in *Lactobacillus plantarum* is inhibited by long-chain unsaturated and cyclopropane fatty acids, whereas saturated fatty acids do not inhibit [100]. This inhibition is reversed by malonate, which may indicate that the effect is due to a feedback inhibition of acetyl-CoA-carboxylase.

Another possibility for controlling the rate of fatty acid synthesis could be the availability of acyl carrier protein. Limitations in the availability of this protein could be produced by turnover of the apoprotein, by the combined action of ACP-hydrolase [101] and ACP-holoprotein-synthetase [102], or by the loading up of available ACP with long-chain fatty acids under conditions where transacylase activity is low, if the rate of chain elongation decreases at longer chain lengths.

Sokawa et al. [103] demonstrated that the rate of fatty acid synthesis decreases by approximately 50% when cultures of *E. coli* are deprived of

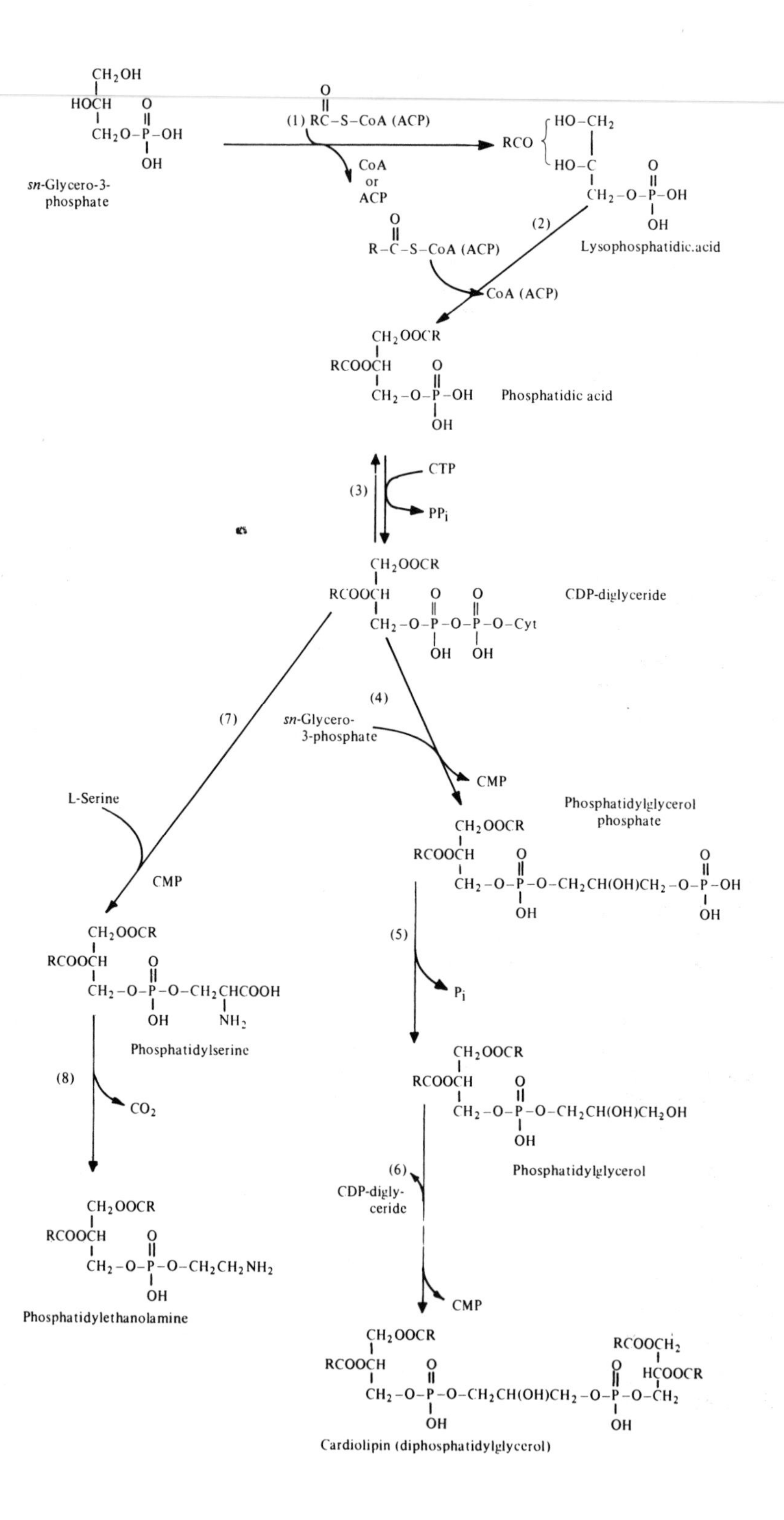

sn-Glycero-3-phosphate
(1) RC-S-CoA (ACP)
CoA or ACP
RCO
Lysophosphatidic acid
(2)
R-C-S-CoA (ACP)
CoA (ACP)
Phosphatidic acid
CTP
(3)
PPi
CDP-diglyceride
(4)
sn-Glycero-3-phosphate
CMP
(7)
L-Serine
CMP
Phosphatidylglycerol phosphate
(5)
Pi
Phosphatidylserine
(8)
CO2
Phosphatidylglycerol
(6)
CDP-diglyceride
CMP
Phosphatidylethanolamine
Cardiolipin (diphosphatidylglycerol)

essential amino acids. They also showed that this effect occurred only in cultures that were "stringent" with respect to amino acid control of the synthesis of ribosomal RNA, a characteristic controlled by the rel gene. Moreover, chloramphenicol which reverses the inhibition of RNA synthesis in stringent cells deprived of amino acids, also reverses the inhibition of fatty acid synthesis.

Stringent strains of E. coli accumulate a guanosine tetraphosphate (ppGpp) and it has been demonstrated [104] that this compound inhibits the specific transcription of ribosomal RNA genes in vitro. It is tempting to speculate that this same compound is responsible for the reduction in the rate of fatty acid synthesis upon amino acid deprival, and that acetyl-CoA-carboxylase, which appears to be an allosteric protein in higher organisms, might be the target molecule for this effector.

Amino acid deprivation also reduces the rate of fatty acid synthesis in B. subtilis [105]. However, chloramphenicol does not reverse this inhibition and actually causes an inhibition of fatty acid synthesis similar to amino acid deprivation. Incubation of B. subtilis in the presence of chloramphenicol or in the absence of required amino acids results in membranes somewhat richer in lipids than normal [63]. However, the imbalance does not approach the twofold difference found when the same strains are prevented from synthesizing lipids and the membrane becomes lipid-poor [63]. Shockman et al. [106] found that cultures of Streptococcus faeculis deprived of amino acids had 36 to 40% lipid in their membranes as compared to 28% in log-phase cells. Kahane and Razin [107] found that membrane lipids in Mycoplasma laidlawii increased in the presence of chloramphenicol. These results show that although lipid synthesis is influenced by protein synthesis, the regulation is only partial.

It has been claimed that fatty acid synthesis in yeast is regulated by the concentration of glycerolphosphate [108]. In B. subtilis and S. aureus it is possible to regulate the supply of glycerol and consequently glycerolphosphate in mutants auxotrophic for glycerol [66, 109]. In this case, diminution in the apparent rate of fatty acid synthesis is observed; however, this diminution is the resultant of a normal rate of fatty acid synthesis and a rapid degradation of the free fatty acids that are formed, since, in the absence of glycerol, the fatty acids do not become esterified to form complex lipids but are found as free fatty acids, which can constitute 10 to 20% of the total fatty acids in the cell [105]. These results indicate that glycerolphosphate does not regulate fatty acid synthesis and there is no

---

FIG. 1. The biosynthesis of the phospholipids of E. coli. ACP = acyl carrier protein. (Reproduced from ref. 16.)

feedback inhibition of fatty acid synthesis by free fatty acids in these organisms. In the same organisms, when fatty acid synthesis is depressed by amino acid deprivation or chloramphenicol, there is no accumulation of free fatty acids. If the inhibition of lipid synthesis were due to an effect on the transacylation step, an accumulation of free fatty acids would be expected; since none was found, the site of inhibition must be in fatty acid synthesis itself.

The situation is somewhat different in two gram negative organisms, E. coli and Rhodopseudomonas capsulata. When glycerol auxotrophs of these organisms are deprived of glycerol, there is very little accumulation of free fatty acids (Mindich [105] and unpublished experiments). It is possible that the fatty acids are made but are incorporated into lipid A of the outer membrane, but it seems more likely that in these cases there is a feedback control of fatty acid synthesis, or, if the fatty acids remain on the acyl carrier protein, it is likely that chain elongation ceases after the fatty acids reach a particular length and total fatty acid synthesis stops due to the lack of free ACP. The apparent lack of regulation in the case of the gram-positive bacteria would then be due to the release of the fatty acids from the ACP, which would allow the use of the latter for further fatty acid synthesis. The ACP derivatives of fatty acids might be more stable in the gram-negative cells than in the gram-positives. The saturation of the ACP with exogenous fatty acids could also explain the inhibition of fatty acid synthesis by exogenous fatty acids, since gram-negative organisms such as E. coli do show a feedback regulation of fatty acid synthesis when exogenous fatty acids are supplied.

Mutants have been isolated that lack the enzyme β-hydroxydecanoyl-thioester dehydrase and consequently cannot make unsaturated fatty acids [83, 110]. When these organisms are deprived of unsaturated fatty acids, the synthesis of total fatty acids increases, suggesting that the incorporation of the exogenous fatty acids depresses the synthesis of total fatty acids. 3-Decynoyl-N-acetylcysteamine inhibits the activity of the above enzyme and when applied to cultures of E. coli, the extent of fatty acid synthesis remains unchanged; however, all the fatty acids are saturated [111].

In a study of the effect of exogenous unsaturated fatty acids on fatty acid synthesis [111a] it was demonstrated that the synthesis of unsaturated fatty acids is reduced, but that the extent of the reduction is strain-dependent. It was also shown that the rate of saturated fatty acid synthesis is reduced in most of these strains. The extent of the regulation was found to vary in the different strains and under different culture conditions [111a, 111b]. The regulation of synthesis by exogenous unsaturated fatty acids is found in both unsaturated fatty acid auxotrophs and wild-type cells [111a, 111b]. When wild-type cells are supplemented with both saturated and unsaturated fatty acids, the rate of fatty acid synthesis can be diminished by 90% [112].

It would appear that the control of fatty acid synthesis is related to both total fatty acid synthesis and to the composition of individual classes. The mechanisms for the control of the types of fatty acids synthesized are not known. Mutants of E. coli that are auxotrophic for unsaturated fatty acids and lacking acyl-CoA-dehydrogenase activity can grow in the presence of various unsaturated fatty acids. These fatty acids are incorporated into phosphatides without degradation. In the presence of long-chain unsaturated fatty acids such as Δ^{11} eicosenoic acid, phospholipids contain a high proportion of myristic ($C_{14:0}$) acid, whereas in the presence of oleic acid this fatty acid constitutes only a few percent of the total [38]. Similar results are found when cells are supplemented with trans-unsaturated fatty acids [38, 113]. There seems to be a mechanism whereby the cell can adjust the saturated fatty acid composition of the phospholipids to maintain fluidity in the membrane, increasing the incorporation of shorter-chain saturated fatty acids when the unsaturated fatty acids have higher melting temperatures than those normally found in E. coli.

When wild-type cells are supplemented with both saturated and unsaturated fatty acids, the synthesis of fatty acids can be reduced by 90%; however, the differences in the ratio of saturated to unsaturated fatty acids that occur at high and low growth temperatures still persist [112]. This selectivity is also found in in vitro preparations utilizing acyl-CoA substrates, indicating that the transacylase might be the determinant of the phospholipid fatty acid composition. Saturated fatty acids are usually found at position 1 of the phospholipid glycerol, unsaturated fatty acids are usually found at position 2. Particulate preparations of E. coli transacylases function in vitro to give this same type of distribution [86]. By studying the thermolability of transacylase activity in mutants with a thermosensitive glycerol-3-phosphate transacylase it has been demonstrated that this enzyme or one of its components is responsible for placing the unsaturated fatty acids at the 2 position or the saturated fatty acids at the 1 position at the first acylation step [112a]. This indicates that the transacylases can function with great specificity and may very well determine the fatty acid composition of the phospholipids. Of course, if the transacylases determine the composition of the phospholipids, there must be some way for the cell to deal with the excess fatty acids that are not used; they must either be gotten rid of or their synthesis must be regulated.

In B. subtilis the relative amounts of branched- and straight-chain fatty acids does not seem to be well regulated and appears to be determined by the availability of precursor molecules [34, 36].

Very little is known about the regulation of the synthesis of the various glycerides in bacteria. It is not certain to what extent this regulation operates through synthesis or turnover, yet there must be specific regulatory mechanisms, since the phospholipid and glycolipid composition of cells is rather characteristic and varies under specific conditions. The various interconversions of the lipids are discussed in Section VI.

VI. TURNOVER OF MEMBRANE COMPONENTS IN BACTERIA

Eukaryotic membranes, especially those of the endoplasmic reticulum, have been shown to have extensive turnover of both lipids and proteins [114, 115]. The rates of degradation vary independently with different proteins in the same membrane. Turnover processes can serve to allow a membrane structure to change composition in the absence of net changes in lipid or protein. The turnover of bacterial proteins is usually very small in actively growing cells, except in special circumstances such as sporulation where a culture is in a limiting medium and must make a morphological transformation in the absence of growth. During sporulation, approximately 8 to 10% of the cell proteins of B. subtilis turn over per hour; however, asporogenic mutants lacking protease do not show turnover [116]. The turnover of membrane proteins has not been studied extensively in bacteria. Membrane proteins in Mycoplasma laidlawaii turn over with a half-life of approximately 3 h [107] under conditions where cytoplasmic proteins are stable.

The turnover of lipid components has been studied more extensively than that for proteins; however, the role of lipid turnover in cell physiology has not been defined. It might be necessary for membrane growth and modification. Although the mechanism of protein turnover is unknown, there are several enzyme activities that have been demonstrated in bacteria that can function in lipid turnover. Phospholipases A and B (or A_1 and A_2), which remove acyl groups from phospholipids and lysophospholipids, respectively, have been demonstrated in E. coli [117]. Lysophospholipids can be acylated by extracts of E. coli [118]. It is not clear whether this activity is independent of the biosynthetic acyltransferase activity described by Hechemy and Goldfine [87].

Phospholipase C has been demonstrated in E. coli [119]. This enzyme forms diglycerides from phospholipids; the resulting diglycerides can be phosphorylated through the action of a 1,2-diglyceride kinase described by Pieringer and Kunnes [120]. The resulting phosphatidic acid could then be transformed into other phospholipids through CDP-diglyceride.

The phosphate group in phosphatidylglycerol was found to turn over with a half-time of about 1 hour in E. coli [90, 121, 122], while the phosphate of phosphatidylethanolamine does not seem to turn over at all. The phosphate group in diphosphatidylglycerol also appears to turn over, but at a slower rate than that of phosphatidylglycerol [121]. In B. subtilis the lipid phosphorous turns over with a half-time of 2 hours [109]; the turnover of individual phospholipids has been studied by Lillich and White [123]. The stability of phospholipids in E. coli varies under various growth conditions [124]. The turnover of phosphatidylglycerol is inhibited by dinitrophenol or anaerobiosis. Phosphatidylethanolamine is degraded as cells approach lysis. When cells are shifted from a rich to a poor growth medium, synthesis of phosphatidylethanolamine seems to be reduced but that of phosphatidylglycerol is stimulated with a concomitant reduction of turnover of phosphate and glycerol in this compound.

Under conditions of growth limitation, the concentration of diphosphatidylglycerol increases dramatically in E. coli membranes [110, 123, 125]. When the cells resume growth, the concentration falls rapidly. An increase in cardiolipin is also seen in abortive infection of E. coli by amber mutants of phage fd [126]. In addition to the activity for synthesizing diphosphatidylglycerol [127] there must be enzymes that degrade the compound as well. Hemophilus influenza has been shown to contain a cardiolipin-specific phospholipase D [128, 129]. The synthesis and turnover of diphosphatidylglycerol may be involved in the observed rapid turnover of the phosphate moiety of phosphatidylglycerol, but the role of these reactions in the cell is not known.

The fatty acid portion of phospholipids seems to turnover at a much slower rate than the polar portion. In Hemophilus parainfluenza [130] the turnover of fatty acids is much slower than that of phosphate. The acylated glycerol in phospholipids also turned over slower than phosphate or the unacylated glycerol of phosphatidylglycerol. In B. subtilis and S. aureus the turnover of fatty acids already incorporated into phospholipids is very slow. An upper limit of 5% turnover per generation can be set [105]. In E. coli the turnover seems to be higher in experiments of the same sort (Mindich, unpublished experiments), but Overath et al. [131] have shown that exchange of fatty acids in phospholipid is extremely low in E. coli. They incorporated cyclopropane fatty acids into the lipids of a fatty acid auxotroph of E. coli labeled with glycerolphosphate and then added unsaturated fatty acids and followed the association of the labeled glycerol with phospholipids containing cyclopropane or unsaturated fatty acids. They found little movement of fatty acids from the original labeled glycerol, indicating either that there is little turnover of the fatty acids or that the glycerol moiety turns over along with the fatty acids.

VII. INCORPORATION OF FUNCTIONAL SYSTEMS IN THE ABSENCE OF LIPID SYNTHESIS

When bacteria such as Rhodopseudomonas are switched from aerobic to semianaerobic growth conditions, the synthesis of the photosynthetic system is induced. Along with the synthesis of the bacteriochlorophyll and carotenoids, there is an increase in lipid synthesis and the fully induced cells have an increased amount of membrane [60]. Anaerobically grown S. aureus does not have respiratory activity or cytochromes. When the cells are aerated they synthesize the respiratory components and also show a stimulation of lipid synthesis [59]. Thus, the synthesis of these two membrane-associated systems is accompanied by increased lipid synthesis. Is concurrent lipid synthesis necessary for the integration of functional systems into the cell membrane?

In B. subtilis glycerol auxotrophs deprived of glycerol, and consequently not increasing their net lipid content, one can find an increase of succinic dehydrogenase [109] and NADH oxidase activity (Mindich, unpublished observation); citrate transport can be induced [65]. In S. aureus the lactose transport system, which functions as a phosphotransferase [56] can be induced in the absence of net lipid synthesis [66]. In these cases there is also an increase in the protein content of the membranes, indicating that activities can increase with an increasing proportion of protein to lipid in the membrane. Although protein turnover might facilitate the incorporation of membrane proteins in the absence of growth, and although turnover might allow for higher activity of these incorporated proteins, it appears that bacteria put new activities into membranes coordinately with new protein synthesis. In E. coli the β-galactoside transport system can be induced in the absence of lipid synthesis for approximately two-thirds of a generation after lipid synthesis is stopped [64]. However, in the case of the induction of the various transport systems, there is a reduced efficiency either from the beginning of the period of no lipid synthesis or there is a progressive loss of efficiency upon induction at later times during this period. In the case of the E. coli system, the presence at normal levels of the transport protein in the membrane can be demonstrated in spite of its low activity by assaying the binding properties of the protein [64]. In the case of the S. aureus system, the presence of normal levels of transport protein in the membrane can be demonstrated by its function as a phosphotransferase [66]. Since the transport proteins are shown to be incorporated into the membrane, it is not obvious as to why the transport activity should be low. It is likely that the new proteins might function at full efficiency only when incorporated at sites of new lipid synthesis or only when complexed with a certain minimum amount of lipid. In the case of the E. coli lactose transport system, the low efficiency of the transport cannot be corrected by subsequent resumption of lipid synthesis [64]. However, in the case of S. aureus, the efficiency of the lactose transport system can be completely recovered by allowing lipid synthesis to resume in the absence of further protein synthesis [66]. The activities of previously induced transport systems in E. coli and S. aureus decrease about 25% during the first hour after the cessation of lipid synthesis is stopped. These results may indicate that there is some mobility of the transport proteins or lipids and that the activity for transport reflects the amount of lipid available to the transport system. It is important to keep in mind that when one is measuring the activity of transport systems one may be looking at only a fraction of the total transport molecules. In the case of the E. coli lac transport system, it seems rather certain that this is the case, since in cells diploid for the lac operon, the activity of the transport system does not increase coordinately with the rest of the activities when compared to the haploid state [132]. The nature of the association of the transport system with the membrane lipids is not clear, but several very interesting experiments have been done by two groups to show that there is an intimate connection between the physical state of lipids and the function of transport systems.

The rate of transport of several sugars was shown to be a function of temperature, and Arrhenius plots of the rate of transport with respect to temperature show an abrupt change in slope at a temperature that is a function of the fatty acid composition of the cell membranes [133, 134]. This temperature corresponds to the temperature for the phase transition from the liquid expanded to a more condensed state found with the isolated lipids of the membrane [135]. Wilson et al. [134] showed that the critical temperature, indicated by this change in slope, varied in an identical manner for two different transport systems when cells auxotrophic for unsaturated fatty acids were grown on different fatty acids. It was also shown that if a culture is induced for the β-glucoside transport system while growing in the presence of oleic acid, and then briefly shifted to medium with linoleic acid and induced for the β-galactoside transport system, the transport activities show critical temperature effects that are characteristic for the fatty acids that were incorporated during their respective inductions [136]. This would indicate that there are separate domains of different types of lipids in the membrane; however, it is possible that these domains refer to lipids that are extremely tightly bound to the transport proteins and the lack of apparent mixing of the lipids reflects more the special relationship of the transport proteins with a small amount of lipid than the general properties of the lipid in the membrane.

Unfortunately, the observations with respect to the temperature dependence of transport activity on the lipid incorporated at the time of induction are in conflict with those of Overath et al. [131] who have found that the "critical temperature" in transport does not reflect the properties of the lipid incorporated during induction but reflects instead the molar proportions of the different lipids in the membrane. They find a gradual change in the "critical temperature" after shifting from one fatty acid supplement to another. These results would indicate that the β-galactoside transport system is not specifically associated with newly synthesized lipid and would be rather consistent with the experiments with glycerol deprivation. Since the experiments performed by the Overath group also included studies with the same strains utilized by the Fox group, the difference in results remains anomalous. Overath et al. [131] also demonstrated that β-galactoside transport activity is inducible in cells that are prevented from synthesizing unsaturated fatty acids by the addition of

3-decynoyl-N-acetylcysteamine. These results are in conflict with those of Fox [132] who reported that β-galactoside transport activity could not be induced in unsupplemented cells of a strain that cannot synthesize unsaturated fatty acids. It appears that transport activity can be induced in the absence of incorporation of unsaturated fatty acids, but the level of induction decreases with increasing times of deprivation.

Another line of evidence indicating the sensitivity of the transport activity to the state of the lipids at the time of the incorporation of permease into the membrane is the finding that the final level of activity of the transport systems is related to whether or not the cells are being induced at a temperature above or below the critical melting temperature for the membrane lipids [137].

Cells grown at 37°C with elaidate as the unsaturated fatty acid supplement show a transition temperature of approximately 30°C. If they are shifted to 25°C and induced for the synthesis of the lac transport system, the components are synthesized but the permease does not function even at high temperatures. If the culture is induced at 37°C, then the permease functions normally. It appears that in order to play its role in transport the permease must interact with the membrane lipids in a specific manner at the time of integration and that lipids below the transition temperature do not allow this process. The observation that the dysfunction cannot be corrected at high temperature indicates that the permease must make this interaction only at the time of incorporation and is in contrast to the previously cited case of the staphylococcal lac permease [66] whose efficiency can be improved subsequent to integration into the membrane.

VIII. DIFFERENTIATION OF BACTERIAL MEMBRANES

In eukaryotic organisms there is no continuum of membranous material; instead there are many different membrane structures. There are membranes that lack enzymes of lipid synthesis and are not associated with ribosomes. The plasma membrane is a good example of such a structure. It must get its lipid and protein from other parts of the cell. The inner membrane of vaccinia virus appears to form with no physical connection to other membrane systems; it may very well be an example of de novo synthesis [138]. In these cases a means of transporting membrane components must exist.

Most membrane in bacteria is continuous. There are, however, several membrane structures in bacteria that seem to differ in composition from the plasma membrane. These systems are now under intensive study, as the mechanisms of their formation might serve as useful models for the construction of differentiated membranes in general. The mesosome is an invagination continuous with the plasma membrane [139]. It is found to be associated with incipient septa and the nucleoplasm in gram-positive cells. It is possible that there are several different classes of mesosomes, perhaps participating in different functions; some mesosomes appear as tubular membranous structures; others appear as lamellar structures [140]. A mesosomal fraction can be separated from the plasma membrane and differences have been noted in the cytochrome content, the activities of certain oxidases, hydrolytic enzymes, and acrylamide gel electrophoresis patterns [141, 142, 143]. The role of mesosomes is not presently known, although it has been suggested that they act in respiration, cell division, lipid synthesis, and exoenzyme transport or function.

Gram-negative bacteria have two examples of differentiated membranes that are separated from the plasma membrane. The outer membrane of the Enterobacteriaceae is clearly different in composition from that of the inner membrane [4, 18] and it lacks the enzymes for its own synthesis. Its components are formed in the protoplast membrane [3, 16, 95] and the interior of the cell. The lipids, proteins, and polysaccharides must be transported from the plasma membrane to the exterior of the cell; the mechanism of this transfer is not known. The development of the lipid-containing phage PM2, which grows in a marine pseudomonad, is a type of membrane differentiation [144], which also involves membrane assembly independent of the plasma membrane. A mechanism must exist to transport lipids from their site of synthesis to the developing virus.

Photosynthetic bacteria have chromatophores that are bounded by membranes; these structures appear to arise from invaginations of the plasma membrane. Chromatophores are differentiated at least in the sense that they have a different composition from the plasma membrane of aerobically grown cells. It is not clear, however, that they have compositional differences from the plasma membrane of cells in which they are found. The question of their physical separation from the plasma membrane is still open [145, 146, 147]. At the end of the logarithmic growth, H. halobium forms a patch of membrane containing a rhodopsin-like protein. Membrane fragments can be isolated that contain only this protein along with phospholipid [21, 148].

Although the nature of the mechanism of the assembly of regions of differentiated membrane has not yet been revealed, we can list the possible ways in which one part of a continuous membrane structure might have a different composition from the rest. The incorporation of specific proteins or lipids might be random, but they could migrate to specific areas of concentration to form the differentiated areas. Alternatively, the specific proteins could be incorporated into specific zones of the membrane. The lipids could also be incorporated into specific zones or they might be incorporated randomly while the proteins are incorporated directly into the differentiated zones. The latter possibility is supported by the finding that animal viruses, the membranes of which are derived from the plasma membrane seem to have virus-specific proteins, but their lipids are similar in composition to those of the plasma membrane [149]. The mechanism by which specific proteins would be placed in specific membrane sites is not known, but could involve such processes as specific transport, specific association with membrane sites of ribosomes making the specific proteins or a process whereby the newly formed proteins associate on the basis of affinity with the differentiated sites. Membrane proteins may also be altered in their primary structure or their association with other subunits during incorporation into their specific sites.

IX. TOPOLOGY OF MEMBRANE SYNTHESIS

The central question in the study of the topology of bacterial membrane synthesis is whether or not there are specific areas of membrane synthesis and whether or not there exist zones of conservation in the growth of membrane. The question can be posed as "Are there such things as new and old membranes in a bacterial cell?" We will concentrate here on the problem of synthesis of the bacterial plasma membrane, a structure which is continuous, and containing enzymes for lipid synthesis and uniform access to the cellular protein synthesis machinery.

Several possibilities can be listed for the manner in which the plasma membrane increases in mass. The simplest would have lipid and protein being synthesized or inserted randomly into the existing structure. Alternatively, either or both the lipid and protein could be inserted at one or a few assembly sites. The material entering the membrane at a given time might stay together for a very long time or there might be mixing in the plane of the membrane or exchange of material through the cytoplasm. Models involving specific growth zones for the membrane have been attractive mainly for models of cell growth and division, Jacob et al. [150] proposed as part of their replicon model, that the bacterial chromosome is attached to the cell membrane and that the growth of membrane between two chromosomes is responsible for the proper apportionment of the chromosomes to daughter cells. Donachie and Begg [151] proposed membrane growth occurs at the poles of slowly growing cells and in the center of more rapidly growing ones. This manner of growth would be consistent with (although not necessary for) the observed pattern of growth of individual E. coli cells. Models including specific zones of membrane growth and conservation can make it easier to approach the problem of localization of the position of septation in dividing cells; for if the membrane is a random, undifferentiated structure, how can the septation site be regulated as carefully as it is?

Although formation of a cell septum involves the apparent growth of membrane at a given region of the cell, it is possible to construct satisfactory models for such localized apparent growth that result from uniform expansion of the membrane. A good example is the formation of the spore septum in Bacillus (Fig. 2). Spore formation involves the growth of a membrane invagination near one of the poles of the cell to segregate a portion of the cell from the rest [152, 152a]. This double membrane continues to grow and the point of attachment moves progressively toward the nearest pole of the cell until finally when the pole itself is reached, the prespore is released from its attachment to the cell membrane. The prespore therefore has a double membrane around its cytoplasm. If one assumes that the spore septal membrane is a specific growth zone, it is difficult to see how the final structure is formed. However, if one assumes that a substance is produced that is incorporated into the membrane at the site of the septal invagination and that this substance causes membrane to become sticky and fold back on itself, then the invagination, which is not really a zone of new synthesis, would increase until a complete septum were formed. The continued production of the hypothetical substance would lead to an increase in the amount of septal membrane and would cause the ring of attachment to the cell membrane to move toward the pole of the cell. In this way a new membrane structure would be formed without the necessity of a specific growth zone in the membrane. A modification of this scheme wherein wall material is synthesized within the invagination would lead to a true division of the cell. Mutants of B. subtilis have been isolated that do not complete the sporulation process. Among the various classes of mutants found are some that do not stop wall growth after the spore septum forms. These mutants form a septum complete with cell wall between the prespore and the rest of the cell [153].

The attempts at demonstrating specific sites of membrane synthesis can be arranged into three classes. They are investigations of (i) the precursor role of mesosomes for plasma membrane lipids and proteins, (ii) the conservation of membrane material, and (iii) pulse-labeling in order to find specific growth zones.

The mesosome has been especially attractive as a possible site for membrane lipid synthesis because of its localization at the sites of septation and its association with the cellular DNA. Fitz-James [154] has claimed that ^{14}C-acetate is incorporated faster into mesosomal lipids than into lipids of the plasma membrane. However, several other studies have not been able to show evidence for a role of mesosomes in lipid synthesis or membrane assembly. Patch and Landman [141] showed that

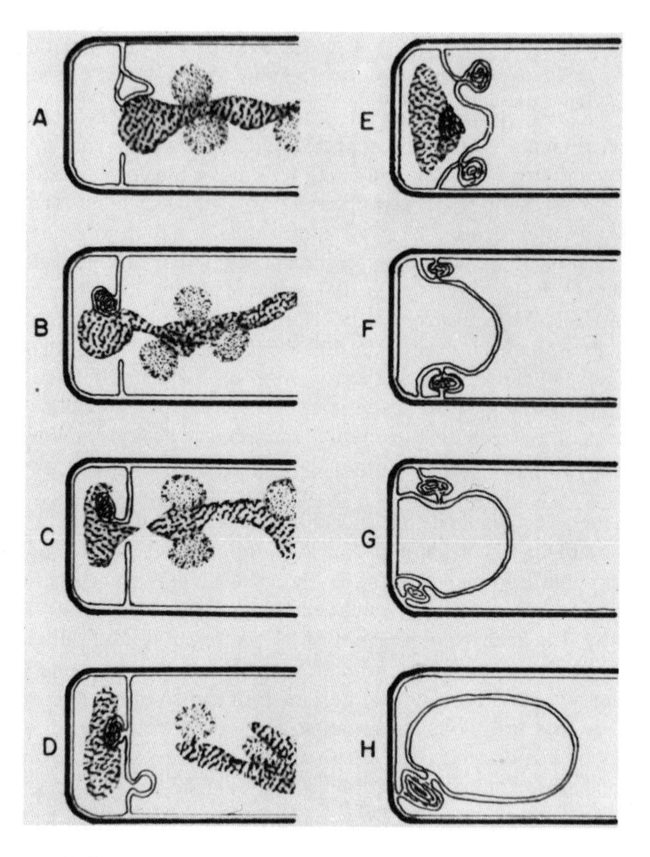

FIG. 2. A diagrammatic interpretation of the electron micrographs of the developing forespore in bacilli. Steps A-D indicate the development of the spore septum which is less well established than steps E-F, the conversion of the spore septum to the forespore membrane. Chromatin and associated lipid granules have been omitted in steps F-H. The fine line represents the plasma membrane; the heavier envelope, the cell wall. (Reproduced from ref. 152a.)

incorporation of radioactive acetate into the lipids of mesosomes was no more rapid than incorporation into the whole plasma membrane. They also found similar results from the incorporation of radioactive sulfate into the proteins of the two structures. Morrison and Morowitz [155] reported similar results in B. megaterium for membrane proteins, and Mindich and Dales [156] found that mesosomal lipids are not labeled faster than those of the plasma membrane in double label experiments. Although these experiments indicate that the mesosome is not the site of membrane synthesis or assembly, they are not conclusive. The preparation of mesosomes takes considerable time, and if redistribution of material takes place after the initial synthesis or assembly, then the higher specific radioactivities expected with specific growth zones would be lost.

Most studies of membrane growth have involved experiments designed to show conservation of membrane materials. Lin, Hirota, and Jacob [157] labeled cells of E. coli with lipid-specific compounds such as glycerol and oleate, but found no evidence for the conservation of large pieces of membrane after growth in nonradioactive media, although by the same technique they were able to show the segregation of thymidine-labeled DNA. The use of whole cell preparations of gram-negative organisms such as E. coli presents a special problem for studies of membrane segregation since these organisms have two lipid containing structures, namely the inner protoplast membrane and the outer membrane. Since both structures contain phospholipids, individual patterns of segregation might be masked by their different behaviors. Morrison and Morowitz [155] used tritium-labeled palmitate to label the phospholipids of B. megaterium. They found that label incorporated into cells over a long period of growth did indeed segregate during subsequent growth in the absence of label. This study would have been a convincing demonstration of the conservation of membrane were it not for several deficiencies in its execution; the most important being that the demonstration of incorporation of label into phospholipid was done at much lower palmitate concentration than that used in the labeling experiments. It is quite possible that much of the label associated with the cells is free palmitic acid and the distributions observed are not those of membrane distribution but of material adsorbed to the outer surfaces of the cells. This interpretation is strengthened by the observation that most of the labeled cells at the end of the labeling period are in small chains, very few are single, yet after growth in unlabeled medium, the chains of cells show only one pole of conservation. This type of distribution is unexpected in any of the models of cell or membrane growth but would be consistent with the desorption of free fatty acid from the labeled cells or poor viability after labeling.

Other studies in search of membrane conservation have involved density shift experiments. Mindich and Dales [156] grew cells of B. subtilis in deuterated media in the presence of ^{3}H-labeled phenylalanine before switching to normal media with ^{14}C-labeled phenylalanine. Membranes isolated one, two or three generations after the shift were examined for heterogeneity of the radioactivity distribution with respect to density. Three generations after the switch there was no significant difference in the distribution of the two labels, indicating that the "older" label was not located preferentially in one part of the membrane. Such an experiment shows that the number of conserved pieces must be greater than six per cell. Tsukagoshi et al. [158] did density shift experiments using bromostearate as a means of increasing the density of E. coli membranes. They found no evidence for conservation of membrane lipid and in further studies [159] in which hybrid membranes have been fragmented, they found no evidence for conservation down to fragments of several hundred angstroms. Another approach to the study of conservation is the use of E. coli mutants that produce minicells. These cells have a division defect that causes the production of small anucleate cells for the ends of cells. Wilson and Fox [160] used this system to study the distribution of membrane proteins in general and two specific transport systems. They found no evidence to indicate conservation of membrane during the production of minicells. Experiments with minicells are open to the criticism that they utilize a system that is clearly abnormal in division, but experiments to demonstrate conservation, if negative, are open to a more serious artifact. Wirtz and Zilversmit [161] and McMurray and Dawson [162] have demonstrated that some tissues of higher organisms have proteins that can facilitate the transfer lipids from one membrane to another. Migration of lipids from one part of the bacterial membrane to another has not been demonstrated but is not unlikely. If migration takes place, the conservation experiments would not reflect the mode of synthesis in assembly of membrane. Rapid diffusion of lipids in the plane of the bilayer has been demonstrated in vesicles of synthetic phospholipids [163] and in functional sarcoplasmic reticulum vesicles [163a]. It is not certain that these results are applicable to bacterial membranes, but if they are they would preclude localization experiments for sites of lipid synthesis. Frye and Edidin [164] have shown that heterokaryons formed between mouse and human cells with Sendai virus have clearly separated areas of mouse-specific and human-specific antigenic composition at the time of fusion. Forty minutes after fusion has occurred, the antigenic composition of the surface of the heterokaryon appears to be homogenous with no specific localization of the antigens. It is not clear from their experiments whether the apparent mixing is due to migration of membrane proteins in the plane of the membrane or to the incorporation of previously formed but cytoplasmically localized membrane protein. However, these studies do raise the question of the possibility that the protein in the bacterial membranes can diffuse, and although this has not been demonstrated, would clearly limit the reliability of conservation experiments.

Pulse-labeling experiments have the advantage that they can demonstrate sites of synthesis even in the case where migration from the site of synthesis takes place, if that rate is slow compared to the rate of synthesis. They have the disadvantage in the case of microbial autoradiography experiments that the resolution is not very great. Morrison and Morowitz [155] found that short periods of labeling with tritium-labeled palmitate resulted in the preferential labeling of one of the poles of B. megaterium cells, a result which is consistent with their finding of conservation in long term labeling with the same compound. Mindich and Dales [156] pulse-labeled cells of B. subtilis with tritiated glycerol and examined the distribution of silver grains over thin sections of cells. They found no preferential site of labeling at the poles of the cells, at the regions of incipient septa, or in the center of the cells. Autoradiography of thin sections allows very good localization of radioactivity [165], but one cannot be sure that lipids are not diffusing in the plane of the membrane during fixation or during the dehydration steps that follow fixation.

The results of the various studies, although not uniform in their conclusions, lend support to the idea that membrane growth does not occur in specific regions and that there are no zones of conservation during membrane growth. As pointed out in this chapter, each type of experiment has its deficiencies; consequently the proof of random, nonconservative membrane growth is not very firm. It should be emphasized that studies in eukaryotic organisms on the growth of membranes in which lipid synthesis occurs yield results that are in agreement with the concept of nonconservative random growth of the membrane. Studies by Leskes et al. [166, 167] on the formation of the endoplasmic reticulum in rat hepatocytes showed uniform incorporation of glucose-6-phosphatase activity into the membranes of individual cells with no evidence of growth zones, and autoradiographic studies by Goldberg and Ohad [168] on the development of thylakoid membranes in the chloroplasts of Chlamydomonas have shown random incorporation of radioactivity in the photosynthetic membranes with no indications of preferential synthesis in any part of the structure. However, these experiments, just as the conservation experiments with bacterial membranes, are also subject to the interpretation that the lipids or proteins are diffusing from one part of the membrane to another.

ACKNOWLEDGMENT

Part of this work was supported by Public Health Service Grant AI-09861.

REFERENCES

1. C. Weibull, J. Bacteriol., 66, 688 and 696 (1953).

2. M. R. J. Salton and J. H. Freer, Biochim. Biophys. Acta, 107, 531 (1965).

3. M. J. Osborn, J. E. Gander, E. Parisi, and J. Carson, J. Biol. Chem., 247, 3962 (1972).

4. T. Miura and S. Mizushima, Biochim. Biophys. Acta, 150, 159 (1968).

5. R. W. Hendler, Physiol. Rev., 51, 66 (1971).

5a. S. J. Singer and G. L. Nicolson, Science, 175, 720 (1972).

6. M. R. J. Salton, Ann. Rev. Microbiol., 21, 417 (1967).

6a. J. Oelze and G. Drews, Biochim. Biophys. Acta, 265, 209 (1972).

7. L. Rothfield and A. Finkelstein, Ann. Rev. Biochem., 37, 463 (1968).

8. G. S. Getz, Advan. Lipid Res., 8, 175 (1970).

8a. S. Razin, Biochim. Biophys. Acta, 265, 241 (1972).

8b. J. E. Cronan, Jr., and P. R. Vagelos, Biochim. Biophys. Acta, 265, 25 (1972).

9. M. L. Vorbeck and G. V. Marinetti, Biochemistry, 4, 296 (1965).

10. J. B. Ward and H. R. Perkins, Biochem. J., 106, 391 (1968).

11. M. Cohen and C. Panos, Biochemistry, 5, 2385 (1966).

12. M. M. Burger and L. Glaser, J. Biol. Chem., 239, 3168 (1964).

13. L. Smith and D. C. White, J. Biol. Chem., 237, 1337 (1962).

14. N. Gellman, M. A. Lukoyanova, and D. N. Ostrovskii, Respiration and Phosphorylation of Bacteria, Plenum Press, New York, 1967.

15. H. R. Kaback, Ann. Rev. Biochem., 39, 561 (1970).

16. R. M. Bell, R. D. Mavis, M. J. Osborn, and P. R. Vagelos, Biochim. Biophys. Acta, 249, 628 (1971).

17. J. S. Anderson, M. Matsuhashi, M. A. Haskin, and J. L. Strominger, Proc. Nat. Acad. Sci. U. S., 53, 881 (1965).

18. C. A. Schnaitman, J. Bacteriol., 104, 882 (1970).

19. R. S. Criddle, R. M. Bock, D. E. Green, and H. D. Tisdale, Biochemistry, 1, 827 (1962).

20. D. Woodward and K. Munkres, Proc. Nat. Acad. Sci. U. S., 55, 872 (1966).

21. D. Oesterhelt and W. Stoeckenius, Nature New Biol., 233, 149 (1971).

22. M. Kates, Advan. Lipid Res., 2, 17 (1964).

23. M. G. MacFarlane, Advan. Lipid Res., 2, 91 (1964).

24. K. Y. Cho and M. R. J. Salton, Biochim. Biophys. Acta, 116, 73 (1966).

25. V. Riazza, A. N. Tucker, and D. C. White, J. Bacteriol., 101, 84 (1970).

26. M. J. Osborn, Ann. Rev. Biochem., 38, 501 (1969).

27. M. R. J. Salton and A. F. M. Ehtisham-Ud-Din, Australian J. Biol. Exp. Med. Sci., 43, 255 (1965).

28. D. C. White and F. E. Frerman, J. Bacteriol., 94, 1854 (1967).

29. J. L. Beebe, J. Bacteriol., 107, 704 (1971).

30. U. Houtsmuller and L. Van Deenen, Biochim. Biophys. Acta, 106, 564 (1965).

31. L. Rothfield and M. Takeshita, Ann. N. Y. Acad. Sci., 133, 384 (1966).

32. A. Endo and L. Rothfield, Biochemistry, 8, 3508 (1969).

33. L. S. Milner and H. R. Kaback, Proc. Nat. Acad. Sci. U. S., 65, 683 (1970).

34. T. Kaneda, Can. J. Microbiol., 12, 501 (1966).

35. T. Kaneda, J. Biol. Chem., 238, 1222 (1963).

36. K. Willecke and A. B. Pardee, J. Biol. Chem., 246, 1032 (1971).

37. C. W. M. Haest, J. DeGier, and L. L. M. Van Deenen, Chem. Phys. Lipids, 3, 413 (1969).

38. M. Esfahani, T. Ioneda, and S. J. Wakil, J. Biol. Chem., 246, 50 (1971).

39. T. Kaneda, Biochem. Biophys. Res. Commun., 43, 298 (1971).

40. J. F. Danielli and H. Davson, J. Cellular Comp. Physiol., 5, 495 (1935).

41. S. Rottem, O. Stein, and S. Razin, Arch. Biochem. Biophys., 125, 46 (1968).

42. H. R. Kaback, J. Biol. Chem., 243, 3711 (1968).

43. F. S. Sjöstrand, J. Ultrastruct. Res., 9, 340 (1963).

44. J. A. Lucy, J. Theoret. Biol., 7, 360 (1964).

45. A. A. Benson, J. Amer. Oil Chem. Soc., 43, 265 (1966).

46. D. E. Green and J. F. Perdue, Proc. Nat. Acad. Sci. U. S., 55, 1295 (1966).

47. D. M. Engelman, J. Mol. Biol., 47, 115 (1970).

48. M. H. F. Wilkins, A. E. Blaurock, and D. M. Engelman, Nature New Biol., 230, 72 (1971).

49. J. M. Steim, M. E. Tourtellotte, J. C. Reinert, R. N. McElhaney, and R. L. Rader, Proc. Nat. Acad. Sci. U. S., 63, 104 (1969).

50. G. B. Ashe and J. M. Steim, Biochim. Biophys. Acta, 233, 810 (1971).

51. W. L. Hubbel and H. M. McConnell, Proc. Nat. Acad. Sci. U. S., 64, 20 (1969).

52. E. Rucker, B. Chance, and D. F. Parsons, Fed. Proc. Fed. Amer. Soc. Exp. Biol., 23, 431 (1964).

53. D. Abram, J. Bacteriol., 89, 855 (1965).

54. E. C. C. Lin, Ann. Rev. Genet., 4, 225 (1970).

55. C. F. Fox, J. R. Carter, and E. P. Kennedy, Proc. Nat. Acad. Sci. U. S., 57, 698 (1967).

56. R. D. Simoni, M. F. Smith, and S. Roseman, Biochem. Biophys. Res. Commun., 31, 804 (1968).

57. D. C. White, J. Bacteriol., 96, 1159 (1968).

58. D. C. White and A. N. Tucker, J. Bacteriol., 97, 199 (1969).

59. F. E. Frerman and D. C. White, J. Bacteriol., 94, 1868 (1967).

60. J. Lascelles and J. F. Szilagyi, J. Gen. Microbiol., 38, 55 (1965).

61. M. Kohiyama, O. Cousin, A. Ryter, and F. Jacob, Ann. Inst. Pasteur, 110, 465 (1966).

62. F. M. Schwartz and N. D. Zinder, Virology, 34, 352 (1967).

63. L. Mindich, J. Mol. Biol., 49, 433 (1970).

64. C. C. Hsu and C. F. Fox, J. Bacteriol., 103, 410 (1970).

65. K. Willecke and L. Mindich, J. Bacteriol., 106, 514 (1971).

66. L. Mindich, Proc. Nat. Acad. Sci. U. S., 68, 420 (1971).

66a. H. Kung and U. Henning, Proc. Nat. Acad. Sci. U. S., 69, 925 (1972).

67. A. Tzagoloff, J. Biol. Chem., 244, 5027 (1969).

68. A. Tzagoloff, J. Biol. Chem., 246, 3050 (1971).

69. C. M. Redman and D. D. Sabatini, Proc. Nat. Acad. Sci. U.S., 56, 608 (1966).

70. M. J. Schlesinger, J. Bacteriol., 96, 727 (1968).

71. S. Razin, H. J. Morowitz, and T. Terry, Proc. Nat. Acad. Sci. U. S., 54, 219 (1965).

72. S. Razin, Z. Ne'Eman, and I. Ohad, Biochim. Biophys. Acta, 193, 277 (1969).

73. T. F. Butler, G. L. Smith, and E. A. Grula, Can. J. Microbiol., 13, 1471 (1967).

74. A. Tzagoloff and D. H. MacLennan, Biochim. Biophys. Acta, 99, 476 (1965).

75. P. K. Stumpf, Ann. Rev. Biochem., 38, 159 (1969).

76. P. W. Majerus and P. R. Vagelos, in Advances in Lipid Research (R. Paoletti and D. Kritchevsky, eds.), Vol. 5, pp. 1-33, Academic Press, New York, 1967.

77. A. W. Alberts and P. R. Vagelos, Proc. Nat. Acad. Sci. U. S., 59, 561 (1968).

78. P. Goldman and P. R. Vagelos, Biochem. Biophys. Res. Commun., 7, 414 (1962).

79. T. Kaneda, J. Biol. Chem., 238, 1229 (1963).

80. P. H. B. Butterworth and K. Bloch, Eur. J. Biochem., 12, 496 (1970).

81. L. L. M. Van Deenen, in Progress in the Chemistry of Fats and Other Lipids (R. T. Holman, ed.), Vol. 8, part 1, p. 1, Pergamon Press, New York, 1965.

82. K. Hoffman, D. B. Henis, and C. Panos, J. Biol. Chem., 228, 349 (1957).

83. D. F. Silbert and P. R. Vagelos, Proc. Nat. Acad. Sci. U. S., 58, 1579 (1967).

84. G. P. Ailhaud and P. R. Vagelos, J. Biol. Chem., 241, 3866 (1966).

85. H. Goldfine, J. Biol. Chem., 241, 3864 (1966).

86. H. Van den Bosch and P. R. Vagelos, Biochim. Biophys. Acta, 218, 233 (1970).

87. K. Hechemy and H. Goldfine, Biochem. Biophys. Res. Commun., 42, 245 (1971).

88. J. E. Cronan, Jr., T. K. Ray, and P. R. Vagelos, Proc. Nat. Acad. Sci. U. S., 65, 737 (1970).

89. H. Van den Bosch, J. R. Williamson, and P. R. Vagelos, Nature (London), 228, 338 (1970).

90. J. N. Kanfer and E. P. Kennedy, J. Biol. Chem., 238, 2919 (1963).

91. J. N. Kanfer and E. P. Kennedy, J. Biol. Chem., 239, 1720 (1964).

92. Y. Y. Chang and E. P. Kennedy, J. Lipid Res., 8, 456 (1967).

93. J. R. Carter, J. Lipid Res., 9, 748 (1968).

94. P. H. Patterson and W. J. Lennarz, J. Biol. Chem., 246, 1062 (1971).

95. D. A. White, F. R. Albright, W. J. Lennarz, and C. A. Schnaitman, Biochim. Biophys. Acta, 249, 636 (1971).

95a. C. R. H. Raetz and E. P. Kennedy, J. Biol. Chem., 247, 2008 (1972).

96. D. B. Martin and P. R. Vagelos, J. Biol. Chem., 237, 1787 (1962).

97. P. R. Vagelos, A. W. Alberts, and D. B. Martin, J. Biol. Chem., 238, 533 (1963).

98. P. W. Majerus, R. Jacobs, M. B. Smith, and H. P. Morris, J. Biol. Chem., 243, 3588 (1968).

99. S. J. Wakil, J. K. Goldman, I. P. Williamson, and R. E. Toomey, Proc. Nat. Acad. Sci. U. S., 55, 880 (1966).

100. T. O. Henderson and J. J. McNeill, Biochem. Biophys. Res. Commun., 25, 662 (1966).

101. P. R. Vagelos and A. R. Larrabee, J. Biol. Chem., 242, 1776 (1967).

102. J. Elovson and P. R. Vagelos, J. Biol. Chem., 243, 3603 (1968).

103. Y. Sakawa, E. Nakao, and Y. Kaziro, Biochem. Biophys. Res. Commun., 33, 108 (1968).

104. A. Travers, R. Kamen, and M. Cashel, Cold Spring Harb. Symp. Quant. Biol., 35, 415 (1970).

105. L. Mindich, J. Bacteriol., 110, 96 (1972).

106. G. D. Shockman, J. J. Kolb, B. Bakay, M. J. Conover, and G. Toennies, J. Bacteriol., 85, 168 (1963).

107. I. Kahane and S. Razin, Biochim. Biophys. Acta, 183, 79 (1969).

108. R. K. Rasmussen and H. P. Klein, J. Gen. Microbiol., 22, 249 (1968).

109. L. Mindich, J. Mol. Biol., 49, 415 (1970).

110. U. Henning, G. Dennert, K. Rehn, and G. Deppe, J. Bacteriol., 98, 784 (1969).

111. L. R. Kass, J. Biol. Chem., 243, 3223 (1968).

111a. D. F. Silbert, M. Cohen, and M. E. Harder, J. Biol. Chem., 247, 1699 (1972).

111b. J. Estroumza and G. Ailhaud, Biochimie, 53, 837 (1971).

112. M. Sinensky, J. Bacteriol., 106, 449 (1971).

112a. T. K. Ray, J. E. Cronan, Jr., R. D. Mavis, and P. R. Vagelos, J. Biol. Chem., 245, 6442 (1970).

113. D. F. Silbert, Biochemistry, 9, 3631 (1970).

114. P. J. Dehlinger and R. T. Schimke, J. Biol. Chem., 246, 2574 (1971).

115. P. Siekevitz, G. E. Palade, G. Dallner, I. Ohad, and T. Omura, in Organizational Biosynthesis (H. J. Vogel, J. D. Lampen, and V. Bryson, eds.), p. 331, Academic Press, New York, 1967.

116. J. Mandelstam and W. M. Waites, Biochem. J., 109, 793 (1968).

117. H. Okuyama and S. Nojima, Biochim. Biophys. Acta, 176, 120 (1969).

118. P. R. Proulx and L. L. M. Van Deenen, Biochim. Biophys. Acta, 125, 591 (1966).

119. P. R. Proulx and L. L. M. Van Deenen, Biochim. Biophys. Acta, 144, 171 (1967).

120. R. A. Pieringer and R. S. Kunnes, J. Biol. Chem., 240, 2833 (1965).

121. Y. Kanemasa, Y. Akamatsu, and S. Nojima, Biochim. Biophys. Acta, 144, 382 (1967).

122. G. F. Ames, J. Bacteriol., 95, 833 (1968).

123. T. T. Lillich and D. C. White, J. Bacteriol., 107, 790 (1971).

124. J. P. G. Ballesta and M. Schaechter, J. Bacteriol., 107, 251 (1971).

125. C. Rampini, E. Barbu and J. Polonovski, Compt. Rend., Acad. Sci., Paris, 270, 882 (1970).

126. Y. Ohnishi, J. Bacteriol., 107, 918 (1971).

127. N. Z. Stanacev, Y. Y. Chang, and E. P. Kennedy, J. Biol. Chem., 242, 3018 (1967).

128. Y. Ono and D. C. White, J. Bacteriol., 103, 111 (1970).

129. Y. Ono and D. C. White, J. Bacteriol., 104, 713 (1970).

130. D. C. White and A. N. Tucker, J. Lipid Res., 10, 220 (1969).

131. P. Overath, F. F. Hill, and I. Lamnek-Hirsch, Nature New Biol., 234, 264 (1971).

132. C. F. Fox, Proc. Nat. Acad. Sci. U. S., 63, 850 (1969).

133. H. U. Schairer and P. Overath, J. Mol. Biol., 44, 209 (1969).

134. G. Wilson, S. P. Rose, and C. F. Fox, Biochem. Biophys. Res. Commun., 38, 617 (1970).

135. P. Overath, H. U. Schairer, and W. Stoffel, Proc. Nat. Acad. Sci. U. S., 67, 606 (1970).

136. G. Wilson and C. F. Fox, J. Mol. Biol., 55, 49 (1971).

137. C. F. Fox, Fed. Proc. Fed. Amer. Soc. Exp. Biol., 30, 1032 (1971).

138. S. Dales and E. H. Mosbach, Virology, 35, 564 (1968).

139. P. Fitz-James, J. Biophys. Biochem. Cytol., 9, 507 (1960).

140. A. Ryter, Bacteriol. Rev., 32, 39 (1968).

141. C. T. Patch and O. E. Landman, J. Bacteriol., 107, 345 (1971).

142. B. Ferrendes, C. Frehel, and P. Chaix, Biochim. Biophys. Acta, 223, 292 (1970).

143. D. A. Reaveley, Biophys. Biochem. Res. Commun., 30, 649 (1968).

144. R. T. Espejo and E. S. Canelo, Virology, 34, 738 (1968).

145. A. L. Tuttle and H. Gest, Proc. Nat. Acad. Sci. U. S., 45, 1261 (1959).

146. K. D. Gibson, Biochemistry, 4, 2027 (1965).

147. A. Gorschein, Proc. Roy. Soc. B., 170, 255 (1968).

148. A. E. Blaurock and W. Stoeckenius, Nature New Biol., 233, 152 (1971).

149. H.-D. Klenk and P. W. Choppin, Virology, 38, 255 (1969).

150. F. Jacob, S. Brenner, and F. Cuzin, Cold Spring Harb. Symp. Quant. Biol., 28, 329 (1963).

151. W. D. Donachie and K. J. Begg, Nature (London), 227, 1220 (1970).

152. D. F. Ohye and W. G. Murrel, J. Cell Biol., 14, 111 (1962).

152a. P. C. Fitz James, J. Biophys. Biochem. Cytol., 8, 507 (1960).

153. T. Yamamoto and G. Balassa, Molec. Gen. Genetics, 106, 1 (1969).

154. P. Fitz-James, in Microbial Protoplasts, Spheroplasts and L-forms (L. B. Guze, ed.), pp. 124-143, The Williams & Wilkins Co., Baltimore, 1968.

155. D. C. Morrison and H. J. Morowitz, J. Mol. Biol., 49, 441 (1970).

156. L. Mindich and S. Dales, J. Cell Biol., 55, 32 (1972).

157. E. C. C. Lin, Y. Hirota, and F. Jacob, J. Bacteriol., 108, 375 (1971).

158. N. Tsukagoshi and C. F. Fox, Fed. Proc. Fed. Amer. Soc. Exp. Biol., 30, 1120 (1971).

159. C. F. Fox, personal communication, 1971.

160. G. Wilson and C. F. Fox, Biochem. Biophys. Res. Commun., 44, 503 (1971).

161. K. W. A. Wirtz and D. B. Zilversmit, FEBS Letters, 7, 44 (1970).

162. W. M. McMurray and R. M. C. Dawson, Biochem. J., 112, 91 (1969).

163. R. D. Kornberg and H. M. McConnell, Proc. Nat. Acad. Sci. U. S., 68, 2564 (1971).

163a. C. J. Scandella, P. Devaux, and H. M. McConnell, Proc. Nat. Acad. Sci. U. S., 69, 2056 (1972).

164. L. D. Frye and M. Edidin, J. Cell Sci., 7, 319 (1970).

165. L. G. Caro and R. P. van Tubergen, J. Cell Biol., 15, 173 (1962).

166. A. Leskes, P. Siekevitz, and G. E. Palade, J. Cell Biol., 49, 264 (1971).

167. A. Leskes, P. Siekevitz, and G. E. Palade, J. Cell Biol., 49, 288 (1971).

168. I. Goldberg and I. Ohad, J. Cell Biol., 44, 572 (1970).

Chapter 2

BIOSYNTHESIS OF PEPTIDOGLYCAN

Jean-Marie Ghuysen and Gerald D. Shockman*

Service de Microbiologie
Faculté de Médecine
Institut de Botanique
Université de Liège
Liège, Belgium

---

*On leave from Department of Microbiology and Immunology, Temple University School of Medicine, Philadelphia, Pennsylvania and "Chaire d'Actualité Scientifique 1971-1972," Faculté de Médecine, Université de Liège, Liège, Belgium.

I. INTRODUCTION

Modern study of the main supporting structure of the bacterial cell started in the 1950's when the external surface structure, i.e., the cell wall, was first isolated from mechanically disrupted bacteria [1]. Walls isolated from cocci were found to be spherical, whereas walls isolated from bacilli were found to be cylindrical. It was soon discovered that the shape and tensile strength of the wall was imparted by a polymer, which, depending upon the bacterial species, represented 5% to perhaps 90% of the isolated walls. In marked contrast with other cell-supporting exostructures that consist of α-cellulose, hemicellulose, glucan, mannan, or chitin, the rigid matrix of the wall was characterized as an heteropolymer consistently composed of two different acetamido sugars and of from four to perhaps as many as eight different amino acid residues. This type of heteropolymer appears to be ubiquitous in the prokaryotic world except for *Mycoplasma* and some L-forms and halophiles. The two acetamido sugars were identified as 2-acetamido-2-deoxy-D-glucose (i.e., N-acetyl-D-glucosamine) and a previously unknown sugar 2-acetamido-2-deoxy-3-O-(D-1-carboxyethyl)-D-glucose (i.e., the 3-O-D-lactic acid ether of N-acetyl-D-glucosamine or N-acetylmuramic acid). In addition to these two acetamido sugars, L-alanine, D-alanine, and D-glutamic acid were always found to be present. There were also other amino acid residues, most often L-lysine or *meso*-diaminopimelic acid. Muramic acid, D-alanine, D-glutamic acid, and *meso*-diaminopimelic acid have been found only in prokaryotic cells. The occurrence of these compounds was thus a very striking feature that showed the uniqueness of bacterial walls. It was also observed that N-acetylglucosamine, N-acetylmuramic acid, L-alanine, D-alanine, D-glutamic acid, and either L-lysine or *meso*-diaminopimelic acid usually occurred in approximately equimolar amounts. This observation led to the conclusion that the assembly of the rigid part of the wall is probably brought about by polymerization of disaccharide-peptide units. The polymer was called peptidoglycan (glycopeptide, mucopeptide, glycosaminopeptide, or murein being synonymous to peptidoglycan).

The essential role played by the peptidoglycan in keeping the cell alive under ordinary hypotonic environmental conditions was established by several techniques. Both hen egg-white lysozyme, when added to resting bacteria, or in some cases to growing bacteria, and penicillin, when added to growing bacteria, were shown to result in dissolution of the cells. Lysis, however, could be prevented at least for a limited period of time, if the external medium contained a solute to which the cell was impermeable, at a concentration that approximately balanced the high osmotic pressure of the cell [2]. Under these conditions, bacteria were seen to undergo transformation into spherical bodies. These physiologically active but osmotically fragile bodies were called protoplasts or spheroplasts [2a, 2b]. Essentially, they are either wall-less bacteria (i.e., protoplasts) or bacteria with some defect in their wall peptidoglycan component (i.e., spheroplasts). The spherical appearance and osmotic fragility of protoplasts and spheroplasts generally has been taken as evidence for the shape-maintaining and osmotically protective nature of the peptidoglycan. The targets of lysozyme and penicillin were thus seemingly the same or similar. In fact, although the overall target, the peptidoglycan, is the same, the mechanism of action of each is entirely different. Lysozyme enzymatically hydrolyzes insoluble peptidoglycans into soluble fragments. Penicillin acts as an inhibitor of the synthesis of insoluble and shape-maintaining peptidoglycan.

Studies of penicillin action on growing cultures of _Staphylococcus aureus_ revealed that the intoxicated bacteria accumulate large amounts of a series of compounds containing uridine diphosphate sugars. The largest of these was later shown [3] to contain muramic acid, L-alanine, D-glutamic acid, L-lysine, and D-alanine in the molar ratio 1:1:1:2. Similar compounds were found in low concentration in many bacterial species. Because of their similarity in composition, these nucleotides appeared to be precursors of the peptidoglycan. Consequently, penicillin was thought to specifically inhibit the biosynthesis of the wall peptidoglycan.

An exhaustive survey of these early studies of the wall was presented by Salton in 1964 [1]. These studies prepared the way for future investigations dealing with the topology of the wall peptidoglycan in the cell envelope, its chemical structure, the reaction sequence and control sites involved in its biosynthesis and mode of growth, the functioning of the membrane in the process and the mechanism of action of several antibiotics. At present, many intriguing and fundamental questions are not yet satisfactorily answered. A number of monographs and recent reviews summarize much of this information [1, 4-14a].

II. STRUCTURE OF PEPTIDOGLYCAN

A. Primary Structure of Peptidoglycan

The wall peptidoglycan can be considered to be a single, enormous macromolecule that forms a more-or-less continuous network around the cellular permeability barrier and provides the cell with a supporting structure of high tensile strength [1, 4, 5, 8, 11]. Basically, the network is composed of glycan strands that are interconnected through peptide chains (Figs. 1 and 2). Despite many variations, there is a remarkable consistency of structure of the wall peptidoglycans throughout the bacterial world.

1. The Glycan Strands

Basically, the glycan moiety consists of linear strands of alternating β-1,4-linked pyranoside-N-acetylglucosamine and N-acetylmuramic acid residues [15, 16] (Fig. 3). It is thus a chitin-like structure in which each alternate N-acetylglucosamine residue is ether-linked at C-3 to a lactyl group that has the D-configuration [17]. The carboxyl of the lactyl groups provides the point to which peptides are amide-linked to the glycan strands. Several variations of the glycan chains are known (Fig. 4). (1) In several Gram-positive bacteria [5] and in at least one Gram-negative species, Proteus vulgaris [18], some of the N-acetylmuramic acid residues are O-acetylated on C-6. (2) In many Gram-positive bacteria, the C-6 of some N-acetylmuramic acid residues is substituted by phosphodiester groups, which covalently link other wall polymers to the peptidoglycan matrix [19, 20, 21]. (3) Muramic acid can occur as N-glycolylmuramic acid, a structural feature, which thus far has been found to be unique to Nocardia and Mycobacterium sp. [22]. (4) In spores of Bacillus subtilis, a high proportion of the muramic acid residues occur in the form of a lactam derivative [23]. The same derivative has been found in trace amounts in Micrococcus lysodeikticus [24]. (5) A recent survey including species of Gram-positive and Gram-negative bacteria from a wide variety of taxonomic groups failed to detect the presence of galactosamine and galactomuramic acid in peptidoglycans [25, 26]. Only glucosamine and gluco-muramic acid were identified in the bacterial species examined. Small amounts of manno-muramic acid, however, were reported to occur along with gluco-muramic acid in Micrococcus lysodeikticus [24]. None of these variations would significantly affect the three-dimensional organization of the glycan chains [see Section II, D below].

2. The Tetrapeptide Units [5]

The D-lactyl groups of the muramic acid residues in the glycan strands, or at least some of them, are substituted by tetrapeptide units. These units have the general sequence L-alanyl-γ-D-glutamyl-L-R_3-D-alanine (Fig. 5). Except for the bond between the D-glutamyl residue and the L-R_3 residue, which is γ-linked, all peptide linkages are α. The L-R_3 residue varies. For example, it may be a neutral amino acid, such

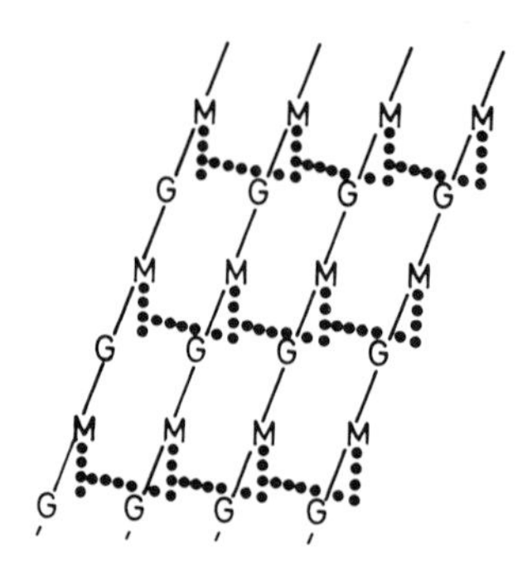

FIG. 1. Schematic representation of a wall peptidoglycan. Glycan chains are composed of N-acetylglucosamine (G) and N-acetylmuramic acid (M). Vertical dots from M represent the amino acids of the tetrapeptide subunits. Horizontal dots represent the peptide cross-linking bridges. Five bridging amino acids are shown, corresponding to the peptide bridges of Staphylococcus aureus presented in Fig. 2. [Reprinted from reference 9 by courtesy of American Elsevier Publishing Co., New York.]

as L-alanine or L-homoserine, or a dicarboxylic amino acid such as L-glutamic acid [26a], or a diamino acid such as L-2, 4-diaminobutyric acid, L-ornithine, L-lysine, L-hydroxylysine, LL-diaminopimelic acid, or meso-diaminopimelic acid. When meso-diaminopimelic acid is at the R_3 position, both the amino group, which is linked to D-glutamic acid, and the carboxyl group, which is linked to D-alanine, are located on the same L-carbon. Variations other than those occurring at the L-R_3 position are known. (1) The α-carboxyl group of D-glutamic acid can be either free, amidated, substituted by a C-terminal glycine or by a glycine amide [26b]. (2) The carboxyl group of diaminopimelic acid not engaged in a peptide bond may be substituted by an amide. (3) threo-3-Hydroxylglutamic acid [26a] can occur instead of glutamic acid. (4) L-Alanine at the N-terminus of the tetrapeptide can be replaced by L-serine or glycine. The aforementioned variations, however, fail to significantly alter the basic structure of the tetrapeptide backbone. Except for the occasional appearance of glycine at the N-terminal position, the backbone of all tetrapeptides exhibits an L-D-L-D sequence.

3. The Interpeptide Bridges

The peptide units belonging to adjacent glycan strands are, in turn, cross-linked through specialized bridges. The C-terminal D-alanine residue of one tetrapeptide unit is always involved in the bridging to a

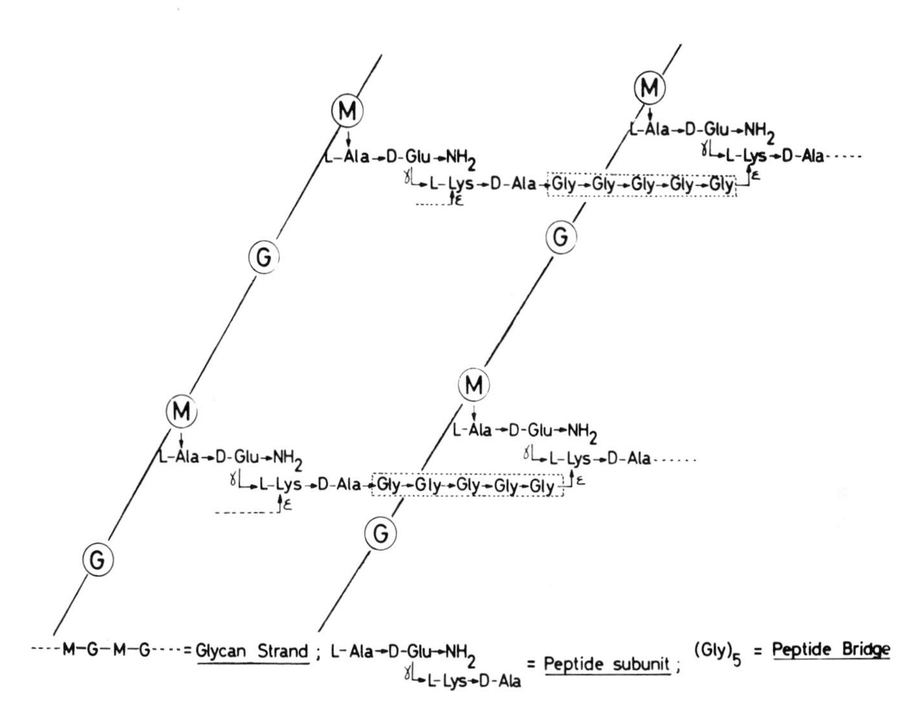

FIG. 2. Primary structure of the wall peptidoglycan in Staphylococcus aureus. In this and all subsequent figures, G = N-acetylglucosamine and M = N-acetylmuramic acid. Arrows indicate the CO→NH direction of the linkages. Usual α-peptide bonds are represented by horizontal arrows; other peptide bonds (e.g., γ or ϵ) are also indicated. The pentaglycine bridges, which extend from the ϵ-amino group of L-lysine on one peptide subunit to the carboxyl group of D-alanine on another, are enclosed by a dashed rectangle.

FIG. 3. A portion of a glycan strand. In the peptidoglycan network, the COOH of the D-lactyl groups is usually peptide substituted.

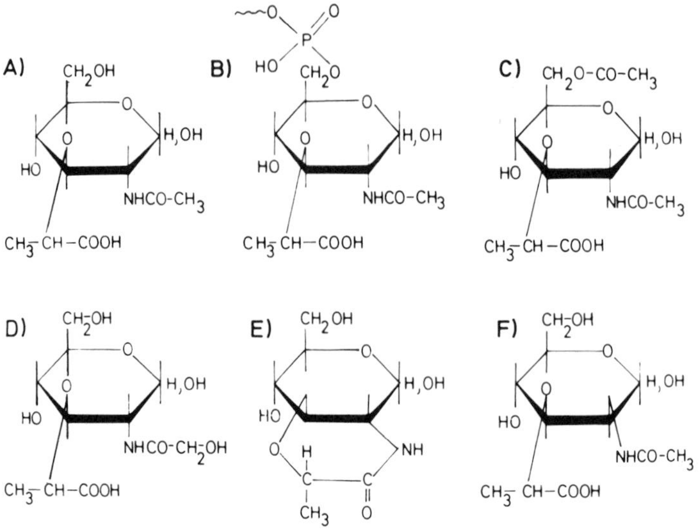

FIG. 4. Known structural variations of the N-acetylmuramic acid residue in glycan chains of various bacterial species. (A) N-acetylmuramic acid; (B) N-acetylmuramic acid with the C-6 substituted by a phosphodiester group; (C) N-acetylmuramic acid with the C-6 substituted by an acetyl group; (D) N-glycolylmuramic acid; (E) muramic lactam; (F) N-acetyl-*manno*-muramic acid.

second tetrapeptide unit. Considerable species variations in composition of the bridges have been observed. The amino acid composition and the location of the bridges have been used to divide bacterial species into chemotypes [5]. This has been considered to be a criterion of taxonomic importance [11, 27-29a].

In chemotypes I, II, and III, the bridges extend from the C-terminal D-alanine of one peptide to the ω-amino group of the diamino acid at the L-R_3 position of another peptide. In chemotype I, the bridging consists of direct N^{ω}-(D-alanyl)-L-R_3 peptide bonds. In many *Bacillaceae* and gram-negative bacteria, the interpeptide bond is a D-alanyl-(D)-*meso*-diaminopimelic acid linkage [5, 30] (Fig. 6). In *Aerococcus* sp. and *Gaffkya homari*, the interpeptide bond is a N^{ϵ}-(D-alanyl)-L-lysine linkage [31] (Fig. 7).

In chemotype II, the bridging is mediated via a single additional amino acid or an intervening short peptide. Variations appear to be endless, only a few examples of which are given in Figs. 8 and 9 [32-41].

```
     (L)          (D)
NH2-CH-CONH-CH-COOH
     |            |
     CH3          CH2
                  |
                  CH2    (L)         (D)
                  |
                  CONH-CH-CONH-CH-COOH
                       |           |
                       X           CH3
```

X	
$-CH_3$	L-Alanine
$-CH_2-CH_2OH$	L-homoSerine
$-CH_2-CH_2-NH_2$	L-diaminobutyric acid
$-CH_2-CH_2-COOH$	L-glutamic acid
$-(CH_2)_2-CH_2-NH_2$	L-Ornithine
$-(CH_2)_3-CH_2-NH_2$	L-Lysine
$-(CH_2)_3-CH(COOH)(NH_2)$ (L)	LL-DAP
$-(CH_2)_3-CH(COOH)(NH_2)$ (D)	meso-DAP

FIG. 5. General structure of tetrapeptide L-alanyl-γ-D-glutamyl-L-R_3-D-alanine subunits. Side chains of amino acids known to occur in the L-R_3 position are shown. DAP = diaminopimelic acid.

In chemotype III, the bridge is composed of one or several peptides, each having the same amino acid sequence as the peptide unit [Fig. 10]. This bridge occurs in M. lysodeikticus and related Micrococcaceae [42-44]. Seemingly, chemotype III is a variant of chemotype II. In fact, this bridge is quite unusual. In M. lysodeikticus many of the N-acetylmuramic acid residues in the glycan strands are not substituted by peptides, as if the peptide units had moved, at a certain stage of the biosynthesis, from these N-acetylmuramic acid residues into a bridging position [5, 42, 43]. The peptidoglycans of other Micrococci belonging to this chemotype III contain more peptide units than disaccharide units, but all of the N-acetylmuramic acid residues are substituted by peptides. It has been hypothesized that after translocation of some peptide units from the glycan to a bridging position, the unsubstituted segments of the glycans were, in turn, removed by excision [44]. Both D-alanyl-L-alanine and N^{ϵ}-(D-alanyl)-L-lysine linkages, sensitive to Myxobacter ALI enzyme and ML endopeptidase, respectively, are involved in the interpeptide bonding in all chemotype III peptidoglycans.

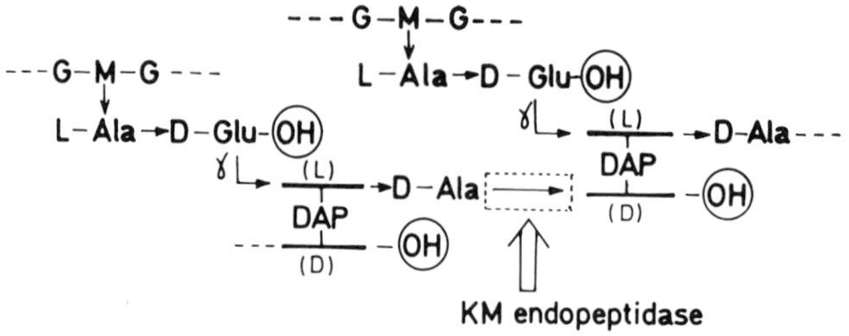

FIG. 6. Peptidoglycan of chemotype I with *meso*-DAP in the L-R_3 position. This structure occurs in the wall of *E. coli*, of probably all other gram-negative bacteria, and of *Bacillus megaterium* KM. Note that the interpeptide linkage, D-alanyl-(D)-*meso*-diaminopimelic acid is at a C-terminal position. The arrow indicates the site of action of the lytic KM endopeptidase (DD-carboxypeptidase) from *Streptomyces* strain *albus* G. In *E. coli* carboxyl groups are not amidated. Peptidoglycan of *Streptomyces* strain R39 has the same primary structure except that the α-carboxyl group of D-glutamic acid is amidated. In some *Bacillaceae*, the α-carboxyl group of D-glutamic acid and/or the carboxyl group of diaminopimelic acid that is not in a peptide bond are also amidated.

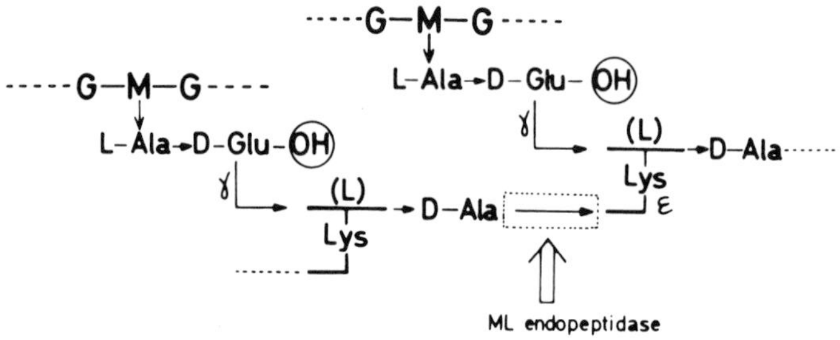

FIG. 7. Peptidoglycan of chemotype I with L-lysine in the L-R_3 position. This structure occurs in the wall of *Aerococcus* sp. and *Gaffkya homari*. Arrow indicates the site of action of the lytic ML endopeptidase from *Streptomyces albus* G.

In chemotype IV, the bridge extends from the C-terminal D-alanine residue of one peptide unit to the α-carboxyl group of D-glutamic acid of another peptide, i.e., between two carboxyl groups. Hence, it necessarily involves a diamino acid residue or a diamino acid-containing peptide. Examples are given in Figs. 11-13 [45, 46, 26a]. This type of bridging occurs between tetrapeptide units containing a neutral amino acid (Fig. 11), a diamino acid (Fig. 12), or a dicarboxylic amino acid (Fig. 13) at the L-R_3 position.

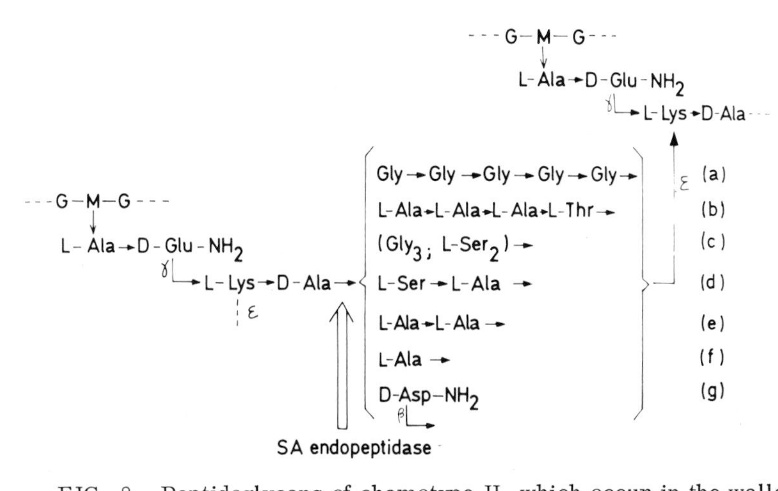

FIG. 8. Peptidoglycans of chemotype II, which occur in the walls of (a) *Staphylococcus aureus* Copenhagen [34, 39-41] ; (b) *Micrococcus roseus* R27 [41] ; (c) *Staphylococcus epidermidis* Texas 26 [33, 34] ; (d) *Lactobacillus viridescens* [38] ; (e) *Streptococcus pyogenes* Group A, type 14 [41] ; (f) *Arthrobacter crystallopoietes* [32, 32a] ; (g) *Streptococcus faecalis* (faecium) ATCC 9790 [35]; and *Lactobacillus casei* RO94 [36]. Arrow indicates the site of action of the lytic SA endopeptidase from *Streptomyces albus* G upon walls (a), (b), (e), and (g). This enzyme has not been tested upon walls (c), (d) and (f).

FIG. 9. Peptidoglycan of chemotype II, which occurs in the walls of *Clostridium perfringens* type A [27], and of *Streptomyces* strains *albus* G, R61, and K11. Arrows indicate the sites of action of *Myxobacter* ALI enzyme. When acting on walls of *C. perfringens*, *Myxobacter* enzyme hydrolyzes both (a) and (b) linkages. When acting on walls of *Streptomyces* sp., *Myxobacter* enzyme hydrolyzes only (a) but not (b) linkages.

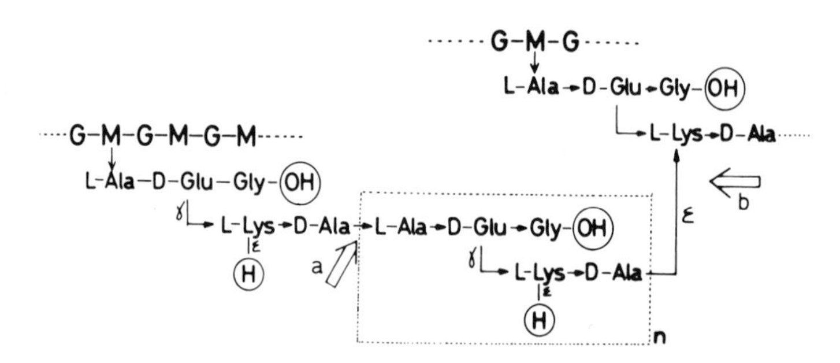

FIG. 10. Peptidoglycan of chemotype III, which occurs in the wall of Micrococcus lysodeikticus. The site of action of Myxobacter ALI enzyme is indicated by arrow a and that of the ML endopeptidase from Streptomyces albus G by arrow b. Note the similarity of action of the latter enzyme upon peptidoglycans of different chemotypes by comparing its action here with that on a peptidoglycan of chemotype I shown in Fig. 7. Also shown are the unsubstituted N-acetylmuramic acid residues that have been found in the wall of M. lysodeikticus, but not in some of the other peptidoglycans of chemotype III.

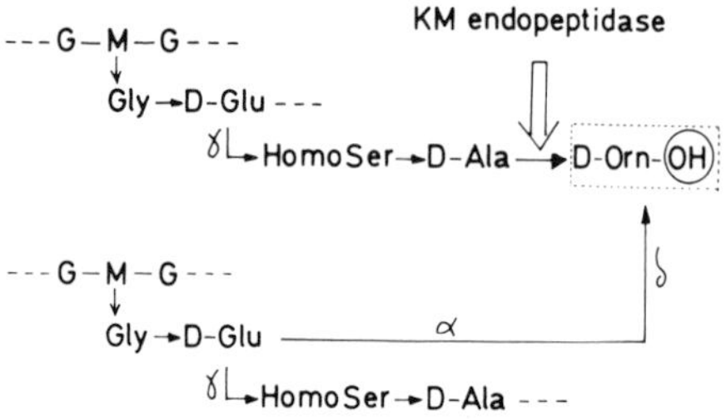

KM endopeptidase
---G-M-G---
Gly-D-Glu ---
γ HomoSer-D-Ala-D-Orn-OH
δ
---G-M-G---
Gly-D-Glu α
γ HomoSer-D-Ala ---

FIG. 11. Peptidoglycan of chemotype IV, which occurs in the walls of some Corynebacteria pathogenic for plants [45]. Arrow indicates the site of action of the KM endopeptidase from Streptomyces albus G (i.e., a lytic DD-carboxypeptidase; cf. Fig. 6).

B. Enzymatic Degradation of Peptidoglycan

The unraveling of the primary structure of the wall peptidoglycan by using nonspecific acidic or basic hydrolyses would have been an impossible task for two reasons. First, this enormous molecule, in contrast with other natural polymers that contain amino acids such as proteins, is built from a limited number of constituents. Second, other polymeric wall

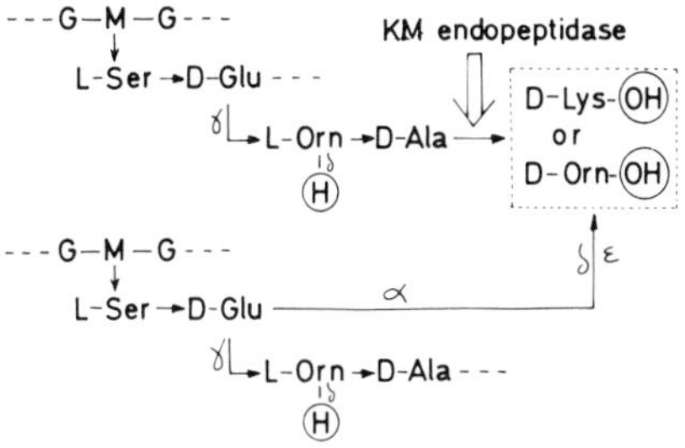

FIG. 12. Peptidoglycan of chemotype IV, which occurs in the wall of Butyribacterium rettgeri (46). Arrow indicates the site of action of the KM endopeptidase from Streptomyces albus G (i.e., a DD-carboxypeptidase that is lytic for walls; cf. Figs. 6 and 11).

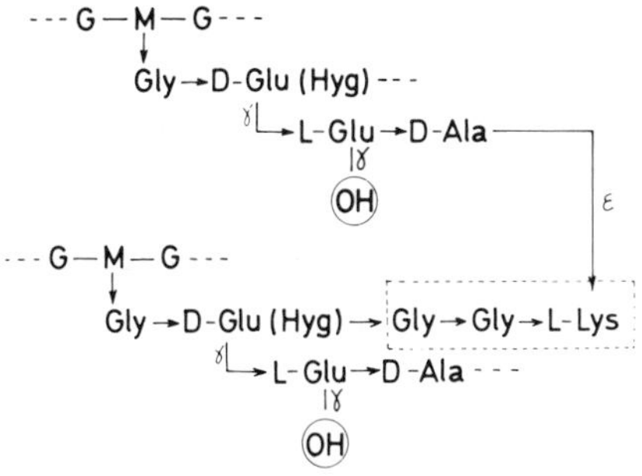

FIG. 13. Peptidoglycan of chemotype IV, which occurs in the wall of Arthrobacter J39 (26a). Hyg = threo-3-hydroxyglutamic acid.

components are often covalently linked to the peptidoglycan, and the resulting association exhibits an exceedingly high degree of complexity. The isolation of a collection of lytic agents selectively active upon different, well-defined linkages within the peptidoglycan was thus essential. Hen egg-white lysozyme, which had been discovered in the 1920's by Fleming, was such a tool. Lysozyme hydrolyzes β-1,4-linkages between N-acetylmuramic acid and N-acetylglucosamine in the glycan strands (Fig. 14, arrow a). The action of lysozyme on walls of M. lysodeikticus, which contain a peptidoglycan of chemotype III (Fig. 10), allowed the isolation and characterization of the disaccharide N-acetylglucosaminyl-N-acetylmuramic acid and of higher oligosaccharides. The choice of M. lysodeikticus for these early studies was rather fortunate. For several reasons, commercially available hen egg-white lysozyme proved to be of extremely limited usefulness for the determination of the chemical

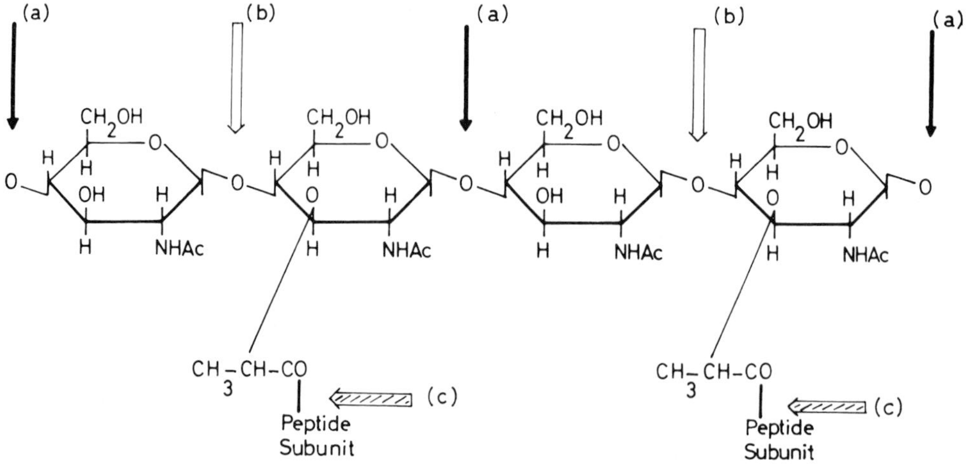

FIG. 14. Sites of action of endo-N-acetylmuramidases at arrows (a), of endo-N-acetylglucosaminidases at arrows (b), and of N-acetylmuramyl-L-alanine amidases at arrows (c). Note that the Streptomyces amidase hydrolyzes N-acetylmuramyl-L-alanine linkages (in many bacterial walls), N-acetylmuramyl-L-serine linkages (in B. rettgeri, cf. Fig. 12), but not N-acetylmuramylglycine linkages (in C. poinsettiae, cf. Fig. 11). Reprinted from reference 5 by courtesy of American Society for Microbiology.

structure of other peptidoglycans. In most cases the degradation products after lysozyme action were highly complex. This was due to a number of factors that include (1) the fact that lysozyme action fails to result in the complete hydrolysis of all sensitive bonds. This appears to be due to both inhibition of the reaction by products and to the transglycosidase activity of lysozyme [47-49] ; (2) inhibition of hydrolysis by the presence of N, O-diacetylmuramic acid residues in the substrate; and (3) since most peptidoglycans contain peptide substituents on virtually all of their N-acetylmuramic acid residues, the action of a glycosidase alone cannot release unsubstituted di- or oligosaccharides.

Therefore a search for other lytic agents was undertaken [50, 51], which resulted in the discovery of three classes of lytic enzymes: glycosidases, N-acetylmuramyl-L-alanine amidases, and endopeptidases. The use of such enzymes in the determination of the structure of the wall peptidoglycans has been reviewed and discussed [5].

i. The glycosidases are either endo-N-acetylmuramidases or endo-N-acetylglucosaminidases (Fig. 14). Endo-N-acetylmuramidases, such as lysozyme, hydrolyze β-1,4-N-acetylmuramyl-N-acetylglucosamine linkages, i.e., they produce fragments with N-acetylmuramic acid at the reducing end. Endo-N-acetylglucosaminidases hydrolyze β-1,4-N-acetylglucosaminyl-N-acetylmuramic acid linkages, i.e., they produce fragments with N-acetylglucosamine at the reducing end. Some of these newly discovered glycosidases exhibit lytic spectra broader than that of lysozyme, and they are not inhibited by the presence of O-acetyl groups on the glycan chains.

ii. N-acetylmuramyl-L-alanine amidases specifically hydrolyze the linkage between the carboxyl of the D-lactyl group of N-acetylmuramic acid and the amino group of the L-alanine residue at the amino terminus of the peptide unit (Fig. 14).

iii. Endopeptidases are of two types: Some hydrolyze peptide linkages in the interior of the peptide bridges of chemotype II peptidoglycans, whereas others specifically hydrolyze those peptide bonds that involve the C-terminal D-alanine residue of the peptide units. Figures 6-12 show the sites of action of some enzymes that hydrolyze D-alanine peptide bonds.

1. When acting on peptidoglycans of chemotype II (Fig. 8) the SA endopeptidase of *Streptomyces albus* G hydrolyzes D-alanyl-glycine, D-alanyl-L-alanine or D-alanyl-D-isoasparaginyl linkages, i.e., bonds located at the N-termini of the peptide bridges and at the C-termini of the peptide side-chain units.

2. The ML endopeptidase from the same strain of *Streptomyces* specifically acts on N^{ϵ}-(D-alanyl)-L-lysine linkages. This enzyme has been used for degrading peptidoglycans of chemotype I from *Aerococcus* sp. and *G. homari* (Fig. 7), as well as peptidoglycans of chemotype III from *M. lysodeikticus* and other *Micrococcaceae* (Fig. 10).

3. The *Myxobacter* ALI enzyme [52] hydrolyzes the D-alanyl-L-alanine linkages, which also occur as interpeptide bonds in peptidoglycans of chemotype III (Fig. 10) and the D-alanyl-glycyl-LL-diaminopimelyl sequence in walls of *C. perfringens* (Fig. 9).

4. The KM endopeptidase, also from *Streptomyces albus* G, hydrolyzes D-alanyl-(D)-*meso*-diaminopimelic acid linkages in peptidoglycans of chemotype I (Fig. 6), as well as other N^{α}-(D-alanyl)-D linkages in peptidoglycans of chemotype IV (Figs. 11 and 12). In such peptidoglycans these bonds occur simultaneously in both an internal and C-terminal position. The KM endopeptidase that hydrolyzes these bonds is in fact a D-alanyl-D-carboxypeptidase.

Figure 15 shows, as an example, how walls of *S. aureus* were enzymatically degraded in a controlled manner into small fragments. The process essentially involved four sequential steps [41].

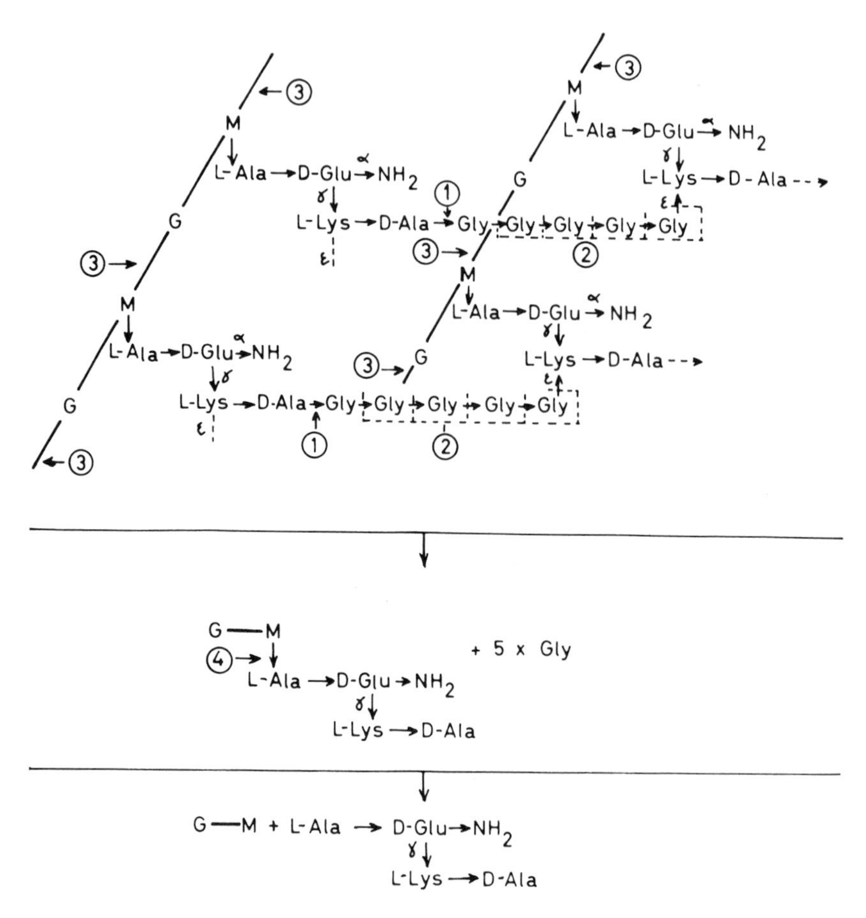

FIG. 15. Degradation sequence for the peptidoglycan of Staphylococcus aureus Copenhagen. Numbered arrows indicate the (1) site of action of SA endopeptidase; (2) degradation of the opened bridges with aminopeptidase; (3) site of action of endo-N-acetylmuramidase; (4) site of action of N-acetylmuramyl-L-alanine amidase.

1. Hydrolysis of the D-alanylglycine linkages by the SA endopeptidase, i.e., the opening of the peptide bridges at their N-termini. As a consequence of such hydrolysis, walls underwent dissolution.

2. Degradation of the opened bridges by aminopeptidase. The glycine residues were sequentially liberated and eventually, the ε-amino groups of L-lysine in the peptide units were exposed.

3. Degradation of the glycan strands into disaccharide units by means of a glycosidase (here an endo-N-acetylmuramidase). At this stage, the disaccharide peptide units were readily isolated and separated from the nonpeptidoglycan polymer of the walls. In S. aureus, this latter polymer is a teichoic acid. It was recovered undegraded, with small peptidoglycan fragments covalently attached to it through phosphodiester bridges [20].

4. Hydrolysis of the disaccharide peptide units into free disaccharides and free peptides, with the help of an N-acetylmuramyl-L-alanine amidase. Disaccharides and peptides were then isolated and characterized.

In the above procedure, an endo-N-acetylmuramidase was used at step 3, yielding β-1,4-N-acetylglucosaminyl-N-acetylmuramic acid disaccharide peptide units. (In the case of S. aureus, some of these disaccharide peptide units contained O-acetyl groups on C-6 of muramic acid). An alternate degradation procedure [53, 54] used an endo-N-acetylglucosaminidase, instead of an endo-N-acetylmuramidase, to degrade the glycan strands, yielding the isomeric β-1,4-N-acetylmuramyl-N-acetylglucosamine peptide disaccharides. Thus, a few enzymes of different specificities were sufficient to quantitatively dismantle the S. aureus peptidoglycan network into its building blocks: the isomeric disaccharides, the tetrapeptide units, and glycine residues originating from the interpeptide bridges. Similar enzymatic degradation techniques were applied to walls of all chemotypes [5]. More recently, partial acid hydrolysis techniques were developed by Kandler and his colleagues [27-29, 38, 42]. Based on the knowledge that all peptidoglycans are built according to the same basic framework, these techniques permitted an exhaustive survey of the peptidoglycan structure in many species, representing a wide variety of taxonomic groups.

C. Size of Peptidoglycans

Solubilization of the wall peptidoglycan can be brought about by the hydrolysis of any single type of linkage, but is achieved only if a sufficient number of bonds are actually hydrolyzed. This property is, of course, imparted by the netlike structure of the peptidoglycan. The organization of the peptidoglycan was initially thought to involve very long glycan strands that extended completely around the cell [8]. Accurate analyses, however, showed that, on the contrary, many terminal groups are present in both the glycan and the peptide moieties of the wall peptidoglycans. These terminal groups presumably reflect the dynamics of the bacterial growth and the active involvement of autolysins (see Sections IV, D and VI, F).

The average length of glycan chains have been estimated for walls of a few species [4, 32, 34, 55, 56]. They vary from 20 to 100 N-acetylhexosamine residues. More detailed analyses showed that the glycan chains are often, if not always, polydisperse. At present, there is no way of determining the actual distribution of chains of different lengths in the different parts of the wall. In several cases, N-acetylglucosamine has been identified at the reducing end of the glycan chains [57-59], suggesting the presence of endo-N-acetylglucosaminidase activity in the autolytic system. There does not appear to be a relationship between cell shape and average glycan chain length.

The average degree of polymerization of the peptide moiety has also been estimated. Escherichia coli has a "loose" peptidoglycan (Fig. 16) [8, 30]. About 50% of the peptide units occur as uncross-linked monomers and the other units as peptide dimers. Peptide oligomers were not detected. In walls from exponential-phase cultures of Lactobacillus acidophilus 63 AM Gasser [60], the average size of the peptide moiety was found to be 2.3 cross-linked peptides with 10% of the peptide units occurring as monomers, 37% as dimers, and 30% as trimers (Fig. 17). One of the most cross-linked peptidoglycans is that found in S. aureus (Figs. 1 and 2). Even in this case, the average size does not exceed 10 cross-linked peptide units [33]. Other examples can be found in reference 11. In S. aureus [39] and L. acidophilus [60], the peptides at the uncross-linked C-termini are pentapeptides (and not tetrapeptides) and end in a D-alanyl-D-alanine sequence (Fig. 17), a structural feature also found in nucleotide peptidoglycan precursors of all bacteria. In most bacteria, however, the uncross-linked C-termini of the peptides have not retained the D-alanine-D-alanine sequence. One of the D-alanine residues, or even both of them, are absent, presumably removed by a carboxypeptidase activity.

The tightness of the peptidoglycan network depends not only upon the frequency with which the peptide units are interlinked, but also upon the frequency with which the glycan chains are peptide substituted. The peptidoglycan of M. lysodeikticus is another example of a "loose" peptidoglycan (Fig. 10) [43]. It is unusual in that many N-acetylmuramic acid residues have unsubstituted D-lactyl groups (see Section II, A).

D. Three-Dimensional Organization of Peptidoglycan

The three-dimensional organization of peptidoglycan is unknown. Molecular models that suggest possible conformations have been constructed. The glycan chains of all of these models are based on that of chitin [61]. Chitin has a linear conformation that is stabilized by hydrogen bonding between the C-3-hydroxyl and the C-1 to C-5 ring oxygen

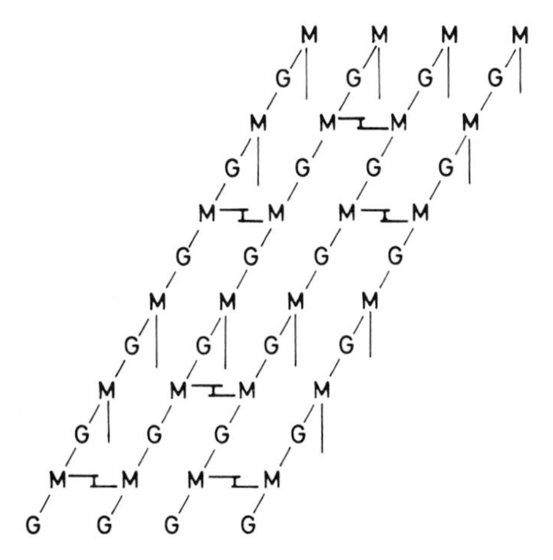

FIG. 16. Schematic representation of the wall peptidoglycan of E. coli. In this relatively loose network, all of the N-acetylmuramic acid residues are substituted, either by uncross-linked peptide monomers or by cross-linking peptide dimers (Fig. 6). Peptide oligomers larger than dimers have not been observed. Reprinted from reference 5 by courtesy of American Society for Microbiology.

of adjacent N-acetylglucosamine residues. As proposed by Tipper [11], a similar conformation is obtained for the glycan strands of the peptidoglycan (Fig. 18), if hydrogen bonding between one N-acetylglucosamine and its two adjacent N-acetylmuramic acids occur through the C-3-hydroxyl of N-acetylglucosamine and the C-1 to C-5 ring oxygen of one N-acetylmuramic acid (as in chitin), and through the C-6-hydroxyl of N-acetylglucosamine and the carbonyl of the lactyl group of the other N-acetylmuramic acid. As pointed out by Tipper [11], this latter hydrogen bond may only be possible with a lactyl group having the D-configuration. In this stabilized chitin-like conformation, the D-lactyl peptide side chains are aligned in parallel along a relatively rigid linear backbone. Moreover, the C-6 of the N-acetylmuramic acid residues are unhindered and hence are readily available for substitution either by acetyl groups or by phosphodiester bridges. It has been emphasized (see Section II, A) that galactosamine, rather than glucosamine, and galactomuramic acid, rather than glucomuramic acid, have not been found to occur naturally in the wall glycans, and that the only variations so far encountered do not alter the basic conformation of the glycan chains. This suggests that the conformation of the glycan is essential for the survival of bacteria, and that any mutation that would alter this conformation is probably lethal.

An α-helical conformation of the peptide units is not likely. Indeed, the glutamyl bond is always γ, and therefore all the carbon atoms of this

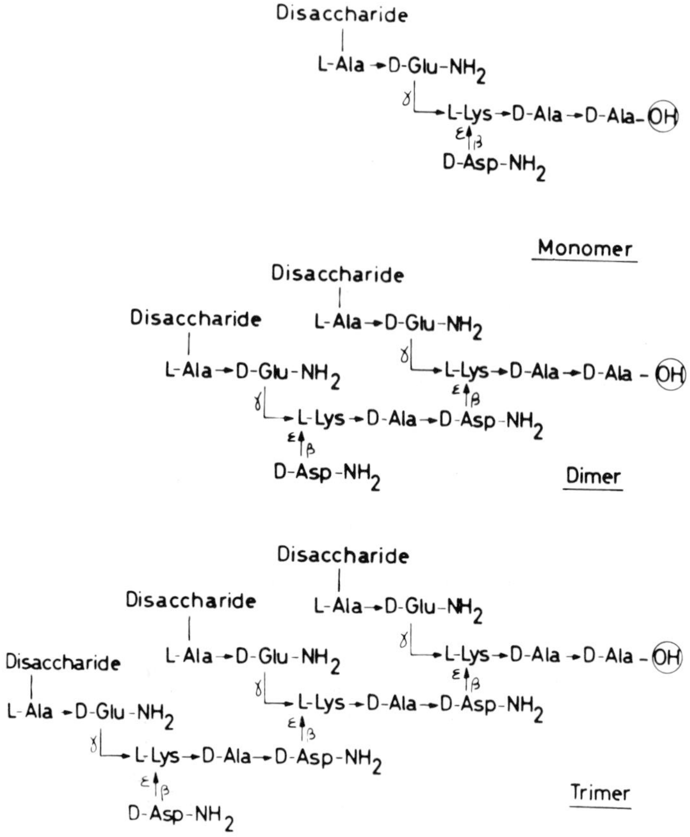

Disaccharide

L-Ala → D-Glu-NH_2

γ → L-Lys → D-Ala → D-Ala-(OH)

ε ↑ β

D-Asp-NH_2

Monomer

Disaccharide

Disaccharide L-Ala → D-Glu-NH_2

L-Ala → D-Glu-NH_2 γ → L-Lys → D-Ala → D-Ala - (OH)

ε ↑ β

γ → L-Lys → D-Ala → D-Asp-NH_2

ε ↑ β

D-Asp-NH_2 Dimer

Disaccharide

Disaccharide L-Ala → D-Glu-NH_2

Disaccharide L-Ala → D-Glu-NH_2 γ → L-Lys → D-Ala → D-Ala - (OH)

ε ↑ β

L-Ala → D-Glu-NH_2 γ → L-Lys → D-Ala → D-Asp-NH_2

ε ↑ β

γ → L-Lys → D-Ala → D-Asp-NH_2

ε ↑ β

D-Asp-NH_2 Trimer

FIG. 17. Disaccharide peptide mono-, di-, and trimer from the wall peptidoglycan of Lactobacillus acidophilus 63 AM Gasser after degradation of the glycan strands into disaccharide units. All the uncross-linked C-termini of the peptides have a D-alanyl-D-alanine sequence. This latter linkage can be hydrolyzed by the DD-carboxypeptidase from Streptomyces albus G (i.e., the KM endopeptidase of Figs. 6, 11, and 12). The action of the SA endopeptidase on this peptidoglycan is the same as that shown in Fig. 8 for chemotype II g.

amino acid residue are in the chain. As shown by Kelemen and Rogers [62], extensive interpeptide hydrogen bonding is possible when the peptides are extended in the pleated sheet (β-like) conformation. A three-dimensional molecular model of the staphylococcal peptidoglycan indicates that the degree of hydrogen bonding between contiguous peptide chains can be as high as 80%.

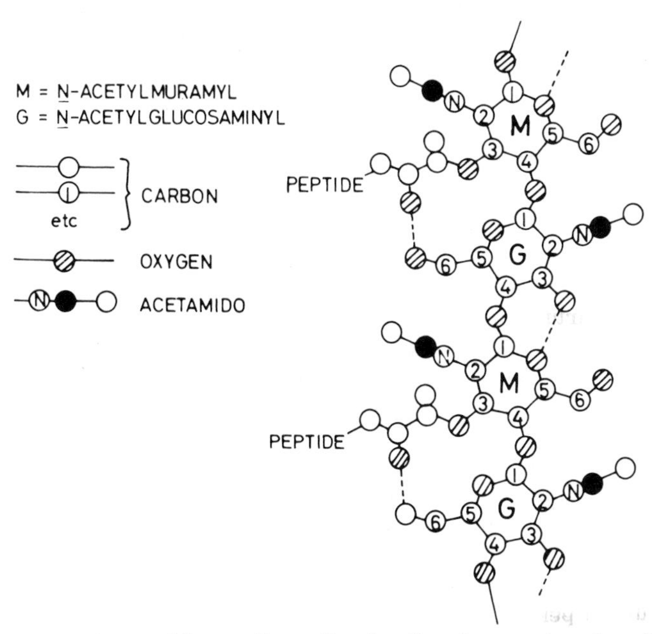

FIG. 18. A possible configuration for the glycan strands. The dashed lines represent hydrogen bonds. Reprinted from reference 11 by courtesy of Iowa State University Press.

An alternative possibility has been suggested by Tipper [11]. It rests upon the flexibility of the peptide chain because its γ-glutamyl linkage allows the chain to double back on itself, and on the similarity between the repeating peptide structure of the peptidoglycan and some peptide antibiotics such as enniatin and the gramicidins. These latter compounds are known for the high stability of their folded conformations and their chelating properties.

A third possible model has been constructed by Oldmixon and Higgins [63, 64]. Although the glycan strands are also based on that of chitin, this model differs from that of Kelemen and Rogers and Tipper in that it involves extensive hydrogen bonding between peptide side chains and the glycan chains. Extending from the lactyl group of N-acetylmuramic acid, the peptide side chains easily fold over one face of the glycan strands in a pattern that results in three hydrogen bonds between D-glutamic acid and D-alanine and the glycan, as well as an additional hydrogen bond between L-alanine and D-glutamic acid within the peptide chain. Because the tetrapeptide side chains consistently contain L-alanyl-γ-D-glutamyl-L-R_3-D-alanine (Section II, A) hydrogen bonding between tetrapeptide and glycan chains is virtually unaffected by the known variations in tetrapeptide sequence. This model results in a somewhat more compact structure than the others.

The models proposed by Tipper, Oldmixon and Higgins, and Kelemen and Rogers, are not only consistent with, but take into account the constancy and variability of peptide structures discussed in Section II, A. It should be emphasized that there is little or no physical evidence to date in support of any of these or other molecular models of the three-dimensional structure of peptidoglycan.

Hydrogen bonding, both in the glycan and the peptide, probably plays a considerable role in defining the conformation and properties of the peptidoglycan. For example, hydrogen bonds may play a role in (1) binding the nascent peptidoglycan to active sites of enzymes involved in its biosynthesis (*vide infra*); (2) aligning this nascent peptidoglycan during wall synthesis and growth; (3) remodeling the shape of the wall during the cell division cycle; and (4) providing a three-dimensional organization compatible with rigidity and tensile strength together with the presence of gaps that are required for the insertion of other wall polymers, diffusion of nutrients and competence.

III. BIOSYNTHESIS OF PEPTIDOGLYCAN AT THE BIOCHEMICAL LEVEL

A. The Three Stages

Peptidoglycan biosynthesis can be divided into three stages, each of which occurs at a different site in the cell: in the cytoplasm, on the membrane, and within the wall itself. Peptidoglycan precursors made on a uridylic acid cytoplasmic carrier (stage 1) are transferred from uridylic acid to an undecaprenyl phosphate membrane carrier (stage 2), and then to a final acceptor in the expanding wall peptidoglycan (stage 3). At some point during the later stages of this process, the nascent peptidoglycan undergoes cross-linking, which is required to make it insoluble, and covalently linked accessory wall polymers are attached. Figure 19 shows the main reactions through which the synthesis of peptidoglycan of chemotype I is carried out. In this case, no additional amino acids are involved in peptide cross-linking. Incorporation of amino acids into interpeptide bridges and addition of specific groups such as amides, in other peptidoglycans, will be discussed in Sections III, F and G below. (For a more comprehensive list of references, the reader is referred to other reviews [9, 11, 65].)

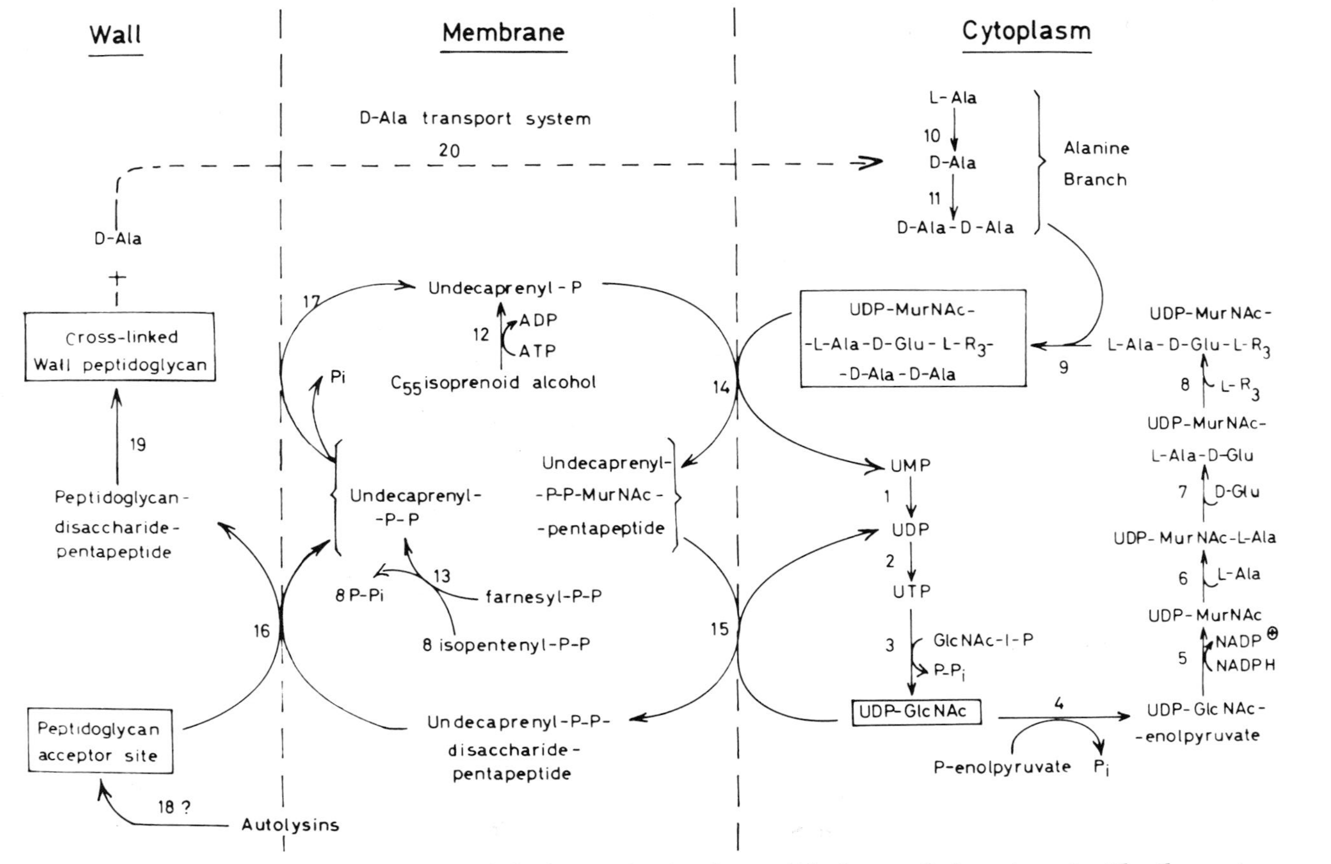

FIG. 19. Schematic representation of the biosynthesis of a peptidoglycan of chemotype I. The three stages — cytoplasmic, membrane-bound, and wall-bound — are separated by the dashed vertical lines. All the reactions are shown in the anabolic sense. GlcNAc = N-acetylglucosamine; MurNAc = N-acetylmuramic acid.

B. Stage 1: Synthesis of Nucleotide Precursors

The first two reactions are phosphorylations of UMP and UDP by ATP to yield UTP. UTP then reacts with α-D-N-acetylglucosamine 1-phosphate to yield UDP-N-acetylglucosamine and inorganic pyrophosphate (reaction 3, Fig. 19). This latter reaction is catalyzed by UDP-N-acetylglucosamine pyrophosphorylase and is analogous to reactions that lead to UDP-glucose and other compounds of this type.

Reaction 4 (Fig. 19) is, in fact, the first step in which the precursor is committed to becoming peptidoglycan. A three-carbon unit is transferred to UDP-N-acetylglucosamine from 2-phosphoenolpyruvate, the glycolytic intermediate. Inorganic phosphate and the pyruvate enol ether of UDP-N-acetylglucosamine are formed. The transferase that catalyzes the reaction in *Enterobacter cloacae* has been purified [66]. The reaction is reversible with a K_m for phosphoenolpyruvate of 3×10^{-5} M and a K_m for UDP-N-acetylglucosamine of 4.6×10^{-4} M. In this reaction, the enolpyruvate structure is preserved, in marked contrast with that observed in many other reactions involving phosphoenolpyruvate. The transferase is irreversibly inactivated by phosphonomycin. Phosphonomycin is believed to be a substrate analog of phosphoenolpyruvate and becomes covalently linked to the enzyme through a sulfhydryl group [67, 68]. The pyruvate UDP-N-acetylglucosamine transferase of *E. cloacae* has been separated from UDP-N-acetylglucosaminepyruvate reductase [66]. This latter enzyme catalyzes the next reaction (5, Fig. 19) in the sequence and yields UDP-N-acetylmuramic acid.

L-Alanine, D-glutamic acid, the L-R_3 residue, and finally, a preformed D-alanyl-D-alanine dipeptide are then added stepwise to UDP-N-acetylmuramic acid (reactions 6-9; Fig. 19). Each step is catalyzed by a separate enzyme requiring ATP and either Mg^{2+} or Mn^{2+}. Unlike the template-directed sequential peptide bond formation in protein synthesis, these reactions totally depend on the specificity of their respective enzymes for their substrates in order to result in a UDP-muramyl-pentapeptide of the correct sequence [9, 65]. The *meso*-diaminopimelic acid-adding enzyme of *E. coli* fails to add L-lysine, and the *S. aureus* L-lysine-adding enzyme fails to add *meso*-diaminopimelic acid to UDP-N-acetylmuramyl-L-alanyl-D-glutamate. A similar difference exists between the vegetative cell and "sporulation specific" L-lysine and *meso*-diaminopimelic acid-adding enzymes of *Bacillus sphaericus* 9602 (see Section IV, C) [69, 70]. In view of the absence of direction by a nucleic acid template, the relative constancy of peptidoglycan composition appears to be especially remarkable. The completed nucleotide is UDP-N-acetylmuramyl-L-alanyl-γ-D-glutamyl-L-R_3-D-alanyl-D-alanine (Fig. 20). Note that D-glutamic acid is γ-linked to the third amino acid, and that the peptide is not a tetrapeptide as is usually found in the completed wall peptidoglycans, but a pentapeptide ending in a D-alanyl-D-alanine sequence.

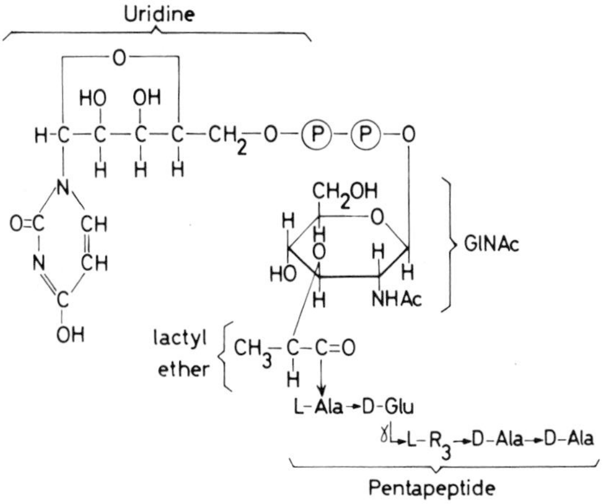

FIG. 20. The completed nucleotide precursor UDP-N-acetylmuramyl-pentapeptide.

C. The Alanine Branch [71, 72]

The synthesis of the dipeptide D-alanyl-D-alanine and its addition to the nucleotide tripeptide UDP-N-acetylmuramyl-L-alanyl-γ-D-glutamyl-L-R_3 have excited considerable interest. D-Alanine antagonists provided tools for studying what is called the alanine branch of peptidoglycan synthesis (Fig. 19). The details of this pathway have been recently reviewed by Neuhaus et al. [71, 72]. Three enzymes are involved in the alanine branch: (1) alanine racemase, (2) D-Ala:D-Ala ligase (ADP) or D-alanyl-D-alanine synthetase, and (3) UDP-MurNAC-L-Ala-γ-D-Glu-L-R_3: D-Ala-D-Ala ligase (ADP) or D-alanyl-D-alanine adding enzyme. Among the D-alanine antagonists, D-cycloserine is of prime importance. It is a competitive inhibitor of both the racemase and synthetase with K_i values considerably smaller than the K_m values for the substrates. D-Cycloserine does not inhibit the D-alanyl-D-alanine adding enzyme.

1. Alanine Racemase

In terms of the synthesis of D-alanyl-D-alanine, alanine racemase (reaction 10; Fig. 19) can function either in an anabolic (2L-Ala → 2D-Ala) or catabolic (2D-Ala → 2L-Ala) sense. The racemases from *S. aureus* [73], *Streptococcus faecalis* [74, 74a], *B. subtilis* [75], and *E. coli* [71] have been isolated and studied.

The S. aureus racemase is competitively inhibited by D-cycloserine when measured in either direction. However, even very high concentrations of L-cycloserine fail to inhibit the reaction in either direction. This observation was puzzling. Indeed, if D-cycloserine competes with D-alanine, L-cycloserine should also be a competitor for L-alanine. In an attempt to explain the lack of inhibition by L-cycloserine, Roze and Strominger [73] postulated the existence of a single site on the racemase on which L- and D-alanine would bind in the same conformation. The conformation specifically recognized by the enzyme would be that found in D- but not in L-cycloserine.

The E. coli racemase exhibits K_m values of 9.7×10^{-4} M for L-Ala and 4.6×10^{-4} M for D-Ala [71]. The V_{max} values (in μmole/h) are 2.22 for the L-Ala → D-Ala reaction and 0.95 for the D-Ala → L-Ala reaction. Hence, the anabolic velocity is larger than the catabolic one, but the apparent affinity for D-Ala is greater than that for L-Ala. Therefore, the Haldane relationship

$$K_{Eq} = \frac{(K_m \text{ D-Ala})\ (V_{max} \text{ L} \rightarrow \text{D})}{(K_m \text{ L-Ala})\ (V_{max} \text{ D} \rightarrow \text{L})}$$

is approximately equal to 1. D-Cycloserine competitively inhibits the E. coli racemase with either D- and L-alanine as substrate, and the K_i value is independent of the configuration of the substrate. In marked contrast with the S. aureus racemase, the E. coli enzyme is also competitively inhibited by L-cycloserine. The ratio K_m(L-Ala)/K_m(D-Ala) is approximately equal to the ratio K_i(L-cycloserine)/K_i(D-cycloserine). Neuhaus et al. [71] believe the above kinetics to be consistent with the existence of separate binding sites on the racemase, for L- and D-alanine. A pronounced substrate inhibition was observed in the D-alanine → L-alanine assay. This was interpreted to mean that, at high concentrations, D-alanine binds to both the L-alanine and D-alanine sites.

A unified model for alanine racemase cannot be proposed. It would appear that the alanine racemase from either S. aureus or S. faecalis is specifically inhibited by D- but not L-cycloserine, and that the alanine racemase from either E. coli or B. subtilis is inhibited by both D- and L-cycloserine. It may be that some racemases have a single site for both D- and L-alanine and that others have two binding sites. A third model has also been proposed [75a] according to which the racemase would have two forms, one which would bind L-alanine and the other D-alanine.

2. D-Alanyl-D-Alanine Synthetase

D-Alanyl-D-alanine synthetase catalyzes the reaction

$$2\text{D-alanine} + \text{ATP} \xrightarrow[K^{+}]{Mg^{2+}} \text{D-alanyl-D-alanine} + \text{ADP} + P_i$$

(reaction 11, Fig. 19). D-Cycloserine inhibits the reaction. It has been proposed by Neuhaus [71, 72, 76, 77] that (in the case of S. faecalis) the primary site of D-cycloserine action is at the donor site of the synthetase, and that the acceptor site of the enzyme is a secondary site of action. D-Alanyl-D-alanine and some analogs of D-alanyl-D-alanine are also inhibitors of the synthetase. The inhibition is specific for dipeptides. Additions to the N-terminal residue decrease the effectiveness of dipeptides as inhibitors. Additions to the C-terminal residue sometimes enhance their effectiveness. For example, D-norvalyl-D-alanine is not an inhibitor, and D-α-amino-n-butyryl-D-alanine is a very poor inhibitor. In contrast, D-alanyl-D-norvaline and D-alanyl-D-α-amino-n-butyric acid are better inhibitors than D-alanyl-D-alanine.

In the presence of D-cycloserine and of Mn^{2+}, the synthetase was found to bind ATP with high affinity. No complex formation was observed when D-cycloserine was replaced by D-alanine in the reaction mixture. D-Alanyl-D-alanine and analogs were tested for their ability to facilitate ATP binding to the enzyme. Two types of responses were observed, which were interpreted as indicating the presence of two types of sites [71]. One site, accessible to D-cycloserine and D-alanyl-D-alanine, facilitates ATP binding. The second site, accessible to D-alanyl-D-alanine, D-alanyl-D-valine and D-alanyl-D-norvaline, does not facilitate ATP binding. Although the inhibitor studies indicated product (or analog) multiple sites, it was not possible to distinguish them from substrate binding sites. It has been suggested that the function of multiple binding sites is to control the rate of dipeptide formation.

3. D-Alanyl-D-Alanine Adding Enzyme

This enzyme catalyzes the addition of D-alanyl-D-alanine dipeptide to UDP-N-acetylmuramyl-L-alanyl-γ-D-glutamyl-L-R_3 in the presence of Mg^{2+} and ATP (reaction 9 in Fig. 19). Its importance lies in its specificity pattern for the addition of dipeptide analogs of D-alanyl-D-alanine [78]. Neuhaus and Struve showed that in S. faecalis the substrate profile of this enzyme is complementary to that of the D-alanyl-D-alanine synthetase. For example, D-alanyl-D-alanine synthetase can incorporate D-norvaline (and D-serine or D-threonine) into the C-terminal end of a dipeptide, but not into the N-terminal end. In contrast, D-norvalyl-D-alanine, but not D-alanyl-D-norvaline, can bind to and inhibit the adding enzyme. This combination of specificities results in the relative accuracy of synthesis of the complete nucleotide peptide and accounts for some of the growth-inhibitory effects of high concentrations of some D-amino acids.

D. Stage 2: The Undecaprenyl Phosphate Membrane Carrier

The membrane lipid carrier is one of the key compounds in peptidoglycan synthesis as it transports the N-acetylmuramyl-pentapeptide and N-acetylglucosamine precursors from the intracellular sites of synthesis through the membrane to the exocellular sites of polymerization. The peptidoglycan lipid carrier from S. aureus [79, 80] was isolated in a very high degree of purity, allowing its structure to be ascertained by mass spectrometry. It is a C_{55}-isoprenoid alcohol containing 11 isoprene units with the chain ending in an alcoholic function (Fig. 21). In S. lactis, a lipid carrier is used for the transport of teichoic acid precursors to the wall [81, 82]. The same peptidoglycan lipid carrier of S. aureus is also effective in the biosynthesis of the O-antigen of Salmonella [83] (i.e., the lipopolysaccharide of the outer membrane). Hence, in a single bacterium, the same lipid may be involved in the synthesis of more than one wall polymer and may be channeled into more than one synthetic route (peptidoglycan and O-antigen, for example). In agreement with this, competition for a limited amount of available lipid carrier was shown to occur in vitro when both precursors and enzymes of two biosynthetic pathways were present [84]. Nucleotide precursors for the synthesis of one polymer inhibit the synthesis of the other polymer. These effects can be enhanced or diminished by preincubation of the enzyme system with appropriate nucleotide precursors [84a]. However, recent evidence suggests that a polyisoprenoid carrier is not involved in teichoic acid synthesis [84b]. In addition, it should be mentioned that the lipid carrier involved in the synthesis of the capsular polysaccharide in Aerobacter [85] can function in mannan synthesis in M. lysodeikticus [86], and that a large number of long-chain isoprenoid alcohols of this type have also been isolated from plants and animals [87, 88]. All are probably relevant to problems of transport through membranes and perhaps to their orientation.

In order to be functional in peptidoglycan synthesis, the C_{55}-isoprenoid alcohol first must be phosphorylated. In S. aureus, phosphorylation is achieved in the presence of ATP by a membrane-bound isoprenyl alcohol phosphokinase [89] (reaction 12; Fig. 19). This enzyme is a lipoprotein and exhibits unusual properties. It is insoluble in water but is soluble and stable in several organic solvents. The enzyme clearly derives from the membrane where it must occur in a lipid phase. The active enzyme is composed of a protein fraction and a phospholipid (phosphatidylglycerol). The protein fraction is inactive and insoluble both in water and in organic

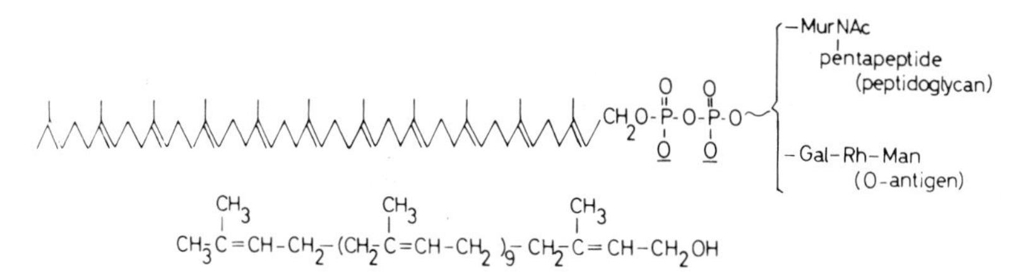

FIG. 21. The C_{55}-isoprenoid alcohol charged with either peptidoglycan or O-antigen precursors, via a pyrophosphate bridge.

solvents. Its activity and solubility in organic solvents are restored by addition of the phospholipid fraction [89, 90].

The presence of an ATP-dependent phosphokinase may be unique to S. aureus. In Salmonella newington, a particulate enzyme catalyzes the synthesis of C_{55}-lipid pyrophosphate (reaction 13; Fig. 19) from farnesyl pyrophosphate and isopentenyl pyrophosphate which, in turn, is probably derived from mevalonate [91, 92]. C_{55}-Lipid pyrophosphate is then dephosphorylated (reaction 17; Fig. 19; vide infra).

E. The Lipid Cycle

The lipid cycle achieves the assembly of the activated precursors into β-1,4-N-acetylglucosaminyl-N-acetylmuramyl-pentapeptide units and carries out the transport of the completed units to the exocellular sites of polymerization. Evidently, this mechanism is responsible for the alternating sequence of N-acetylglucosamine and N-acetylmuramic acid residues found in the glycan strands of the completed wall. The first two reactions (14 and 15; Fig. 19) consist of transferring the precursors from the hydrophilic environment of the cytoplasm to the hydrophobic environment of the membrane in "an interchange of carriers that are compatible with the particular microenvironments" [93]. Such processes are tranlocations, and the enzymes which catalyze them have been called translocases. First (reaction 14), phospho-N-acetylmuramyl pentapeptide is translocated from UDP-N-acetylmuramyl-pentapeptide to the membrane undecaprenyl carrier with formation of UMP and undecaprenyl-PP-N-acetylmuramyl-pentapeptide. N-Acetylglucosamine is then translocated from UDP-N-acetylglucosamine to undecaprenyl-PP-N-acetylmuramyl-pentapeptide with formation of UDP and undecaprenyl-PP-disaccharide-pentapeptide (reaction 15). Note that in this reaction, only the N-acetylglucosamine and not the phosphate is transferred. In order to complete the lipid cycle a third reaction (reaction 16; Fig. 19) transfers the disaccharide-pentapeptide unit from the membrane carrier into the expanding wall peptidoglycan.

1. Translocation of N-Acetylmuramyl-Pentapeptide [93, 94]

This translocation (reaction 14; Fig. 19) is a transphosphorylation reaction which proceeds without loss of energy: UDP-MurNAC (pentapeptide) + undecaprenyl-P $\rightleftharpoons$ undecaprenyl-PP-MurNAC (pentapeptide) + UMP. One phosphate group of the pyrophosphate is derived from UDP and the other from the phosphorylated lipid carrier. Forward and reverse reactions require Mg^{2+}. The transfer reaction is stimulated by K^+ and other monovalent cations. UMP is a very effective inhibitor of the forward transfer. The equilibrium may be reached by either route and the K_{Eq} value is about 0.25. The translocase has been solubilized from membrane fragments by a variety of diverse reagents into two major active fractions with apparent molecular weights of 2×10^6 and 100,000 to 200,000 [95]. Further purification has not yet been achieved.

It has been proposed by Neuhaus [93-95] that an enzyme intermediate enzyme-P-N-acetylmuramyl (pentapeptide) is transitorily formed according to the reactions

Enzyme + UMPP-N-acetylmuramyl (pentapeptide) $\rightleftharpoons$ ...
... $\rightleftharpoons$ enzyme P-N-acetylmuramyl (pentapeptide) + UMP (14a)

Enzyme-P-N-acetylmuramyl (pentapeptide) + undecaprenyl-P $\rightleftharpoons$ ...
... $\rightleftharpoons$ enzyme + undecaprenyl-PP-N-acetylmuramyl (pentapeptide) (14b)

The enzyme intermediate has not been isolated, but several experimental evidences support the proposed mechanism. (1) The rate of exchange that occurs in reaction 14a largely exceeds the rate of transfer. (2) Kinetics of product formation showed that UMP continues to be formed after the lipid product undecaprenyl-PP-N-acetylmuramyl (pentapeptide) has attained a steady-state level. (3) After a lag period, phospho-N-acetylmuramyl (pentapeptide) is found in the reaction products. This product could arise through hydrolysis of the postulated enzyme-P-N-acetylmuramyl (pentapeptide) complex. (4) A small part (5-10%) of the maximal exchange activity was found to be sufficient to ensure a major part (about 50%) of the transfer activity. (5) The K_m values for K^+ ions are 1×10^{-2} M in the exchange assay and 2.4×10^{-3} M in the transfer assay. (6) At selected concentrations, certain detergents stimulate the transfer reaction as if they enhanced the ability of the enzyme to bind a recognized acceptor (i.e., the undecaprenyl-P), whereas they inhibit the exchange assay. Vancomycin stimulates the transfer reaction and inhibits the exchange reaction at low concentrations, but higher concentrations inhibit both reactions. These results have been interpreted to be an indication that vancomycin acts as a surfactant in the transfer assay and inhibits the exchange assay by forming an adduct with the terminal D-alanyl-D-alanine of the pentapeptide substrate [93].

The specificity profiles [93] of reactions 14a and b (Fig. 19) are also of interest with respect to their ability to complement those of the earlier reactions. The translocase exhibits a high specificity for the uracil moiety, suggesting a key role for pyrimidine in the reaction. While all of the biosynthetic enzymes will utilize the fluoro-substituted nucleotide, the translocase does not. Thus, fluorouracil causes great accumulations of FUDP-N-acetylmuramyl pentapeptide when various bacteria are grown in the presence of this analog [93, 93a]. Although complete studies of the use of analogs by the translocase have not yet been reported, it is known that a C-terminal D-alanyl-D-alanine sequence is an essential feature of the substrate. The utilization of UDP-N-acetylmuramyl pentapeptide with substituents on the ε-amino group of L-lysine (e.g., pentaglycine, L-alanine, or L-serine) in the transfer reaction by preparations from various microorganisms has been reported. An interesting example of the complementary specificities of the various reactions is the case of O-carbamyl-D-serine in S. faecalis [74]. The analog UDP-N-acetylmuramyl-L-alanyl-γ-D-glutamyl-L-lysyl-D-alanyl-O-carbamyl-D-serine is as effective as the normal UDP-N-acetylmuramyl pentapeptide in the exchange reaction (reaction 14a). O-Carbamyl-D-serine is used as a substrate by the D-alanyl-D-alanine synthetase to form D-alanyl-O-carbamyl-D-serine, which in turn can be incorporated into UDP-N-acetylmuramyl-pentapeptide. However, the analog is an effective inhibitor of alanine racemase (reaction 10) and so would prevent the initiation *in vivo* of a series of reactions resulting in an altered peptidoglycan.

2. Translocation of N-Acetylglucosamine [12, 96-98]

This translocation is a transglycosidation reaction (reaction 15; Fig. 19). N-Acetylglucosamine is transferred from UDP-N-acetylglucosamine to undecaprenyl-PP-N-acetylmuramyl (pentapeptide) carrier. The reaction results in the generation of UDP and of undecaprenyl-PP-disaccharide (pentapeptide). The disaccharide is β-1,4-N-acetylglucosaminyl-N-acetylmuramic acid and is linked to the pyrophosphate through the C-1 of N-acetylmuramic acid. This reaction differs from reaction 14 (Fig. 19) in that only N-acetylglucosamine, and not the terminal phosphate of UDP, is transferred to the lipid intermediate.

3. Translocation of Disaccharide-Pentapeptide into the Wall

This reaction (16; Fig. 19), which occurs at the interface of the membrane and wall, completes the lipid cycle. It is also the first of the series of reactions of stage 3 that occur in the wall itself, and is discussed in Section H.

F. Incorporation of Interpeptide Chains

In all peptidoglycans other than those of chemotype I, the bridges that extend between the peptide units consist of one or several additional amino acid residues. The incorporation of these bridging amino acid residues onto the pentapeptide units takes place before the transfer of disaccharide peptide units from the lipid carrier to the expanding wall. In the case of the peptidoglycans of chemotype II, the process of incorporation essentially consists of the extension of the length of the side chain of the L-R_3 residue through the substitution of the ω-amino group by one or several amino acid residues. Often, but not always, these amino acids are activated as aminoacyl-transfer RNA derivatives and are then transferred to the peptidoglycan unit on the lipid intermediate. Most of the reaction sequences investigated are catalyzed by particulate enzymes.

The synthesis of the N^{ϵ}-pentaglycyl-L-lysine sequence in S. aureus (Fig. 8a) has been studied in detail [102-104]. Five glycine residues from glycyl-tRNA are sequentially added to the ϵ-amino group of L-lysine on the undecaprenyl-PP-disaccharide-peptide intermediate. Four species of $tRNA^{Gly}$ exist in S. aureus [105]. All four support the incorporation of glycine into the peptidoglycan unit. Three of these $tRNAs^{Gly}$ participate in template-directed polypeptide synthesis. One of them, however, is inactive in polynucleotide-directed polypeptide synthesis and does not bind to ribosomes in the presence of any of the glycine codons. This apparently peptidoglycan-specific $tRNA^{Gly}$ might be a unique gene product. It has been extensively purified, shown to lack all of the minor bases except for a single thiouridine residue, but its sequence has not yet been published [12, 106]. Staphylococcus aureus contains only a single $tRNA^{Gly}$ synthetase which has been isolated and purified [107]. The synthetase functions in catalyzing the formation of glycyl-tRNA's used for both peptidoglycan and protein biosynthesis. In the presence of ATP and Mg^{2+}, it can also charge the tRNA's of other bacteria and yeast with glycine.

The synthesis of the N^{ϵ}-L-alanyl-L-lysine sequence in Arthrobacter crystallopoietes [108] (Fig. 8f) is carried out through a mechanism similar to that described for the synthesis of N^{ϵ}-pentaglycyl-L-lysine in S. aureus. L-Alanine is incorporated by particulate enzymes into the peptidoglycan unit at the level of the lipid intermediate from L-alanyl-tRNA. L-Alanyl-$tRNA^{Cyst}$ was prepared by reduction of L-cysteinyl-$tRNA^{Cyst}$ and was shown not to be utilized in peptidoglycan synthesis. This suggested that tRNA did not act simply as a carrier of activated L-alanine, but that some specific feature of this tRNA-Ala molecule was recognized by the peptidoglycan-synthesizing system.

In Staphylococcus epidermidis [109], four different pentapeptide chains have been found as substituents on the ϵ-amino group of L-lysine (Fig. 8c). All four contain three glycine residues and two L-serine residues, but each has a different sequence. Particulate enzymes, glycyl-tRNA, and L-seryl-tRNA carry out the incorporation of glycine and L-serine at the level of the lipid intermediate. Maximum L-serine incorporation requires the simultaneous incorporation of glycine, but glycine incorporation is independent of L-serine incorporation. Four $tRNAs^{Ser}$ were isolated. All participate equally well in peptidoglycan synthesis. Two of them bind to E. coli ribosomes in the presence of either of two serine codons, UCA or UCG. A third binds in the presence of either of two other serine codons, AGU or AGC. The fourth does not bind in the presence of any of the known serine triplets. Hence, if tRNA is specifically recognized by the peptidoglycan-synthesizing system, the tRNA specificity is unlikely to be at the anticodon region. Whether or not each of the four $tRNAs^{Ser}$ is able to catalyze only one type of bridge is an interesting problem that has not yet been solved.

The incorporation of L-alanine to the ϵ-amino group of L-lysine in the peptidoglycan unit of Lactobacillus viridescens [110] (Fig. 8d) is unusual in that L-alanine is transferred from L-alanyl-tRNA to cytoplasmic UDP-N-acetylmuramyl-pentapeptide and not to the membrane-bound lipid intermediate. Moreover, the transfer is catalyzed by a soluble enzyme that exhibits a low specificity for the amino acid and a lack of specificity for tRNA. This transferase catalyzes the transfer not only of L-alanine from L-alanyl-tRNA but also, although with a lower efficiency, of L-serine, L-cysteine, and probably glycine, from their corresponding specific tRNA's, to UDP-N-acetylmuramyl-pentapeptide. Moreover, the transferase is able to utilize L-alanyl-$tRNA^{Cyst}$ in peptidoglycan synthesis.

In all the aforementioned examples of incorporation of interpeptide chains, the amino acids that are added to the ω-amino group of the peptidoglycan unit are glycine or L-amino acid residues or both. The involvement of tRNA in the process seems to be general, although as yet a unified model cannot be proposed. The mechanism completely differs from mRNA-coded protein synthesis on ribosomes. However, there exist both in bacteria and in animal tissues, several soluble aminoacyl-tRNA transferases which catalyze the addition of certain amino acids to the N-terminus of proteins in the absence of ribosomes [111-113]. It should also be noted that a requirement for tRNA has also been demonstrated in the synthesis of aminoacylphosphatidylglycerols by bacterial enzymes [114].

In many Lactobacilli other than L. viridescens, the interpeptide bridge consists of a single isoasparaginyl residue that has the D-configuration (Fig. 8g). The syntheses of the N^{ϵ}-(D-isoasparaginyl)-L-lysine sequence in S. faecalis and L. casei have been studied. In these cases [115, 116] tRNAs do not participate. On the contrary, D-aspartic acid is activated in the form of a β-D-aspartylphosphate by an enzyme that can be released from the membrane by high salt concentrations. D-Aspartic acid is then transferred to the ϵ-amino group of L-lysine of the lipid intermediate, and finally its α-carboxyl group is amidated (in the presence of NH_3 and ATP).

By analogy with the foregoing, one can postulate that the synthesis of the peptide bridges in the peptidoglycans of chemotype IV (Figs. 11-13) also occur via the addition of one or several amino acid residues to the pentapeptide units. In this latter case, however, the extension of the pentapeptide must occur from the α-carboxyl group of D-glutamic acid. The interpeptide bridges of chemotype IV always contain a diaminoacid residue, which frequently has the D configuration. The mechanism by which these amino acids are incorporated is completely unknown. The peptide bridges in the peptidoglycans of chemotype III (Fig. 10), consist of one or several peptides, each having the same amino acid sequence as the peptide unit. It seems very unlikely that, in this latter case, bridge synthesis would occur through stepwise addition of amino acid residues. It has been hypothesized (Fig. 22) that some of the completed peptide units are translocated from N-acetylmuramic acid to a bridging position through the successive, alternating action of transpeptidase(s) and N-acetylmuramyl-L-alanine-amidase [43].

G. Further Alterations of Peptidoglycan Units

Variations occur both in the glycan and peptide portions of the peptidoglycans of many bacterial species (see Section II, A above). In terms of the general structure of the polymers, they are of minor importance. In most cases, the mechanisms involved are not understood. It is known, however, that in *S. aureus* [117], the conversion of D-glutamic acid of the pentapeptide unit into D-isoglutamine is carried out by amidation of the α-carboxyl group in the presence of ATP and ammonium ions at the level of the lipid intermediate. In *M. lysodeikticus* [118] the substitution of the same α-carboxyl group of D-glutamic acid by a glycine residue is also carried out in the presence of ATP at the lipid level. It should be noted that this glycine residue remains unsubstituted in the completed peptidoglycan and is not utilized as an interpeptide bridge. Another variation that has been described recently is the occurrence of small amounts of pentapeptides ending in either a glycyl-D-alanine or D-alanyl-glycine sequence, instead of the usual D-alanyl-D-alanine sequence, in the nucleotide precursors of some species of *Arthrobacter* and *Corynebacteria* [119, 120]. These alterations were observed only when inhibitory concentrations of glycine were present in the media. It is of interest to note here that the specificity profiles of the D-alanyl-D-alanine synthetase and the D-alanyl-D-alanine adding enzymes of *S. faecalis* would permit the substitution of glycine for D-alanine at the penultimate position of the UDP-N-acetylmuramyl pentapeptide [78].

The peptidoglycan of the cortex of bacterial spores is different from the wall peptidoglycan of the corresponding vegetative cells, although

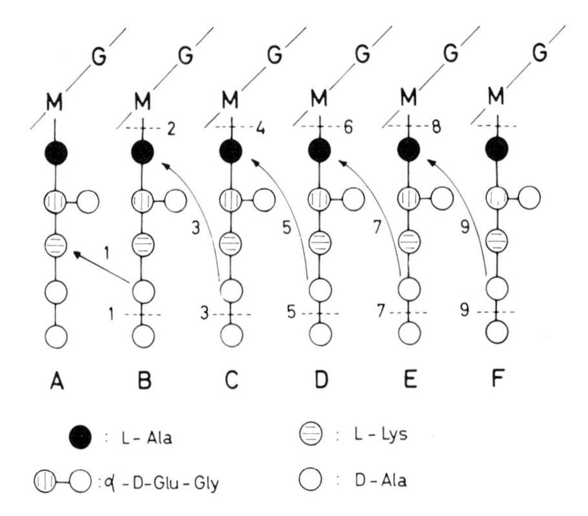

FIG. 22. Proposed biosynthetic sequence of the modification of the peptidoglycan in Micrococcus lysodeikticus. Reactions 2, 4, 6, and 8 are the result of hydrolysis by an N-acetylmuramyl-L-alanine amidase. Reactions 1, 3, 5, 7, and 9 are transpeptidations. Reprinted from reference 43 by courtesy of the American Chemical Society.

the two peptidoglycans are basically composed of the same constituents (see Section IV, C). It has been suggested that different segments of the genome are involved in the synthesis of the two peptidoglycans. Some of the enzymes that are responsible for the unique features of the cortex have been studied. Most of them, however, remain to be investigated. Similarly, the factors that must control the repression and expression of the various enzymes involved in the differentiation process are unknown.

H. Stage 3: Translocation of Disaccharide-Peptide Units into the Wall

The transfer of the disaccharide-peptide units from the membrane carrier into the expanding wall peptidoglycan should not be regarded as having been fully established. It is likely that each unit is transferred from the lipid intermediate to an endogenous wall acceptor, resulting in the addition of a disaccharide-peptide unit to the growing peptidoglycan (reaction 16; Fig. 19). In several cases the cell-free reaction product has been found to be solubilized by an endo-N-acetylmuramidase, suggesting the formation of glycosidic rather than peptide bonds. The formation of lipid-linked disaccharide-peptide oligomers, however, (such as those found in O-antigen synthesis in gram-negative bacteria [99]) has not been eliminated as an intermediate step. Whatever the mechanism, the reaction generates undecaprenyl pyrophosphate, which in turn, is dephosphorylated (reaction 17; Fig. 19). This dephosphorylation reaction yields inorganic phosphate and the initial C_{55}-isoprenoid alcohol phosphate carrier, which then can begin a new cycle. Bacitracin [100, 101] was found to be an inhibitor of the phosphatase, which catalyzes the dephosphorylation of the undecaprenyl pyrophosphate.

The wall acceptors, which are assumed to be used for the insertion of the activated disaccharide-pentapeptide units, must be nonreducing termini of N-acetylglucosamine. Such termini can be generated by an endo-N-acetylmuramidase autolysin (reaction 18; Fig. 19; see also Section IV, D). There is no definite proof, however, that incorporation is actually carried out through such a mechanism. The suggestion has already been made (see Section II, D) that hydrogen bonding may also play an important role in aligning the nascent peptidoglycan in the expanding wall. It must also be remembered that somewhere in or after the peptidoglycan biosynthetic sequence, the addition of covalently linked accessory wall polymers, such as teichoic acids, must occur.

I. Transpeptidation Reaction

The insertion of newly synthesized, but as yet uncross-linked disaccharide peptide units into the growing wall peptidoglycan must be followed by closure of the bridges between the peptide units if the process is to yield an insoluble network. It may well be, however, that insertion and peptide cross-linking occur simultaneously. The mechanism for peptide cross-linking was first revealed by a study of the effects of penicillin on S. aureus [121-123]. Penicillin G, when added at sublethal doses to growing staphylococcal cells, was shown to reduce the extent of peptide cross-linking. Walls isolated from cells grown in the presence of low concentrations of the antibiotic contained higher amounts of uncross-linked peptide units than did walls isolated from cells grown in the absence of penicillin. Moreover, the peptide units that had not undergone transpeptidation because of the presence of penicillin, retained the C-terminal D-alanyl-D-alanine sequence of the peptidoglycan precursors and remained uncross-linked after subsequent growth in the absence of penicillin [123]. It was therefore proposed that the closure of bridges in S. aureus was achieved through a transpeptidation reaction, and that the transpeptidase, which catalyzes the reaction, was the target of the penicillin molecule. The mechanism of the transpeptidation reaction in S. aureus is shown in Fig. 23. The penultimate C-terminal D-alanine of a peptide donor is transferred to the N-terminal glycine of a peptide acceptor. Interpeptide bonds are formed, and equivalent amounts of D-alanine residues are liberated from the donor peptides.

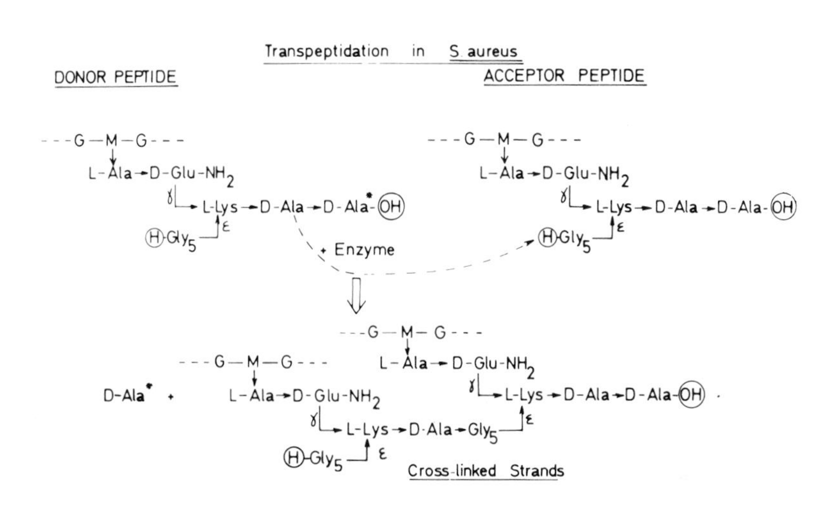

FIG. 23. Proposed transpeptidation reaction in Staphylococcus aureus Copenhagen. The completed, cross-linked peptidoglycan of S. aureus is shown in Fig. 2.

The use of the terminology for "donor" and "acceptor" here, and subsequently, is the same as that used in protein and polypeptide antibiotic biosynthesis [123a, 123b].

A cell-free system, which would carry out the reaction *in vitro*, has not yet been isolated from *S. aureus*. Fortunately, gram-negative bacteria allowed the study of the transpeptidation reaction in a more direct way. Particulate preparations [124-128] were obtained from *E. coli* and *Salmonella*, which performed the bridge-closure reaction (Fig. 24). In this case, the amino group involved in the reaction is located on the D-carbon of the *meso*-diaminopimelic acid residue of the peptide acceptor, and the peptide bond formed is D-alanyl-D-*meso*-diaminopimelic acid. The *in vitro* reaction was also shown to be inhibited by penicillin. Efforts of several laboratories to extend these studies to microorganisms other than *E. coli* and *Salmonella* have not yet achieved this goal. Transpeptidation, however, is thought to be ubiquitous among bacteria for several reasons. (1) It effectively explains why a single D-alanine residue is involved in cross-linking the peptide units in all bacterial peptidoglycans, whereas the largest peptides of the nucleotide precursors end in a C-terminal D-alanyl-D-alanine sequence. (2) Transpeptidation does not require an input of energy and therefore can result in peptide-bond formation at exocellular sites where ATP is not available. This is a particularly useful feature when one of the substrates is not only exocellular but may be insoluble. (3) Treatment with penicillin or related antibiotics results in lysis of bacteria from a wide variety of taxonomic groups.

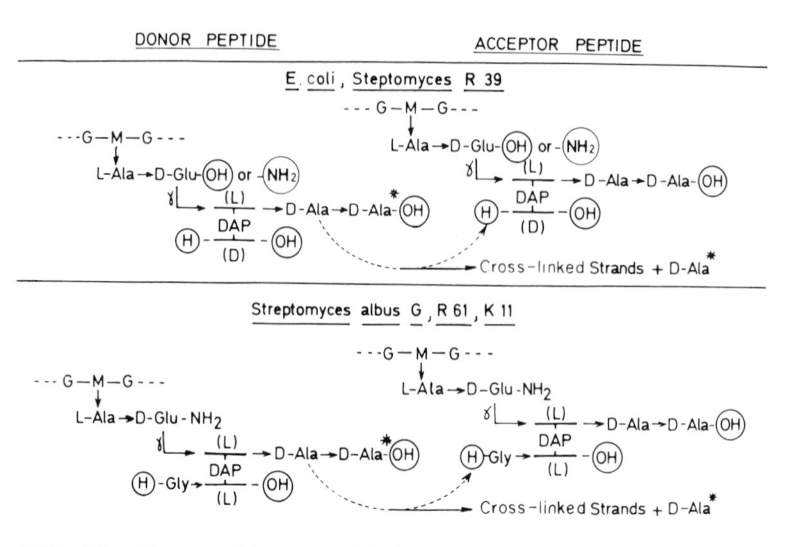

FIG. 24. Proposed transpeptidation reactions in E. coli and several Streptomyces sp. The structure of the cross-linked strands is shown in Fig. 6 for E. coli and Streptomyces R39, and in Fig. 9 for Streptomyces albus G, R61, and K11. Note that the proposed cross-linking reactions in Streptomyces sp. are based on the structure of the completed wall peptidoglycans.

The direction of growth of the peptide moiety is unknown, i.e., it is not known whether the D-alanyl-D-alanine donor group belongs to the newly transported disaccharide peptide unit or to the wall peptide oligomer. Whatever the mechanism, the D-alanine residue that is released can be reutilized by the cell either for peptidoglycan synthesis or for teichoic acid synthesis, by means of a specific transport system (reaction 20; Fig. 19). In E. coli the transport of D-alanine, L-alanine, and glycine has been partially resolved [71, 129]. It is mediated by at least two systems. The systems for D-alanine and glycine are related and are different from that for L-alanine. Study of some mutants resistant to D-cycloserine has shown that D-cycloserine is accumulated in sensitive cells through at least some part of the D-alanine/glycine transport system.

J. DD-Carboxypeptidase Transpeptidase

No bacterial membrane-bound transpeptidase has yet been released from the membrane, purified and characterized. Streptomyces sp., however, excrete into the culture medium enzymes that can function as carboxypeptidases (hydrolytic enzymes) or as transpeptidases (synthesizing enzymes), depending upon the availability of nucleophilic acceptor

(H_2O or NH_2-R). Such enzymes produced by several strains of Streptomyces were isolated and purified [130-135a]. They act on peptides ending in a C-terminal D-alanyl-D-alanine sequence and can differ in function since, after elimination of the C-terminal D-alanine residue of a peptide donor, the peptidyl-enzyme complex can be attacked by water (leading to simple hydrolysis; D-alanyl-D carboxypeptidase activity) or by a recognizable amino group (leading to peptide bond formation, transpeptidase activity) (Fig. 25). The two steps involved in the mechanism may occur either in a sequential manner, i.e., formation of a peptide-enzyme acyl intermediate and subsequent attack by a nucleophilic group with regeneration of the free enzyme, or in a concerted manner, i.e., via the formation of a donor-acceptor-enzyme tertiary complex. Structural studies of the wall peptidoglycan in several strains of Streptomyces suggest that the transpeptidation reaction in these organisms must be that shown in Fig. 24.

The substrate requirements for carboxypeptidase and transpeptidase activity of several purified enzymes isolated from different strains of Streptomyces were studied with the help of synthetic and natural peptides. The enzymes produced by strains R39, R61, K11, and albus G had high hydrolytic activity on the synthetic tripeptide N^{α}, N^{ϵ}-diacetyl-L-lysyl-D-alanyl-D-alanine from which they release the C-terminal D-alanine residue. The discovery of these enzymes allowed the study of their carboxypeptidase action on peptides possessing the general structure acetyl-L-R_3-R_2-R_1 (OH) (Tables I-III). The series of enzymes differed in their K_m and V_{max} values with various peptides, but all of them exhibited the same general substrate profile [131-134], which was characterized by considerable specificity for the presences of (1) a D-amino acid residue, often preferentially D-alanine, at the C-terminal or R_1 position, (2) solely D-alanine at the penultimate R_2 position, and (3) a relatively long side chain on the L-R_3 residue (for further details, see reference 135a). This profile closely resembles the structure of the peptide units of the nascent Streptomyces peptidoglycans which undergo transpeptidation (Fig. 24). In fact, a C-terminal D-alanyl-D-alanine sequence occurs in all nascent peptidoglycans and an L-R_3 group with the same type of profile as above is ubiquitous in peptidoglycans of chemotypes I and II.

Evidence that these Streptomyces enzymes are the transpeptidases that effect the closure of the bridges in the biosynthesis of the wall peptidoglycans includes the following.

1. In the presence of a suitable carboxyl donor, such as N^{α}, N^{ϵ}-diacetyl-L-lysyl-D-alanyl-D-alanine, and a proper amino acceptor, the purified enzymes from Streptomyces R61, K11, and R39 were found to be able to catalyze transpeptidation with the concomitant release of the terminal D-alanine of the donor peptide [135]. The reactions occur in the absence of an exogenous input of energy. With either [^{14}C]-D-Ala, [^{14}C]-Gly, or [^{3}H]-meso-diaminopimelic acid as acceptor, these three

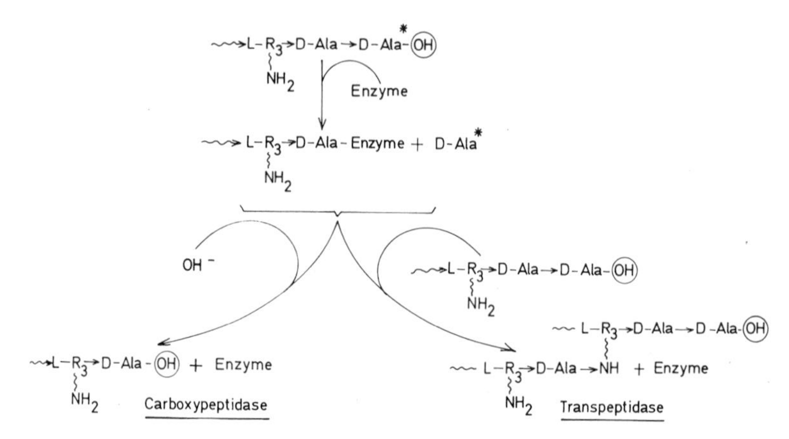

FIG. 25. Possible mechanism of hydrolysis and transpeptidation by "DD carboxypeptidase-transpeptidase" enzymes (see text).

enzymes catalyzed the formation of either diacetyl-L-lysyl-D-alanyl-[^{14}C]-D-alanine, diacetyl-L-lysyl-D-alanyl-[^{14}C]-glycine, or diacetyl-L-lysyl-D-alanyl-D-[^{3}H]-meso-diaminopimelic acid. In the presence of saturating acceptor concentrations, the time course of transpeptidation paralleled the time course of hydrolysis of the donor peptide when no acceptor other than water was present.

2. In some cases, transpeptidation was shown to occur at a molar ratio of amino acceptor:water as low as $1{:}1.8 \times 10^6$ (assuming that the concentration of water in the active site region of the enzyme is 55 M), demonstrating the exceedingly high efficiency of the enzymes as transpeptidases [135].

3. The Streptomyces enzymes exhibited differences with respect to their requirements for peptide acceptors and these differences reflected the differences in structure of the wall peptidoglycans of the corresponding strains.

As shown in Figs. 9 and 24, the interpeptide bond in strains R61 and K11 is a D-alanyl-glycyl linkage, which is in an endoposition, whereas the interpeptide bond in strain R39 is a D-alanyl-(D)-meso-diaminopimelic acid linkage, which is in an α-position to a free carboxyl group. As shown in Table IV, the R39 transpeptidase only catalyzes the synthesis of interpeptide bonds that are α to a free carboxyl group, whereas both R61 and K11 enzymes are able to catalyze the synthesis of peptide bonds either in an endo- or at a C-terminal position.

DD-Carboxypeptidase activity (as shown by the release of the C-terminal D-alanine residue from UDP-N-acetylmuramyl-pentapeptide precursors) also exists in gram-negative bacteria (E. coli; Salmonella) [125, 127], in gram-positive bacteria (particularly in several Bacilli) [136-139], and blue-green algae (Anaboena variabilis) [140]. Only cell-bound activity has been found so that carboxypeptidase-transpeptidase excretion may be a property unique to Streptomyces. In E. coli, the DD-carboxypeptidase is located in the inner plasma membrane of the cell envelope and has been solubilized by sonication of the cells [127] and by detergent treatment of the membrane [141]. The specificity of the soluble DD-carboxypeptidase from E. coli has not yet been studied, and so far an attempt to test it in a transpeptidase assay with defined peptide substrates has not been reported. In contrast, the crude "particulate" preparation obtained from E. coli exhibited transpeptidase activity as well as carboxypeptidase activity, in assays that involved concomitant peptidoglycan synthesis from UDP-N-acetylglucosamine and UDP-N-acetylmuramyl-pentapeptide precursors [128]. It is possible that the E. coli enzyme requires a polymerized (at the glycan level) nascent peptidoglycan in order to function as a transpeptidase. It is also possible that E. coli, as well as other bacteria, contain a protein that exhibits only a carboxypeptidase activity, in addition to the carboxypeptidase-transpeptidase system such as that described for some Streptomyces enzymes.

K. Inhibition of the Transpeptidation Reaction by Penicillin

It is certain that an early and important step in the lethal action of penicillin G (and other penicillins and cephalosporins) upon bacteria is the abolition of, or reduction in efficiency of, the transpeptidase involved in peptide cross-linking. The purified Streptomyces R61, K11, and R39 enzymes were found to be inhibited by several penicillins and cephalosporins. With each antibiotic, inhibition of both carboxypeptidase and transpeptidase activity occurred at the same concentrations [135 and unpublished results]. This was additional evidence that at least in these Streptomyces sp. it is one and the same enzyme that is responsible for both activities.

The mechanism of inhibition by penicillin of hydrolysis of the peptide donor was studied in the absence of amino acceptor. Penicillin appeared to combine with the R61 enzyme at a site that was not identical with the substrate binding site. Kinetically, the inhibition of the R61 enzyme was competitive [133]. However, competitive kinetics do not necessarily exclude an inhibitory mechanism other than a direct competition between substrate and inhibitor for the same site on the

TABLE 1

Substrate Requirements of *Streptomyces* DD-Carboxypeptidases-Transpeptidases[a,b]

Ac → L-Lys → R_2 ↓→ R_1(OH) (↑ε Ac on L-Lys)	R61			K11			R39			albus G		
	K_m	V_{max}	Efficiency	K_m	V_{max}	Efficiency	K_m	V_{max}	Efficiency	K_m	V_{max}	Efficiency
Ac → L-Lys → (↑ε Ac) D-Ala → D-Ala	12	890	72	11	2,000	182	0.8	330	410	0.33	100	300
D-Ala → D-Lys	13	80	7							0.80	85	106
D-Ala → D-Leu	10	50	5	10	160	16	0.7	230	320	0.33	33	100
D-Ala → Gly	36	200	6	12	220	18	2.5	100	40	2.50	60	24
D-Ala → L-Ala	Virtually no hydrolysis			Virtually no hydrolysis			No hydrolysis			No hydrolysis		
Gly → D-Ala	16	1.7	0.1	14	10	0.7	No hydrolysis			15.0	107	7
D-Leu → D-Ala	10	10	1	13	38	2.9	No hydrolysis			No hydrolysis		
L-Ala → D-Ala	Virtually no hydrolysis			Virtually no hydrolysis			No hydrolysis			No hydrolysis		

[a] Release of the C-terminal residue. Influence exerted by the C-terminal $R_2 \rightarrow R_1$ dipeptide sequence.

[b] K_m: m*M*; V_{max}: μmoles/mg/hour; Efficiency: V_{max}/K_m

TABLE 2

Substrate Requirements of *Streptomyces* DD-Carboxypeptidases-Transpeptidases[a]

~~→ L-R_3 → D-Ala ↓→ D-Ala (OH)		Specific activity[b] R39	K11	R61	*albus* G
Ac → L-Ala →	D-Ala → D-Ala	200	1,100	650	300
R_1 → L-homoSer →	D-Ala → D-Ala	6,000	1,600	1,400	2,000
N^{α}, N^{γ}-bisAc → L-DAB →	D-Ala → D-Ala	14,000	6,500	4,000	22,000
N^{α}, N^{δ}-bisAc → L-Orn →	D-Ala → D-Ala	75,000	50,000	21,000	22,000
N^{α}, N^{ϵ}-bisAc → L-Lys →	D-Ala → D-Ala	90,000	90,000	47,000	40,000

[a] Release of the C-terminal residue. Influence of the length of the side-chain of the L-R_3 residue that precedes the C-terminal D-alanyl-D-alanine sequence.

[b] mμEq of linkage hydrolyzed/mg/h; R_1: UDP-MurNAC-Gly-γ-D-Glu.

TABLE 3

Substrate Requirements of *Streptomyces* DD-Carboxypeptidases-Transpeptidases[a,b]

~~→ (L) CH → D-Ala ↓→ D-Ala (OH); $(CH_2)_3$; R'-NH-CH-R"	R61			K11			R39			*albus* G		
	K_m	V_{max}	Effi-ciency	K_m	V_{max}	Effi-ciency	K_m	V_{max}	Effi-ciency	K_m	V_{max}	Effi-ciency
(1) Ac → ⊤ → D-Ala → D-Ala; $CH_3CONH-CH_2$	12	890	72	11	2,000	182	0.80	330	410	0.33	100	300
(2) Ac → ⊤ →; NH_2-CH_2	15	4	0.3	30	9	0.3	0.20	600	3,000	6.0	20	3
(3) R_1 → ⊤ →; NH_2-CH-COOH (D)	11	8	0.3	17	18	1.06	0.25	400	1,600	0.4	10	25
(4) R_2 → ⊤ →; $NH_2-Gly_5-NH-CH_2$	14	800	57	11	1,050	95	0.30	420	1,400	0.28	9	52

[a] Release of the C-terminal residue. Influence exerted by, the presence of charged groups at the end of the side chain of the L-R_3 residue that precedes the C-terminal D-alanyl-D-alanine sequence. The side chains are those of the following amino acids. (1) N^{ϵ}-acetyl-L-lysine; (2) L-lysine; (3) *meso*-diaminopimelic acid; (4) N^{ϵ}-pentaglycyl-L-lysine.

[b] K_m mM; V_{max}, μmole/mg/h; efficiency, V_{max}/k_m. R_1, UDP-MurNAC-L-Ala-D-Glu-γ-; R_2, [N^{α}-Disacch-L-Ala-D-Glu(NH_2)-γ].

free enzyme. Binding of penicillin G to the R61 enzyme was found to cause quenching of its fluorescence. The dissociation constant of the enzyme-penicillin complex as measured by fluorescence quenching (10^{-8} M) was in close agreement with the K_i value as determined by kinetic measurements. Both acceptor and donor peptide did not decrease the affinity of the R61 enzyme for penicillin [141a]. The penicillin molecule has a highly reactive CO-N bond in its lactam ring causing it to be a powerful acylating agent. The inhibition of the Streptomyces R61 carboxypeptidase-transpeptidase by penicillin G was found to be reversible and to occur in the absence of detectable acylation of the protein by the antibiotic.

The transpeptidase activity of the particulate preparation from E. coli [128], the carboxypeptidase activity of the soluble preparation obtained by sonication of the same organism [127] as well as the carboxypeptidase activity of the particulate preparations from Bacilli [136, 137] and Anaboena [140] were also found to be inhibited by low doses of penicillin. The E. coli soluble DD-carboxypeptidase was competitively [127], and most probably, reversibly inhibited by penicillin. With some of the other enzyme preparations, irreversible fixation of penicillin was observed, but a direct demonstration that enzyme inactivation via penicilloylation occurred is lacking. It is possible, however, that different carboxypeptidase-transpeptidases from different organisms may be inhibited or inactivated in different ways.

Despite a number of hypotheses and experiments, even today little is known about the precise molecular basis of penicillin action [142]. The previously proposed ideas that penicillin could be a structural analog of the nascent peptidoglycan either at the level of N-acetylmuramic acid [143] or at the level of the D-alanyl-D-alanine donor site [122] were based on too few experiments and were probably premature. The resemblance between penicillin and D-alanyl-D-alanine is not close in two respects. The peptide bond between the two D-alanine residues is about 25% longer than the corresponding bond in the β-lactam ring of penicillin. Moreover, the angle around the D-alanyl-D-alanine peptide bond (normally 180 deg) is considerably different from the angle around the corresponding bond in the β-lactam ring (135.7 deg). It has been proposed [144] that cleavage and formation of peptide bond by the transpeptidase occurs via a transition state during which the D-alanyl-D-alanine peptide bond would be distorted to an angle of 135.7 deg. Such a distortion would be energetically unfavorable and the driving force required for it would be produced by the energy of binding to

TABLE 4

Substrate Requirements of Three _Streptomyces_ DD-Carboxypeptidases-Transpeptidases Acting as Transpeptidases [a, c]

Acceptor used	Product of transpeptidation	Yield of product with enzymes [b] R61 or K11	R39
Gly-Gly	Ac_2-L-lys-D-Ala-Gly-Gly	+ + +	0
Gly-Gly-Gly	Ac_2-L-lys-D-Ala-Gly-Gly-Gly	+ +	0
Gly-L-Ala	Ac_2-L-lys-D-Ala-Gly-L-Ala	+ + +	0
Gly- L Ac- L	Ac_2-L-lys-D-Ala-Gly- L Ac- L	+ +	0
Gly-D-Ala	Ac_2-L-lys-D-Ala-Gly-D-Ala	+	0
D-Ala-Gly	Ac_2-L-lys-D-Ala-D-Ala-Gly	+	0
D-Ala-L-Ala	Ac_2-L-lys-D-Ala-D-Ala-L-Ala	+	0
L-Ala-L-Ala		0	0
L-Ala-D-Glu L -D-Ala D - (OH)	L-Ala-D-Glu -D-Ala Ac_2-L-lys-D-Ala - (OH)	+ +	+ +

L-Ala-D-Glu-[L/D: -D-Ala, -(amide)] L-Ala-D-Glu-[-D-Ala, -(amide)], Ac_2-L-lys-D-Ala	+ +	0

[a] Transfer of diacetyl-L-lysyl-D-alanyl from diacetyl-L-lysyl-D-alanyl-D-alanine tripeptide donor to various peptides acceptors.

L–D (I-shaped symbol) = meso-diaminopimelic acid; L–L (I-shaped symbol) = LL-diaminopimelic acid.

[b] + + +, + +, +: high, good, and low yield of product of transpeptidation, respectively. 0 : no detectable product.

[c] Unpublished results from H. R. Perkins, M. Nieto, J. M. Ghuysen, J. M. Frère, M. Leyh-Bouille, R. Moreno, J. Dusart, and A. Marquet.

the protein. Since the penicillin molecule is already suitably distorted, its binding to the transpeptidase would be much more favorable. It has also been assumed that cleavage of the β-lactam amide bond would then follow with concomitant acylation and irreversible inactivation of the enzyme. Hence, according to this hypothesis, the molecular basis for penicillin action would rest upon a combination of two features: the resemblance to the transition state and the acylating capacity of the penicillin molecule. Experiments with the DD-carboxypeptidase-transpeptidase from Streptomyces R61 however, are in open conflict with this idea. Penicillin inhibition of this carboxypeptidase-transpeptidase does not rely on structural analogy as shown by fluorescent and peptide analog inhibition [141a, 144a] studies. In addition, penicillin inhibition was not accompanied by detectable acylation of the enzyme protein ([134] and unpublished results). The penicillin-treated enzyme could be recovered with full activity. Moreover, it has been recently found [144b] that the 6-methyl derivative of the methyl ester of penicillin which is a structurally closer analog of D-alanyl-D-alanine than the methyl ester of penicillin (since it has a methyl group in the position where penicillin has a hydrogen atom) is in fact completely inactive as an antibiotic. It is hoped that the Streptomyces transpeptidases and other soluble transpeptidases, which probably will be isolated from other bacteria in the near future, will provide accurate models for the study of penicillin-receptor interactions at the molecular level.

Much also remains to be done before the mechanism of action of penicillin at the cellular level is clearly understood. The problem is complicated by the fact that (1) on binding to living cells, irreversible fixation of the great majority of penicillin molecules takes place on sites that may not be concerned with peptidoglycan synthesis and that are probably irrelevant to the killing action of penicillin [145, 146], (2) more than one transpeptidase (for example, one for cell elongation and one for septum formation) may be present in the same bacterium, (3) transpeptidase may not be the only target of penicillin. It has been suggested that an autolytic glycosidase activity of E. coli was inhibited by very high doses of penicillin [147, 148]. Recent data on this topic can be found in reference 148a and 148b.

L. Mechanism of Action of Vancomycin [93, 149-154]

Vancomycin and ristocetin inhibit peptidoglycan synthesis. Vancomycin binds in vitro to wall peptidoglycan precursor nucleotides and other peptides terminating in R-D-alanyl-D-alanine or related structures in which both terminal residues are glycine or have the D-configuration. Thus, complex formation is thought to be the basis for the lethal action of this antibiotic. Experiments with radioactive iodovancomycin [155] showed that very little passed into the cytoplasm of M. lysodeikticus cells, although some was found in the membrane fraction. This suggests that vancomycin cannot reach and bind to the nucleotide precursors in living cells. It may complex with the lipid intermediate that is about to be polymerized on the membrane surface (Section III, E). It may also interfere with later stages of maturation of the peptidoglycan involving transfer from the lipid intermediate into the wall and/or peptide cross-linking by transpeptidation. Reversal of vancomycin action has been obtained with the soluble peptide diacetyl-L-diaminobutyryl-D-alanyl-D-alanine [156]. This peptide binds vancomycin and counteracts both its inhibition of peptidoglycan synthesis and its enhancement of lipid intermediate accumulation. When the tripeptide was added to a culture of Bacillus megaterium at various intervals after vancomycin treatment, growth inhibition caused by the antibiotic was reversed.

IV. PEPTIDOGLYCAN AS PART OF THE CELL ENVELOPE

A. Location of the Peptidoglycan

In the bacterial cell, the peptidoglycan is located outside of the ultimate permeability barrier (protoplast membrane). In fact, the bacterial cell wall can be defined as that cellular organelle which is located outside of the protoplast membrane and contains peptidoglycan. Thus, the wall is an exocellular structure.

An important difference between the Gram-negative and Gram-positive species is that the envelope of Gram-negative bacteria contains an outer membrane very similar in appearance in thin sections to the inner plasma membrane. This outer membrane, sometimes called the L membrane [157], exhibits different permeability properties and contains lipopolysaccharides and lipoproteins. In a variety of Gram-negative species, a number of proteins and enzymes have been found to be released from the cells by various procedures involving osmotic shock, while other (intracellular) proteins are completely retained by the cells.

The complex layering in Gram-negative bacteria is clearly seen in thin sections [18, 157-158] (Figs. 26 and 27). An electron-dense layer G2 is separated from the outer L triple-layered membrane unit by a transparent layer G1, and from the inner triple-layered plasma membrane by another transparent layer M (the terminology is that proposed by De Petris [157]). The peptidoglycan has been located within the G2 layer, a dense line 1.5-2.0 nm thick, at the inner boundary of the wall part of the envelope. The peptidoglycan of E. coli was the first one to be isolated in the form of a rigid layer [8]. Granules of lipoprotein were shown to be covalently linked to it and they could be removed with proteolytic enzymes. In this organism a lysylarginine dipeptide extends from some of the meso-diaminopimelic acid residues of the peptidoglycan and covalently connects the structural lipoprotein to the peptidoglycan. On the average, one lipoprotein molecule

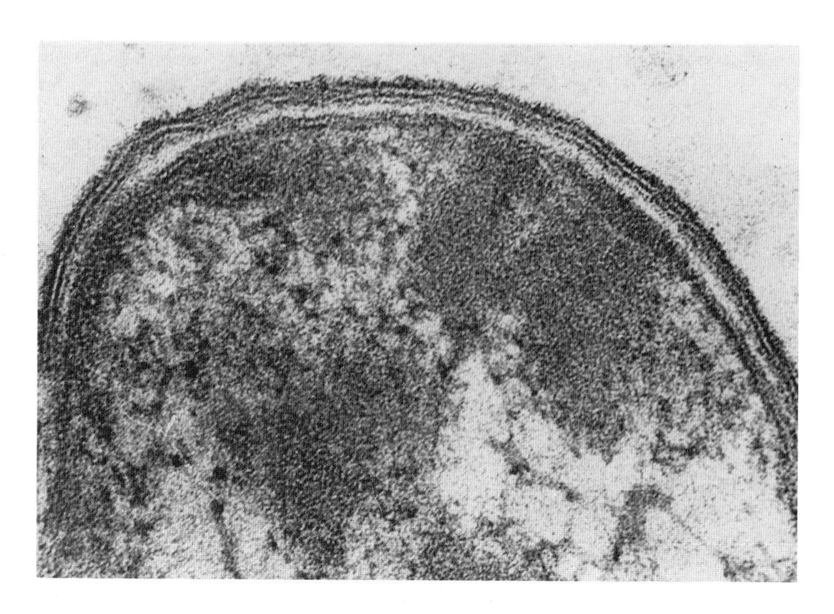

FIG. 26. Electron micrograph of a section of Proteus vulgaris P18 heated (5 minutes at 80°C) showing the complex multilayered structure of the cell envelope. Magnification x129,500. Reprinted from reference 18 by courtesy of Elsevier Publishing Co., Amsterdam.

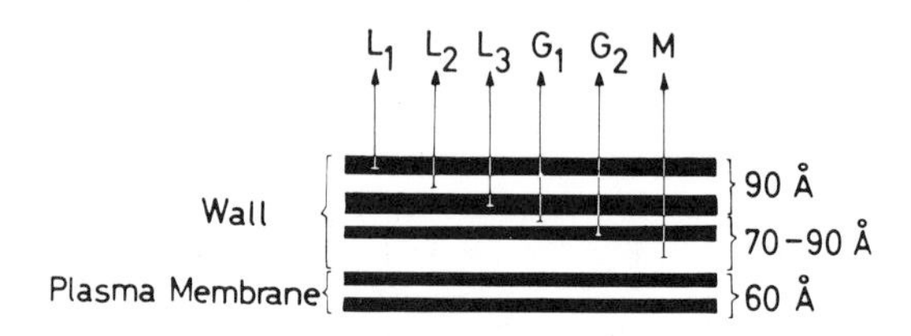

FIG. 27. Schematic representation of the structure of the cell envelope of Proteus vulgaris P18. The terminology is that used by De Petris [158] to describe the cell envelope of E. coli. The G_2 layer is thought to contain peptidoglycan. Reprinted from reference [18] by courtesy of Elsevier Publishing Co., Amsterdam.

would be linked to every tenth repeating peptide unit in the peptidoglycan [159]. In Gram-negative bacteria, this appears to be the only envelope polymer covalently linked to the peptidoglycan. Physical association between the plasma membrane and the peptidoglycan-containing G2 layer is not clear. It is thought that "adhesion sites" are present within the M region [160]. Techniques are available which, after enzymatic degradation of the G2 layer, allow the separation of the inner membrane from the outer membrane and the purification of each of these organelles [161, 162].

In general, the cell envelope of Gram-positive species lacks the outer L membrane [163] (Fig. 28). The wall, which surrounds the membrane, is considerably thicker (15-50 nm) than the equivalent G2 layer of the Gram-negative bacteria. It consists of 40-90% of peptidoglycan. An almost endless variety of polysaccharides that are frequently negatively charged and of polyolphosphate polymers that are collectively called teichoic acid [21] are covalently linked to the peptidoglycan and constitute the nonpeptidoglycan part of the wall. These latter polymers are highly antigenic. One, but probably not the only link between peptidoglycan and other wall polymers is via phosphodiester bridges from C-6 of N-acetylmuramic acid (see Section II, A above). By providing the chemical groups through which these polymers are anchored into the wall, the peptidoglycan contributes to the surface properties of the cell. In *S. aureus* [164, 165], for example, the N-acetylglucosamine residues of the teichoic acid are essential for phage fixation but, in order to be operative, these groups must possess a definite orientation that is imparted by the supporting insoluble peptidoglycan.

Truly layered wall profiles have been observed only infrequently in Gram-positive bacteria. One example is the layers seen after wall thickening had occurred during stationary-phase growth [166]. A second example is the highly ordered layer found at the surface of some species of *Bacilli* and *Clostridia*. For example, Nermut and Murray [167] suggest that (1) in *Bacillus polymyxa* the close-packed array (RS) of 7 nm units with a repeating frequency of 10 nm, which is visible on the outer surface, is a protein, (2) the intermediate layer is probably polysaccharide, and (3) the innermost layer is the site for teichoic acid and peptidoglycan.

B. Interrelationship between Wall and Plasma Membrane

The surface antigens of Gram-positive bacteria have two remarkable properties, i.e., their "mobility" and their functional importance in concentrating Mg^{2+} ions at the plasma membrane. Growth environment conditions exert deep influences on the composition of the wall anionic polymers in altering the relative proportion of polysaccharides and teichoic acids [168]. As noted by Rogers [14], "providing either one or another of the groups of negatively charged polymers is present on its surface, the microorganism seems content." This mobility occurs despite the fact that polysaccharides and teichoic acids are covalently linked to the peptidoglycan and thus implies the presence of an active system for turnover. The physiological importance of this phenomenon is revealed by the recent demonstration

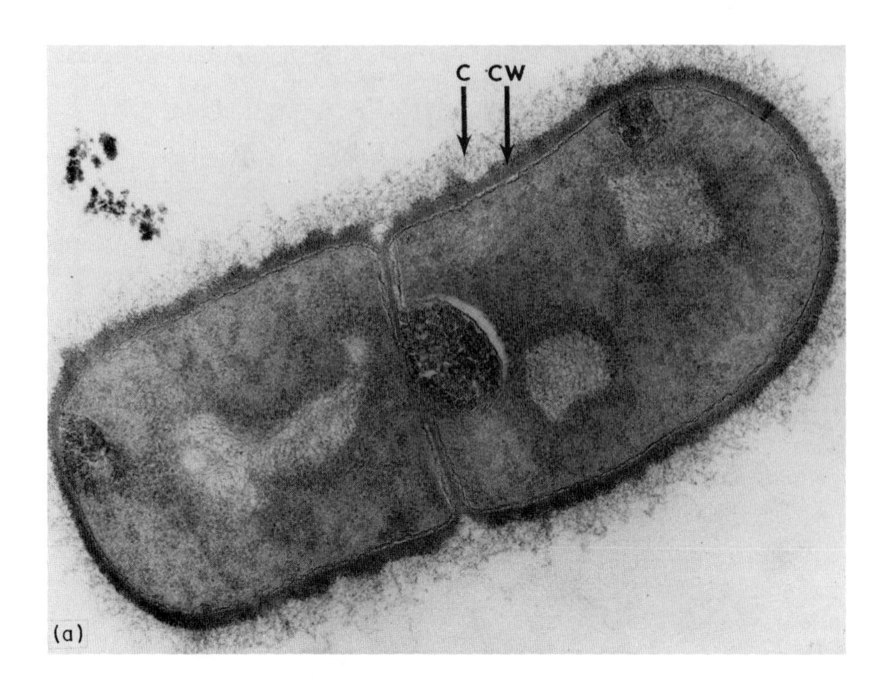
C
CW
(a)

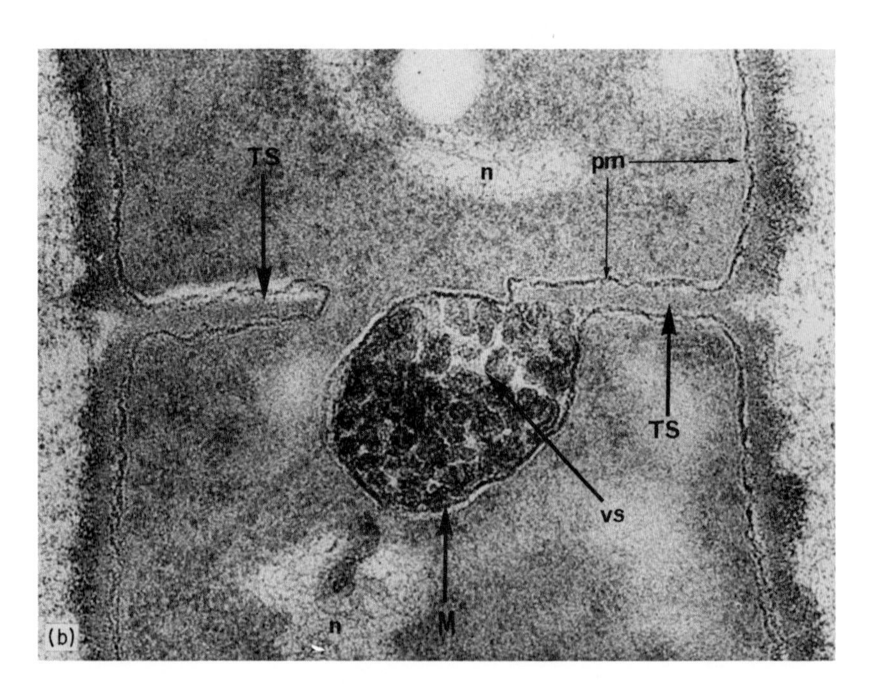
TS
n
pm
TS
vs
M
n
(b)

that teichoic acids seem to be essential for normal cellular activity and stability of the membrane [169]. In fact, two types of teichoic acids are involved. The wall teichoic acid is responsible for the capacity of isolated walls to bind Mg^{2+}. In addition to it, bacteria also contain a polyglycerol phosphate-teichoic acid, which is located in the periplasmic region of the cell, i.e., between the wall and the membrane. According to the model proposed by Baddiley and his colleagues, these two regions of anionic polymer would function as an integrated cation-exchange system between the exterior of the cell and the membrane, insuring to the latter the high concentration of Mg^{2+} required for many processes. One may assume that polymers other than teichoic acids have a similar function in Gram-negative bacteria.

C. Peptidoglycan in Resting and Germinating Spores

The peptidoglycan of the cortex in resting spores is extensively different from the wall peptidoglycan of vegetative cells. About 50% of the muramic acid residues in the spore peptidoglycan occur as nonacetylated internal amides [23]. The rest occurs as N-acetylmuramic acid, of which about half is substituted by only a C-terminal L-alanine residue, whereas the remainder is substituted by the usual type of peptide, which has a very low extent of cross-linking. Such profound alterations of the peptidoglycan suggest that the polymer fulfills an altered function in the spore cortex [69, 70]. The occurrence of muramic acid residues in the form of the δ-lactam prevents the lactyl groups from being involved in hydrogen bonding (see Section II, D). A likely consequence of this and of the low degree of peptide cross-linking, is a high flexibility of the polymer, which is, perhaps, better suited to its function in the spore cortex [11].

The envelope of resting spores has an exceedingly complex organization. Proceeding inward from the surface, thin sections of resting spores of *B. polymyxa* [170] reveal (1) the outermost sculptured surface of the sporangium, (2) the various laminae of the spore coats, (3) a zone of low

---

FIG. 28. Electron micrograph of a section of *Bacillus megaterium* ATCC 19213, taken from a synchronously dividing population at 3 hours. Note the absence of a distinctly layered appearance of the wall (CW) as compared with that usually seen in Gram-negative species (e.g., see Fig. 26). Also shown in the electron micrograph are the surrounding capsular material (C), the nascent transverse septum (TS), plasma membrane (pm), mesosomes (M), mesosomal vesicles (vs), and nuclear material (n). Note the close association of the nascent septum with a mesosome. Reprinted from reference 163 by courtesy of the American Society for Microbiology.

scattering corresponding to the cortex, (4) the wall primordium located at the inner boundary of the cortex in close vicinity to the plasma membrane, and (5) the spore body. The wall primordium presumably consists wholly or in part of peptidoglycan. The possession of a tripartite wall by the vegetative cells of B. polymyxa (see Section IV, A) implies that considerable change occurs to the wall primordium during spore germination. These changes involve elaboration of a new peptidoglycan structure and incorporation of other new wall components. They were followed by electron microscopy throughout germination [170] (Figs. 29 and 30). Rupture of the sporangial envelope and disintegration of the cortex were paralleled by a thickening of the wall primordium. This thickening was then reduced as cracks appeared in the laminae of the spore coats and the cell increased in size. At this stage of germination, patches of additional wall components composed of repeating elements, appeared under the cracks of the coat. The tripartite wall then enlarged from these centers and eventually covered the entire surface of the new cell. This remarkable study of the sequence of events in germination has made it possible to "see" a rearrangement and/or de novo synthesis of an apparent wall peptidoglycan and the delayed synthesis and assembly of an additional wall component, leading to a mature multilayered vegetative wall.

The presence of different peptidoglycan chemical structures in spores and vegetative cells offers interesting possibilities to study differentiation mechanisms at the biochemical and genetic levels. The peptidoglycan of vegetative cells of B. sphaericus 9602 contains L-lysine and D-isoasparagine (a peptidoglycan of chemotype II), whereas the spore peptidoglycan contains meso-diaminopimelic acid but not L-lysine or D-isoasparagine (peptidoglycan of chemotype I) [69, 70]. Both peptidoglycans, however, have a common precursor, uridine-diphospho-N-acetylmuramyl-L-alanyl-D-glutamic acid, which can accept either L-lysine or meso-diaminopimelic acid. Tipper and Pratt [70] found that as sporulation proceeds, the L-lysine-adding activity of the vegetative cells decays to a level that is maintained until meso-diaminopimelic acid-adding activity is detectable. The remaining L-lysine-adding activity then declines rapidly, whereas meso-diaminopimelic acid-adding activity increases rapidly. This process, as well as an increase in dipicolinate synthetase, was dependent on continued RNA and protein synthesis, and necessarily involves transcription and translation of at least one "sporulation-specific gene."

D. Autolytic Enzymes

Many bacterial species possess enzymes capable of hydrolyzing their own peptidoglycan. When such enzymes are permitted to act, the cells lose their osmotic protection, and autolyze. The presence and activity of such autolytic systems can be observed under a variety of conditions,

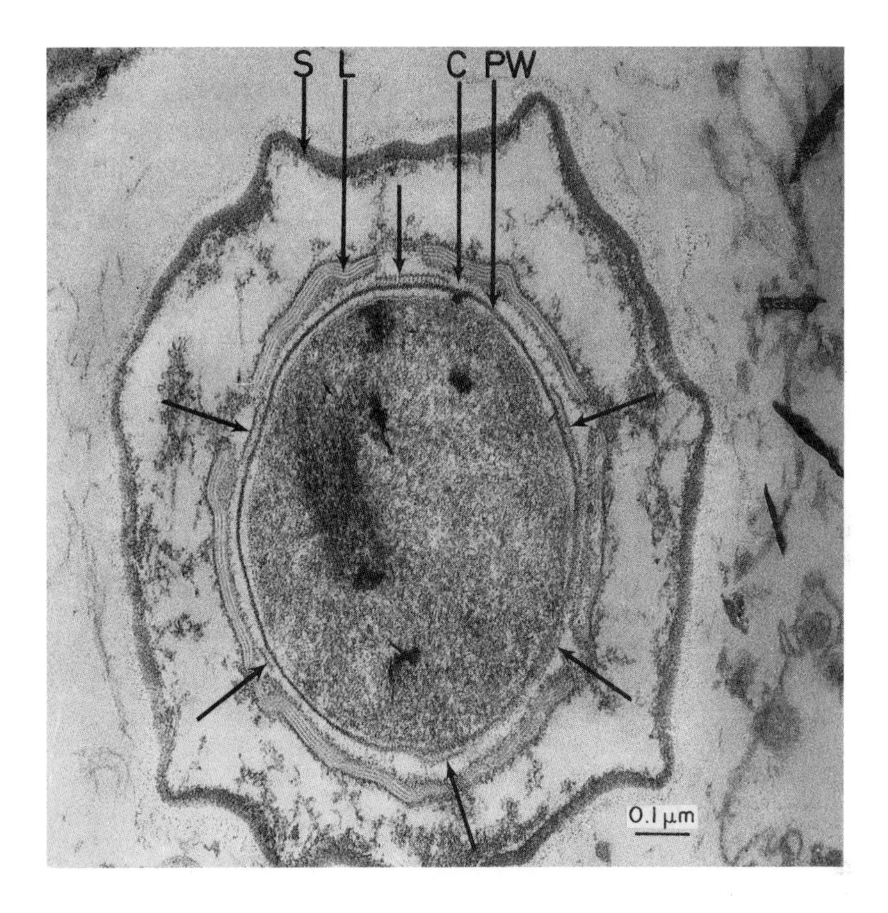

FIG. 29. Section of a germinating spore of Bacillus polymyxa. The exterior parts of the spore, i.e., the outermost surface of the sporangium (S) and the laminae of the spore coats (L), are disintegrating. The cortex (C) has all but disappeared. The primordial cell wall (PW), which is of slight density in the resting spore, now consists of a single highly scattering component. As shown by the six smaller arrows, the multilayered spore coats have cracked in several places and under the cracks, areas of the germ-cell wall show additional layers. The bar equals 0.1 μm. Reprinted from reference 170 by courtesy of the National Research Council of Canada.

many of which involve the inhibition of continued peptidoglycan synthesis. For example, the addition of specific inhibitors of peptidoglycan biosynthesis such as the penicillins or D-cycloserine, or the deprivation of nutrionally required peptidoglycan precursors such as glucose, glucosamine,

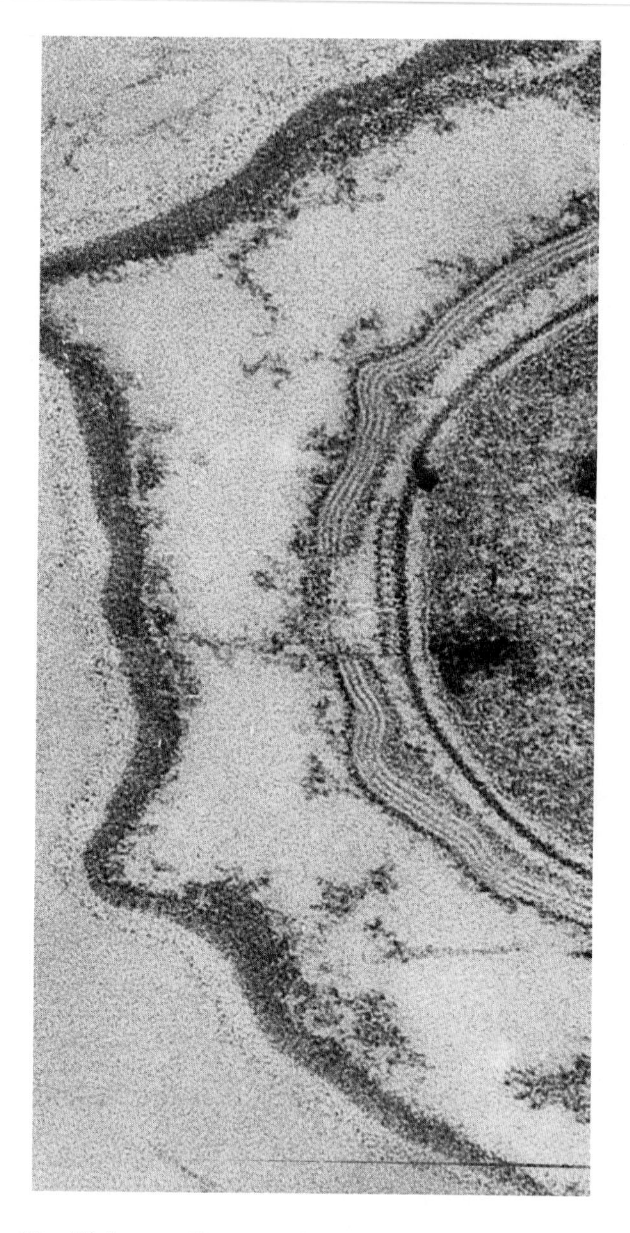

FIG. 30. High magnification of a portion of Fig. 29 to show detail of an early area of complete wall assembly during germination of spore of Bacillus polymyxa. Reprinted from reference 170 by courtesy of the National Research Council of Canada.

lysine, diaminopimelic acid, alanine, or glutamic acid, may result in cellular autolysis when the conditions of pH, salt concentration, etc., are suitable [171].

The relationship of inhibition of peptidoglycan synthesis to cellular autolysis is not due simply to the failure of wall growth to keep up with continued synthesis of protoplasmic constituents. This has been demonstrated in two ways. First, in S. faecalis ATCC 9790 (S. faecium) and other species, deprivation of lysine results in an inhibition of a net increase in both cellular protein and peptidoglycan, and, in cellular lysis. Second, both penicillin and D-cycloserine induce lysis of S. faecalis under conditions in which the cells are not significantly increasing their surface area but are engaged in wall thickening [172]. An increase in penicillin and cycloserine resistance occurs as the ability of the cells to autolyze decreases, and, as the walls thicken. An additional implication of these results is that inhibition of synthesis of the complete cross-linked peptidoglycan does not in itself result in cellular lysis, but that the active participation of one or more peptidoglycan hydrolases is required. A recent report of Tomasz et al. [173] showing that defects in the ability of Diplococcus pneumoniae to undergo cellular autolysis, induced nutritionally or via mutation, were accompanied by resistance to penicillin-induced lysis, serves to reinforce this concept.

The specificities of the autolytic enzymes found in various species correspond to the four described in Section II, B (N-acetylmuramidase, N-acetylglucosaminidase, N-acetylmuramyl-L-alanine amidase, and peptidases). Some have been found to be closely associated with the cell wall while others were found excreted in the growth medium, associated with the membrane or in the cytoplasm. In S. faecalis 9790 [174] and L. acidophilus [175] strain 63 AM Gasser, the only activity detectable in the wall fraction was an N-acetylmuramidase. In D. pneumoniae [176] various Bacillaceace [177] and Clostridia [177a, b] and in Listeria monocytogenes [177c], the major activity appears to be an N-acetylmuramyl-L-alanine amidase. Recently, however, the activity of a hexosaminidase of an as yet undetermined specificity, has been detected in a spore$^-$, proteinase$^-$ mutant of B. subtilis 168 [177]. It was postulated that the hexosaminidase activity may also be present in the parent Bacillus but was lost during preparative procedures, perhaps by proteolysis. Other species, such as E. coli [8] and S. aureus [57, 178-181], possess a variety of activities including N-acetylglucosaminidases, N-acetylmuramyl-L-alanine amidases, and cross-bridge-splitting peptidases. Evidence of an N-acetylmuramidase in S. aureus has not been obtained.

The presence of autolysins in rapidly growing and dividing cells led to the idea that such potentially dangerous enzymes may play an important role in bacterial growth [8, 14a, 182-185] as well as in other cellular functions such as sporulation, the ability of cells to become competent for transformation and the excretion of toxins and exoenzymes. Peptidoglycan hydrolases of any specificity could effectively function for the latter processes.

Several potential roles for autolytic peptidoglycan hydrolases in cell wall growth and cell division have been proposed [5, 14, 14a, 64]. These include (1) hydrolytic action to provide new acceptor sites for the addition of peptidoglycan precursors, (2) the hydrolysis of bonds in selected areas of the wall so that changes in cell shape may occur ("remodeling" function), (3) in cell division, that is compartmentalization into two new cell units separated by wall or membrane or both, (4) cell separation, and (5) the potential biosynthetic capacity of a hydrolytic activity.

1. Provision of New Acceptor Sites

A number of years ago it was thought that the number of acceptor sites could be a limiting factor for the expansion of the cell surface [8]. As discussed in Section II, C, this does not seem to be the case for peptidoglycans of most species. In any event, from what is known of the final steps of peptidoglycan biosynthesis (Section III), the only peptidoglycan hydrolase that would provide suitable additional "ends" (nonreducing N-acetylglucosamine) would be an N-acetylmuramidase. Species thus far found to have such an activity include S. faecalis [174], L. acidophilus [175], Arthobacter crystallopoietes [186] and Bacillus thuringiensis [187]. It is possible that a bridge-splitting peptidase could also function in a similar way, if disaccharide-peptide monomers can be added to the growing peptidoglycan via transpeptidation before the glycosidic bonds are formed. It also remains possible that nascent peptidoglycan units are formed from small oligomers (e.g., on a lipid carrier as in O-antigen biosynthesis [99]) and that oligomers are added to the growing wall by transpeptidation after preparation of suitable sites via peptidase action [187a]. It seems certain that the frequently found N-acetylmuramyl-L-alanine amidase and N-acetylglucosaminidase activities cannot produce suitable acceptor sites. In fact, in some organisms that produce relative large amounts of amidase, such as S. aureus and Bacilli spp., evidence for the presence of substantial quantities of the products of amidase action, i.e., unsubstituted N-acetylmuramic acid and N-terminal L-alanine peptide residues in their peptidoglycan is lacking. For example, in S. aureus the number of N-terminal L-alanine residues is no more than 6% of the L-alanine in the wall [57].

2. Remodeling Function of Autolysins

To our knowledge, this idea was first proposed by Rogers [14]. There is no doubt that bacterial cells change their shape and surface area-to-volume ratio throughout the cell division cycle. This is true not only for rod-shaped organisms but also for cocci. Hydrolysis of a few bonds at selected specific topological sites could permit a rearrangement of both the covalent and hydrogen bonding of the peptidoglycan network. Peptidoglycan hydrolase of any of the known specificities may serve this function.

Perhaps the most striking morphological change that has been observed to be accompanied by a change in level of autolytic enzyme activity, and a corresponding change in the structure of the resulting peptidoglycan, is the conversion of rods to spheres in *A. crystallopoietes*. A transient increase in N-acetylmuramidase activity was found to occur during rod-to-sphere conversion [186]. In addition, the average glycan chain length of the rods is 126, while that of the spheres is 34 hexosamine residues [32], which is consistent with the activity of a glycanase in morphogenesis. Similarly, the conversion of rod-shaped vegetative to spherical microcysts of *Myxococcus xanthus* was found to be accompanied by a transient increase in N-terminal *meso*-diaminopimelic acid residues [188], possibly the result of increased endopeptidase or decreased transpeptidase activity.

3. Cell Division

It is often hard to separate this function from cell separation (Fig. 31) since it is often technically difficult to determine the presence of a cross-wall, membranous septum, or some other type of physiological separation, especially in some rod-shaped species. The question then arises: does the division of one cell into two, or the fragmentation of a filamentous form, involve cross-wall growth, or merely the separation of preexisting cross-walls? To compound this problem, the time between initiation of nascent cross walls and their completion appears to be rather short, at least in some species. Also, the formation of a cross wall and its separation to become two new poles, results in subtle, sequential, and definable changes in cell shape and surface area-to-volume ratio, overlapping with the remodeling function discussed previously.

4. Cell Separation

Autolysins have been shown to play a role in cell separation in two *Bacillus* spp. [189-191], and *D. pneumoniae* [192], all of which possess potent amidase activities. For example, mutants of *Bacillus licheniformis* (lyt^-), deficient in ability to autolyze, grew at rates very similar to that of the wild type, but two of the least lytic mutants grew as very long chains of unseparated bacilli [189]. The wall chemistry of these two mutants also differed from that of the wild type [189].

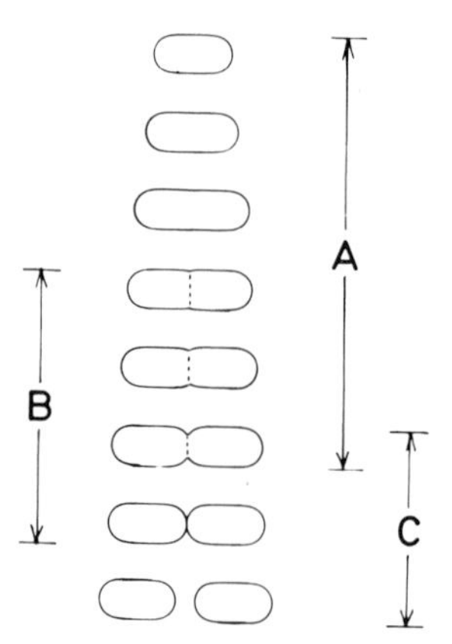

FIG. 31. Diagrammatic representation of the cell surface during the cell division cycle of a rod-shaped species. For simplicity the poles are drawn as hemispheres. Note the sequential changes in shape and surface area-to-volume ratio during the cell cycle. The overlapping elongation (A), division (B), and cell separation (C) sequence is also indicated.

5. Potential Biosynthetic Capacity of a Hydrolytic Activity

Some carboxypeptidases have been recently shown to be able to carry out transpeptidation reactions (Sections III, I and III, J. Endo-N-Acetyl-muramidases, such as hen egg-white lysozyme, are known to be trans-glycosidases [47-49]. Transglycosidases could be useful for linking together short glycan chains, perhaps even at the level of the lipid intermediate, in a manner similar to that proposed for cross-bridge formation via transpeptidation (Section III, I).

6. Multiple Roles for Autolysins

As pointed out by Rogers [14], autolytic enzymes may have multiple functions. Mutants defective in ability to autolyze, as well as other types of mutation which affect wall chemistry, morphology, or function, are frequently pleiotropic. Without detailed genetic analysis, it is often difficult to differentiate between multiple mutagenic events and the multiple effects of a single mutation. However, in most cases, the defects in mutants investigated all seem to be related to surface function.

The change in wall chemistry accompanying a defect in cell separation and low autolytic activity in lyt^- mutants of *B. licheniformis* is mentioned in the preceding text. In addition, a temperature-sensitive mutant of

B. subtilis deficient in autolytic activity grew somewhat more slowly than the wild type, and had an altered morphology, at the nonpermissive temperature [191]. The cells were irregular in width and had many bends. Normal morphology and an increased growth rate occurred upon the addition of either B. subtilis autolysin or hen egg-white lysozyme.

Two mutants of S. faecalis with defects in their autolytic system were partially characterized [193]. Compared to the wild type, both mutants autolyzed more slowly, had walls of very similar chemical composition, and grew at nearly the same rate. One mutant, E81, grew in long chains, contained low levels of both the active and proteinase-activatable latent forms of the autolysin, had walls that were less cross-linked and more sensitive to added isolated autolysin. The other mutant, E71, contained low levels of active but high levels of latent autolysin, and had walls of a similar degree of cross-linking and sensitivity to added enzyme as the wild type. In addition, both mutants showed altered morphology, being longer and having thicker walls than the wild type.

In view of the problem of distinguishing multiple from single mutations, the pleiotropic effects of replacement of choline by ethanolamine in walls of D. pneumoniae are particularly significant [173, 176, 192]. Choline is required for growth and is a constituent of the wall teichoic acid of this organism. Replacement of choline by ethanolanine results in (1) loss of the ability of the cells to autolyze, (2) resistance of isolated walls to the action of the autolytic amidase, (3) physical association of daughter cells in long chains, (4) loss of the capacity to undergo transformation, and (5) resistance to dissolution by deoxycholate and to penicillin, cycloserine, and phosphonomycin-induced lysis. Since this change appears to be a single one that affects only the composition of a nonpeptidoglycan polymer of the wall, it would seem that similar pleiotropic effects could arise from a single mutation affecting surface growth or composition.

7. Relationship of Autolytic Activity to In Vivo Peptidoglycan Synthesis

There does not appear to be a requirement for autolytic activity for peptidoglycan synthesis per se. For example, inhibition of growth of S. faecalis via an inhibition of protein synthesis either nutritionally (valine or threonine deprivation), or by means of the addition of antibiotics (chloramphenicol or tetracycline), results in a rapid decrease in ability of cells to autolyze, but in the continued synthesis of peptidoglycan and the other wall polymers [64, 171, 184, 185, 194-196a]. However, instead of enlarging the surface, most of the newly made wall polymers go into a thickening of the wall [196]. The result is more wall and peptidoglycan per unit mass, and per cell. Wall thickening (Fig. 32), which is a normal part of wall growth [197], occurred over the entire

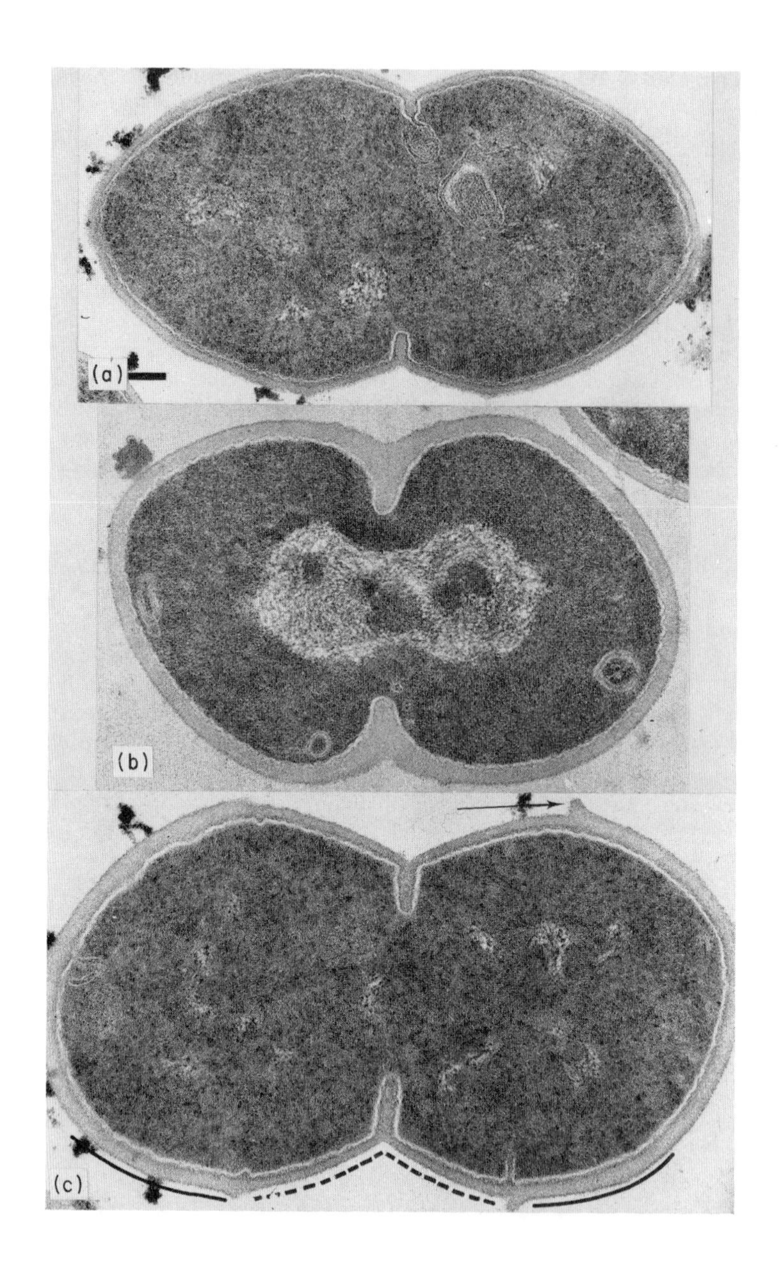
(a)
(b)
(c)

coccal surface [196], while it could be shown via electron microscopy (Fig. 33) that cellular autolysis was initiated at the tip and sides of nascent cross walls [198]. Using selective isotope-labeling techniques, it could be shown that the peptidoglycan made during chloramphenicol-induced wall thickening was not associated with autolytic enzyme activity [196a]. Thus, at least in this species it would appear that the autolytic N-acetylmuramidase can be associated only with peptidoglycan synthesis, which results in an increase in cellular surface area, a process which has also been shown to occur in the region of nascent cross walls [197]. A model for wall growth in this species has been presented recently [64, 197]. This model is consistent with multiple roles for the autolysin (remodeling, division, separation) and accounts for its continued action throughout most of the cell division cycle (see Section V, C).

This negative correlation between peptidoglycan synthesis per se and autolytic activity is not limited to S. faecalis. Similar though less complete information is available for other species including L. acidophilus [199], B. subtilis [166, 200] and B. megaterium [200]. In all cases thickening of the entire wall surface after an inhibition of protein synthesis was observed and correlated with a decrease in ability of cells to autolyze.

E. Heterolytic Enzymes

Some bacteria produce "heterolytic" enzymes, i.e., enzymes that are lytic only upon bacteria of other species. Both the ML endopeptidase, which acts on N^{ϵ}-(D-alanyl)-L-lysine linkages (Figs. 7 and 10), and the

---

FIG. 32. Electron micrographs of sections of Streptococcus faecalis. (A) Central longitudinal section of a cell from an exponentially growing culture. Arrow shows one of the wall bands that separate the equatorial wall made during the most recent generation from that made during a previous generation. Mesosomal membrane associated with the nascent cross wall can also be seen. (B) A central section of a cell from a culture deprived of threonine for 20 hr. Note the extensive thickening of the cell wall. (C) A central section of a cell from a culture recovering from 10 hr of threonine starvation. The culture had regrown for 80 min and had undergone 1.28 doubling in mass. Note the conservation of polar wall thickened during threonine starvation, which is now separated by thinner equatorial wall, synthesized during the 80-min period of regrowth. An increasing gradient of wall thickness from the nascent cross wall to the wall bands can be seen here as well as in part A of this figure. The bar equals 0.1 μm and applies to all three micrographs. Reprinted from references 196 and 227 by courtesy of the American Society for Microbiology.

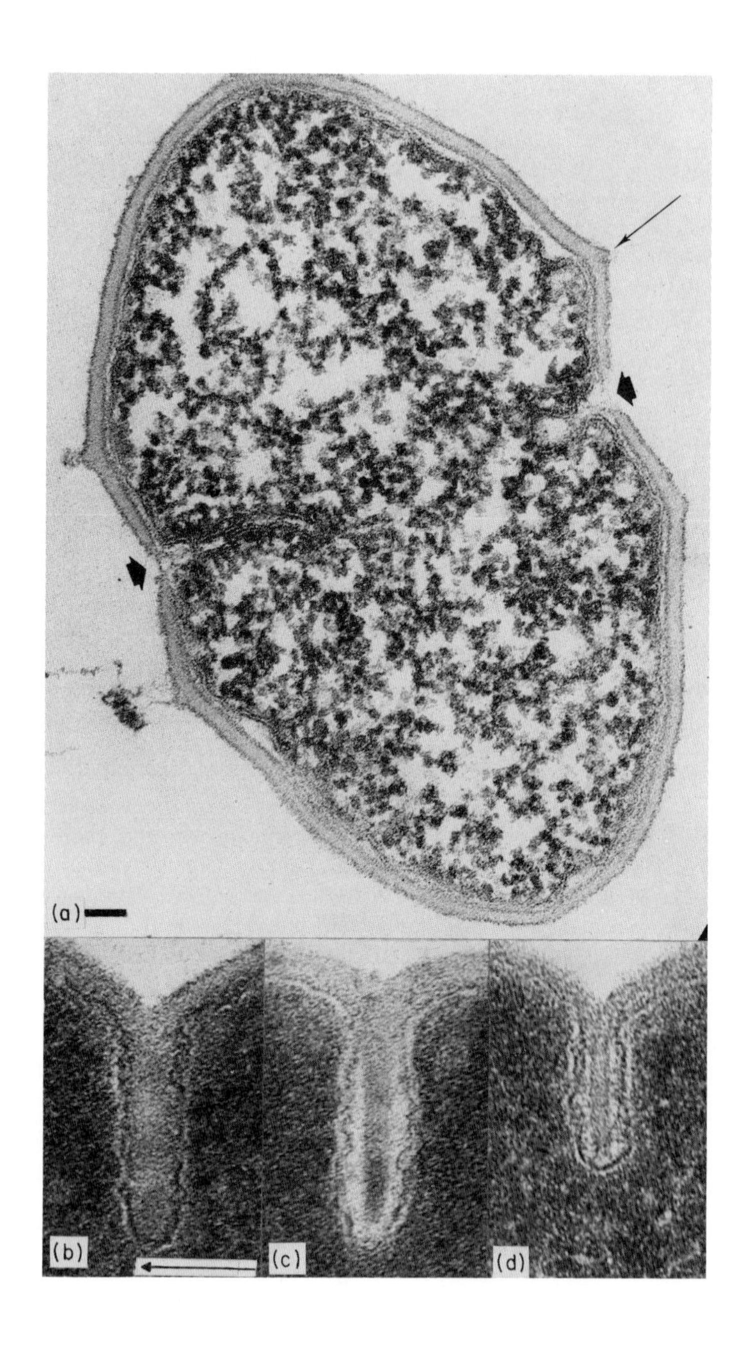
(a)
(b)
(c)
(d)

KM endopeptidase, which acts on C-terminal D-alanyl-D linkages (Figs. 6, 11, and 12), are produced by Streptomyces albus G. Linkages sensitive to these enzymes do not occur in the Streptomyces peptidoglycan. It may be that heterolytic enzymes are weapons devised by bacteria in order to enhance their possibilities of survival in a natural niche. It may also be that the lytic activities of these enzymes are "accidental" and that their real function in the bacterial cell that produces them is to achieve a specific function unrelated to their ability to lyse other species. The KM endopeptidase from Streptomyces albus G is of particular interest since, in fact, it is D-alanyl-D-carboxypeptidase, and D-alanyl-D-carboxypeptidases are known to play an important role in peptidoglycan biosynthesis (see Sections III, I and III, J.).

F. Peptidoglycan Turnover

Not only do the nature and relative amounts of acidic accessory wall polymers change during shifts in nutritional conditions (Section IV, B), but turnover of the peptidoglycan portion of the wall has been observed in some species but not in others. Pitel and Gilvarg [201] saw no evidence of peptidoglycan turnover in a diaminopimelic acid-and lysine-requiring mutant of B. megaterium during exponential growth or the early part of the stationary phase. No loss of diaminopimelic acid from the peptidoglycan of E. coli during excretion of a lipopolysaccharide-lipoprotein complex was observed [201a]. Peptidoglycan turnover in S. faecalis appears to be below a detectable level during (1) exponential growth, (2) wall thickening during valine or threonine starvation, (3) growth reinitiation after amino acid starvation, (4) growth in the presence of a just subgrowth inhibitory concentration of penicillin, and (5) recovery from a period of wall damage inflicted by lysozyme treatment [202].

On the other hand, relatively rapid rates of turnover of peptidoglycan and other wall polymers have been observed in other bacterial species

---

FIG. 33. Autolysis of Streptococcus faecalis. (A) Central, longitudinal section of a cell of Streptococcus faecalis 60 min after the beginning of autolysis (at 74% of the initial turbidity). The peripheral wall seems to be intact, but the entire cross wall is no longer visible (large arrows). The wall bands (small arrows) mark the separation of "new" equatorial, peripheral wall from "old" polar wall. Stages of cross wall dissolution. (B) Control, i.e., nonautolyzing cross wall. (C and D) Primary autolytic attack at the leading edges of the cross wall. The base of the cross wall remains intact and the septal membrane maintains its original invaginated position. Bar equals 0.1 μm. Reprinted from reference 198 by courtesy of the American Society for Microbiology.

[202, 203, 203a, 203b]. Peptidoglycan and wall teichoic acid were shown to turn over at the same rate in B. subtilis W23 [203]. The rate of peptidoglycan turnover was equivalent to a loss of almost 50% of wall material per generation in B. subtilis W23, and to about 30% per generation in B. megaterium KM. For B. subtilis the products of wall turnover were shown to result from the action of an N-acetylmuramyl-L-alanine amidase, and the turnover rate was decreased but not eliminated in the presence of 10 μg/ml of actinomycin D.

The peptidoglycan of L. acidophilus strain 63 AM Gasser was shown to turn over rapidly during exponential growth and upon recovery from amino acid starvation [199, 202]. The rate of turnover during exponential growth (generation times of 52-152 min) was approximately proportional to growth rate and 22-44% of the peptidoglycan was lost during each generation time. In this case, turnover does not appear to be related to autolytic activity since a slowly autolyzing mutant showed a turnover rate comparable to the controls. Peptidoglycan turnover slowed and stopped as the stationary growth phase was approached and was virtually completely prevented by chloramphenicol. A large number of anomalies of this turnover were noted of which some were as follows. (1) A lag of 0.8 to 2 generation times before turnover commenced even in random cultures that had been labeled for over six generations. (2) After pulse-labeling of less than 0.2 generation times, turnover was not observed while longer pulses resulted in turnover after the "normal" lag period. (3) Turnover was followed for many generations in a relatively undisturbed culture, maintained at a constant growth rate. After each doubling of mass, the culture was diluted (1:1) with fresh medium. After a lag of about one generation time, turnover occurred, with about a 40% loss of peptidoglycan per generation. Strikingly, however, peptidoglycan turnover stopped when about 10-20% of the initial labeled peptidoglycan was still present, suggesting that a significant fraction is immune to turnover. The observations do not appear to be either artifacts of the method of growth or of measurement of turnover, since, in L. acidophilus, significant turnover of protein was not observed, nor was detectable peptidoglycan (or protein) turnover seen in S. faecalis, grown and measured in the same way. A similar fraction of peptidoglycan immune to turnover was observed in experiments with B. megaterium, which turns over its peptidoglycan at a rate of about 15% per generation [203a].

At this time, it is difficult, if not impossible, to interpret the role of peptidoglycan turnover in the process of wall growth, in those species in which it occurs. Apparently, peptidoglycan turnover is somehow related to an increase in cell surface area and not to wall thickening. On the other hand, since the absence of detectable peptidoglycan turnover has been observed during growth of E. coli, S. faecalis, and a mutant of B. megaterium, turnover does not seem to be an essential feature of wall growth. Its occurrence can, however, complicate the interpretation of several types of experimental approaches.

V. BIOSYNTHESIS OF PEPTIDOGLYCAN AT THE CELLULAR LEVEL

A. Localization of Peptidoglycan Synthesis

It seems clear that membrane-bound enzymes effect the final steps of peptidoglycan synthesis and that the final wall acceptor must be in very intimate association with the membrane sites. It remains to be determined whether or not there are specific topological sites at which this occurs and/or specific units of membrane that carry out an integrated process [204]. Some consideration has been given to the mesosome as the site of the membrane-bound peptidoglycan-synthesizing apparatus. Two types of evidence point to the presence of this system in the entire membrane surface. First, the undecaprenol lipid carrier has been found to occur in plasma as well as mesosomal membrane of L. casei [205]. Although this compound may be used by the cell for other purposes, we only know of its functions as a carrier in the synthesis, and perhaps transport, of precursors of exocellular substances (see Section III, D) such as peptidoglycan and O-antigens. Second, even in an organism such as S. faecalis, for which it is known that wall synthesis which results in an enlargement of the surface area, occurs at a rather restricted portion surface (in the vicinity of nascent cross walls), peptidoglycan biosynthesis, which results in wall thickening, not only can, but does occur over virtually the entire coccal surface (Fig. 32) even during rapid exponential growth [64, 196, 197].

Such findings are surely related to the regulation of peptidoglycan biosynthesis more in terms of exactly where on the cell surface and in what direction peptidoglycan (and wall) assembly occurs, rather than in terms of factors affecting the overall rate of peptidoglycan synthesis. Direction of peptidoglycan (and wall) assembly are not only important in the consideration of wall thickening versus enlargement of the surface but also for cross-wall versus peripheral-wall formation. These last factors are crucial in the determination of cell division, cell size, and morphogenetic changes during the cell division cycle, which, in turn, are related to the synthesis and regulation of synthesis of informational macromolecules (DNA, RNA, and protein). Only the peripheral wall elongation and cross-wall formation processes, and not the wall-thickening process, can be coupled to the synthesis of informational macromolecules. After the cessation of protein synthesis, peptidoglycan synthesis proceeds at a virtually unchanged rate for a considerable period of time, thus emphasizing the complexity of the problem and the necessity to consider the topological location and direction of synthesis of the product [64, 196a, 206].

B. Relationship to Surface Enlargement

In cells of S. faecalis undergoing enlargement, newly synthesized peptidoglycan is closely associated with the active form of the autolysin [207], which has been in turn shown to be localized at nascent cross walls (Fig. 33) [198]. Electron microscopy (Fig. 32) showed that the wall elongation process also originates at nascent cross-walls [197]. Thus, in this case, and because of the covalent linkage of accessory wall polymers to the peptidoglycan, in probably all Gram-positive species, elongation of peptidoglycan is likely to be synonymous with elongation of wall. A limited number of discrete wall elongation sites have been seen in many cocci, which divide in one plane, by immunofluorescent or ferritin-labeled antibody or radioautographic techniques [208-213]. In all cases, antibodies to polymers other than peptidoglycan were used.

With rod-shaped organisms, results are conflicting. Hughes and Stokes [214] used an antipeptidoglycan-specific antiserum and were able to visualize a limited number of discrete sites of wall growth in a lyt^- strain of B. licheniformis, which exhibited very little peptidoglycan turnover. These investigators pointed out the complications of interpretation of wall-growth localization data in the presence of turnover. Chung et al. [210, 215] also observed a limited number of wall-growth sites in B. cereus, B. megaterium and E. coli using an immunofluorescence technique. Other data including the use of immunofluorescence, electron microscopic, and radioautographic techniques, are consistent with a large number of wall-growth sites in a variety of Bacillaceae and enteric bacteria [200, 209, 216-219a].

The use of penicillin to inhibit transpeptidation and thereby produce localized weakening in the protective wall has also produced conflicting results. In some cases, localized bulges in the wall were seen to develop [147, 220]. In others, multiple random weakened sites on the surface were seen [221].

For all of these techniques, there are a number of technical and other problems that are sometimes overlooked. The problem of wall turnover has been mentioned previously. In addition, there are the following problems. (1) Resolution and specificity of the techniques employed are limited. (2) Wall thickening has been shown to occur in a variety of bacterial species. In S. faecalis, it is a normal part of the wall elongation process during rapid exponential growth. Thickening occurs over a significant portion of the wall surface, but mostly near the zone engaged in wall elongation. This could be a more significant problem in other bacterial species, especially with techniques that are unable to clearly differentiate wall enlargement from wall-thickening sites. (3) As pointed out by Donachie and Begg [222] for E. coli, the number, location, and direction of growth zones may depend on growth rate and cell size (which are related). (4) Lytic, or morphological, effects induced by penicillin may well depend on the conditions used.

Penicillin can undoubtedly inhibit the large number of sites of wall biosynthesis that are engaged in wall thickening as well as those that are engaged in wall elongation and/or cross-wall formation. The ability of penicillin to induce cellular lysis under conditions where wall thickening rather than surface expansion is occurring [171, 172, 185] clearly shows an effect of the antibiotic on wall thickening. The electron micrographs of Murray et al. [223] suggest that penicillin inhibits wall thickening as well as cross-wall formation in S. aureus. Penicillin also inhibits the incorporation of precursors into peptidoglycan in "resting cells" of S. aureus [224], and lysis also occurs.

The morphological consequences of penicillin treatment of S. faecalis appear to depend on the physiological state of the cells at the time of antibiotic addition [225]. Exposure of exponentially growing streptococci to 0.4 μg/ml of penicillin, a concentration just below that which affects the rate of growth, resulted in the appearance of mostly swollen, balloon-shaped, or lemon-shaped cells. The rate of peptidoglycan synthesis was the same as the control. It would appear that penicillin inhibition of peptidoglycan transpeptidation at sites engaged in wall thickening as well as at sites engaged in wall enlargement, resulted in a modified product over most of the coccal surface. In contrast, the addition of the same concentration of penicillin to approximately the same number of cells emerging from 3 hr of threonine starvation, at any time up to about 35 min after inoculation into fresh medium, resulted in the appearance of rod-shaped cells (about one-half of the population) at 100-130 min. In this latter case, it would seem that penicillin treatment was primarily modifying the peptidoglycan product at the midpoint of the cells. In the absence of penicillin, cells recovering from 3 hr of threonine starvation undergo two synchronized divisions, the first at 40-45 min, much the same as that reported for another strain of S. faecalis starved and regrown similarly [226].

A combination of factors may contribute to the appearance of a significant number of rod-shaped cells. The thickened walls of the threonine-starved cells are conserved to become the poles of the first generation of daughter cells [227]. Such thickened poles would tend to maintain the lateral dimensions of the cells. Also, since the polar wall has already thickened during starvation, little peptidoglycan synthesis is presumed to occur at these sites. The inhibitory effect of penicillin would therefore be directed primarily toward peptidoglycan synthesis occurring at nascent cross walls and involved in wall enlargement. The localized effect of penicillin seen during regrowth of threonine-starved S. faecalis is similar to the localized effects observed during growth of an overnight culture of E. coli [147]. However, it should be noted that a similar degree of cross-linking in both the peptidoglycan of the control and penicillin-treated E. coli cultures was observed [147], consistent with similar observations on Proteus mirabilis and its peptidoglycan-containing, penicillin-induced L-form [228].

C. Relationship to Morphogenesis and Division

From the above and considerable additional information, it seems clear that factors that control the site, direction of assembly, and perhaps subsequent modifications of the peptidoglycan are those that govern surface enlargement, sites of division, shape, and changes in shape of the cell during the cell division cycle. At present, we know little about these factors except that they must be integrated with the other cellular processes. Elsewhere, a model for wall growth of cocci that divide in a single plane has been presented in some detail [64]. This model is based on considerable experimental evidence. The same article also presents a highly speculative model for the surface growth of rod-shaped bacteria, based on considerably less, and more conflicting information. These models were constructed on the principle that cocci are more primitive, in that they synthesize only cross walls (which become poles), while rods, in addition to making cross walls in order to septate the cytoplasm, form a cylindrical section of surface between cross walls. For increasing the surface area of rods, cylindrical extension is more economical in terms of surface area to volume ratio than is cross wall formation plus separation [64, 229]. Rods need to form and separate cross walls only to divide and increase in cell number. For several reasons, it has been assumed that rod-shaped species possess some additional genetic information, lacking in at least some cocci, for the production of a cylindrical wall section.

For both rods and cocci, cross-wall formation must be oriented not only topologically but also in direction of assembly. In virtually all cases examined, cross-wall formation proceeds centripetally from the cell surface, forming a continuously closing annular ring. Thus, even in rod-shaped species that appear to have a large number of peptidoglycan growth sites, those sites involved in cross-wall formation and cell division must be localized. The location of these cross-wall sites relative to the poles, and the relative rates of elongation of cross-wall and peripheral wall, govern the ultimate length (shape) of the cells as well as the shape changes that occur during the division cycle. When the size and shape of rod-shaped bacteria, such as E. coli or Salmonella, grown at different growth rates were compared, it was found that the faster-growing population had a greater mean cell volume [229]. The larger cell volume is due not only to an elongation of the cylinders but also to an enlargement of cell diameters. Thus, when cells are shifted up from a poor to a rich medium they increase not only in length but also in width, by a mechanism that is not yet understood.

The interrelationship of the processes of cross-wall closure and peripheral-wall enlargement is perhaps most simply illustrated by the streptococcal wall-growth model [64], where both processes are known to occur at the same sites (Fig. 34). In this model, peptidoglycan, which is synthesized and assembled at, or near nascent cross-walls, can find its way into peripheral wall, cross wall, or thickened wall. The fraction of peptidoglycan precursors that finds its way into each of the various products is influenced by growth conditions, and appears to depend upon the relative rates of three processes (diagramed in Fig. 34): (1) the rate at which precursors go into the wall-elongation process at the leading edges of the cross wall (R_1 in Fig. 34); (2) the rate at which precursors go into the wall thickening process which, even in rapidly growing cells, takes place at a much larger number of sites covering a larger portion of the surface and; (3) the rate at which newly elongated wall peels apart at the base of the cross wall into two layers of peripheral walls (R_2 in Fig. 34).

During rapid exponential growth, most peptidoglycan synthesis appears to occur at the more limited number of sites at the nascent cross wall that are engaged in wall elongation (R_1 in Fig. 34). However, a significant portion of synthesis results in a continuously decreasing gradient of the rate of the wall thickening, extending from the tip of the nascent septum (most rapid) to the poles (slowest). The size of the resulting daughter cells as well as their exact shape and the shape changes that occur during the division cycle, are, according to this model, governed by the relative rate at which peeling apart of the two new wall layers occurs at the base of the nascent cross wall (R_2 in Fig. 34). Relatively rapid peeling apart results in longer peripheral walls and shorter cross walls, i.e., longer cells. Relatively slow peeling apart (R_2 in Fig. 34) results in shorter peripheral wall and more cross wall, i.e., shorter and smaller cells. The timing and degree of the peeling apart process are also essential for cell separation. At present, little is known about the factors that govern the relative rates of these processes or if and when they vary in a cell division cycle. It is thought that the activity of the autolytic N-acetylmuramidase of S. faecalis plays a role, especially with regard to the rate of R_2 in Fig. 34. Factors thought to be involved in the regulation of surface enlargement are discussed in Section VI, E.

The streptococcal model is intended to suggest the basis for a generalized mechanism for cross-wall growth in bacteria of a variety of shapes

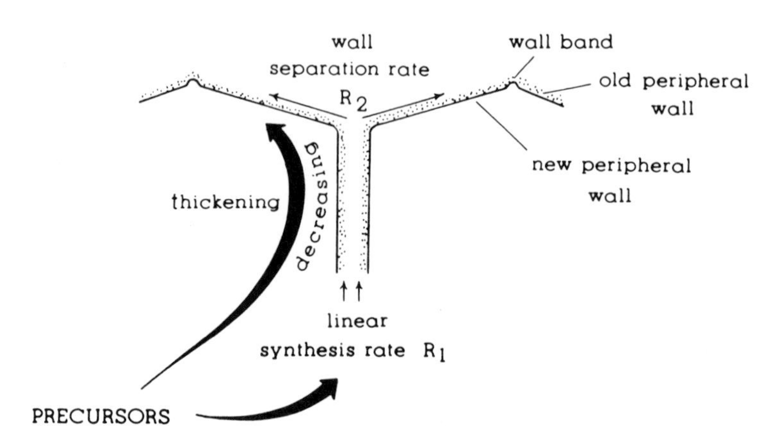

FIG. 34. Model of wall growth for Streptococcus faecalis. During exponential growth most of the wall precursors are fed into the leading edges of the growing cross wall. This forms two layers of cross-wall that separate into two layers of peripheral wall. By controlling the rate (R_2) at which the two layers of cross wall peel apart in relation to the rate of linear extension (R_1), the shape of the cell can be controlled. When the rate of separation (R_2) is approximately equivalent to the rate of linear extension (R_1), peripheral wall synthesis is favored and the cell increases in length experiencing a large increase in surface area. By inhibiting the rate of separation (R_2) relative to the rate of wall extension (R_1), the cross wall migrates centripetally into the cytoplasm to septate the cell. On completion of the septation process, the two layers in the cross wall complete separation and the two daughter cells then split apart. Besides linear wall extension, wall precursors are also used to thicken the cross wall to about twice the thickness of nearby peripheral wall and to continuously thicken peripheral wall. Most of the thickening occurs at the cross-wall tip, and it appears that the rate of wall thickening decays with wall age (as the wall is found further away from the cross-wall tip). The wall bands appear to mark the boundary between equatorial wall produced in the current generation and polar wall produced in a previous generation. Entrance of a culture into the stationary phase of growth interrupts the process of linear extension. During the early stages of this period of growth (1) precursors are preponderantly diverted into an overall thickening of the wall surface at the expense of linear extension; and (2) the rate of wall separation (R_2) is greatly inhibited. Reprinted from reference 64 by courtesy of the Chemical Rubber Co., Cleveland, Ohio.

and planes of division. Superimposed on the model, would be other factors such as cylinder formation by rod-shaped species. On the other hand, rod-shaped bacteria may use entirely different mechanisms. It also seems possible, both for this model and others, that changes occur in the primary, secondary or tertiary structure of the peptidoglycan subsequent to its initial incorporation into the wall. Such changes could be important in shape determination and morphogenesis and may be governed by other factors such as those discussed by Dr. Henning and Dr. Schwarz in Chapter 9.

VI. REGULATION OF PEPTIDOGLYCAN SYNTHESIS

A. General Considerations

The peptidoglycan is the most rigid part and one of the largest molecules of the bacterial cell. At the same time, it is a dynamic structure. This enormous molecule completely surrounds the cell, exhibits a high degree of mechanical and tensile strength, maintains shape, and yet undergoes constant remodeling during cell expansion and division. The peptidoglycan and the other wall polymers sometimes share a common constituent. D-Alanine and N-acetylglucosamine, for example, are constituents of the peptidoglycan and of some teichoic acids [21]. A given intermediate can participate in the synthesis of more than one wall polymer. The C_{55}-polyisoprenoid alcohol, for example, is utilized as a membrane carrier in peptidoglycan, and O-antigen syntheses [169]. Many factors including changes in cell size and shape with growth rate suggest that the peptidoglycan covering not only expands in order to keep up with an increase in cell volume but that the quantity, topological localization, and direction of peptidoglycan synthesis is, in a largely unknown way, regulated and integrated with the synthesis of other macromolecules and structures. These considerations emphasize the importance of the regulation of enzyme activity, formation and localization for peptidoglycan synthesis itself and its interrelationship with the synthesis of the rest of the cell.

B. At the Level of Nucleotide Precursors

Osmotically fragile temperature-sensitive mutants of *E. coli* K12 impaired at the levels of UDP-N-acetylglucosamine enolpyruvate reductase (proposed genetic symbol: *MurB*), of L-alanine-adding enzyme (*MurC*), of diaminopimelic acid-adding enzyme (*MurE*) and of D-alanyl-D-alanine-adding enzyme (*MurF*), respectively, were isolated and studied [230-235]. Mutants which would be impaired at the level of UDP-N-acetylglucosamine-2 phosphoenolpyruvate transferase (proposed genetic symbol: *MurA* [231]) and at the level of D-glutamic acid-adding enzyme (*MurD*) were not obtained. The *MurC*, *E*, and *F* genes were located extremely close to each other (between *Leu* and *azi*: 1 - 1, 5 min.) on the *E. coli* chromosome so that they may form or be part of an operon. In contrast, the MurB gene was located at 77 min. Mutants with impaired L-alanine racemase (alr, at 3 min) and D-alanyl-D-alanine ligase (ddl, at 17 min) activities, respectively, were also obtained.

A number of mutants probably altered in one of the membrane-bound enzymes were also isolated. The exact lesions or genetic loci have not yet been determined. Strikingly, accumulation of the completed UDP-N-acetylmuramyl-pentapeptide precursor could not be detected in any of these mutants [231]. Similarly, the amount of UDP-N-acetylmuramyl-pentapeptide in the K12 parental strain was not increased significantly by penicillin or vancomycin treatment, whereas in contrast, the amount of UDP-N-acetylmuramyl-tripeptide could be enormously increased by D-cycloserine. As suggested by Lugtenberg, the apparent difficulty in accumulation of large amounts of UDP-N-acetylmuramyl-pentapeptide in K12 strains, either under the action of antibiotics or by mutation, strongly suggests that UDP-N-acetylmuramyl-pentapeptide regulates its own biosynthesis by feedback inhibition [231]. It would appear that the choice of *S. aureus* in the early studies of penicillin action was rather fortunate since the accumulation of UDP-N-acetylmuramyl-pentapeptide may be limited to perhaps relatively few species that lack regulation at this level.

C. At the Level of the Alanine Branch

Regulation of the amount of D-alanyl-D-alanine available for condensation with the UDP-N-acetylmuramyl-tripeptide has been studied. The mechanism is complex and involves (1) regulation of the size of the intracellular pools of D- and L-alanine: Two unrelated transport systems, one for each isomer, have been identified [129]; (2) competition for available D-alanine by the several reactions that utilize this compound: A D-alanine membrane acceptor (ADP), which participates in the incorporation of D-alanine into membrane teichoic acid has been purified from *L. casei* [236]. A specific transaminase, which utilizes D-alanine as an amino donor for D-glutamic acid synthesis, has been purified from *B. subtilis* [237]; (3) D-alanine regulates its own biosynthesis. By acting as a specific inducer of L-alanine dehydrogenase in *B. subtilis*, D-alanine limits the amount of L-alanine available to the alanine racemase [238]. In *Pseudomonas aeruginosa*, D-alanine (as well as other D-amino acids) induces a D-amino acid dehydrogenase which directly limits the level of D-alanine [239]; (4) a high intracellular concentration of alanine apparently represses the synthesis of alanine racemase in *E. coli* [71]; (5) the intracellular ratio of L- to D-alanine influences the anabolic activity (L → D) of the racemase: This is an intrinsic property of the racemases of both *E. coli* and *S. faecalis*. The anabolic (L → D) velocity is higher than the catabolic (D → L) velocity and the K_m value for L-alanine is higher than for D-alanine (see Section

III, C). The Haldane relationship K_{Eq} is one. Thus, the intracellular L-alanine pool must be larger than the D-alanine pool in order to insure the necessary anabolic velocity; (6) The D-alanyl-D-alanine synthetase contains several product-binding sites, some of which probably function in the control of enzyme activity, which results in D-alanyl-D-alanine formation (see Section III, C). Since K_{Eq} of the racemase is one, it follows that inhibition of the D-alanyl-D-alanine synthetase by its product would, in fact, control utilization of L-alanine by the racemase.

D. Of Peptide Cross-linking

Whether it occurs in single or multiple enzyme forms, the DD-carboxypeptidase-transpeptidase system is likely to play a role of prime importance in controlling the degree of peptide cross-linking of wall peptidoglycans. Studies of several *Streptomyces* enzymes which perform both carboxypeptidase and transpeptidase activities, indicated the presence of several sites or subsites in the active region of the protein involved in substrate binding, or inducing a correct alignment of the catalytic groups, or both. These multiple sites may control the actual functioning of the system either in an anabolic (transpeptidase activity) or catabolic (carboxypeptidase activity) sense. The direction of the reaction could also depend upon the exact location of the enzyme within the plasma membrane, the availability of a nucleophilic group (water or a recognizable amino group), and upon the structure of the peptides.

The efficiency of the transpeptidation reaction is obviously determined by the intrinsic properties of the enzyme that catalyzes it. Hence, efficiency is likely to vary with bacterial species (and possibly, in the course of the cell division, or more complex life cycle of a given cell). In *S. aureus* and in *L. acidophilus* 63 AM Gasser, for example, the residual noncross-linked C-termini of the wall peptide moieties have retained the D-alanyl-D-alanine sequence of the nucleotide precursors (Fig. 17). Hence, in these cases, peptide hydrolases and particularly carboxypeptidase activity cannot be involved in the control of the size of the peptide moieties. Estimations of the extent of peptide cross-linking in the isolated walls indicate that the transpeptidation reaction is very efficient in *S. aureus* (80%) but poorly efficient in *L. acidophilus* (30%).

In most bacteria, the residual noncross-linked C-termini of the wall peptides have not retained the D-alanyl-D-alanine sequence of the nucleotide precursor. Elimination of the donor site involved in transpeptidation necessarily limits the size of the peptide moiety of the wall peptidoglycan. This can be carried out by carboxypeptidase activity upon the nascent disaccharide peptide units (during or after wall insertion), or at the level of cross-linked peptide oligomers within the wall itself. Note that carboxypeptidase action at the UDP-N-acetylmuramyl-pentapeptide level would prevent transfer of this unit to the lipid intermediate. Indeed, the translocase has a considerable specificity for C-terminal D-alanyl-D-alanine (see Section III, E). Consequently, removal of the C-terminal D-alanine residue from the precursor would prevent, or at least interrupt, peptidoglycan synthesis.

Limitation of the size of the peptide moiety could also result from the hydrolysis of D-alanyl interpeptide bonds in the completed peptide moiety of the walls through the action of endopeptidase autolysins. The C-termini appearing as a consequence of endopeptidase action would contain a single D-alanine residue. Such a reaction also could be carried out by carboxypeptidase activity at least when the interpeptide bonds are mediated through C-terminal D-alanyl-D linkages. When these interpeptide bonds are in a different position, other endopeptidases (see Section II, B) would be required to fulfill this function. The endopeptidase activity of the DD-carboxypeptidases is explained by the fact that these enzymes do not require a specific side chain of the C-terminal residue, providing that it is located on a D-amino acid center. When acting on peptide oligomers in which the interpeptide bonds are C-terminal D-alanyl-D-linkages (Figs. 6, 11, and 13; for a more complete list, see Chart I in reference 130), the carboxypeptidases hydrolyze the interpeptide bonds and degrade these oligomers into monomeric peptides.

E. At the Cellular Level

As mentioned in Sections V, B and V, C, not only is it necessary to regulate the activity and biosynthesis of the many enzymes involved in peptidoglycan biosynthesis, but it is also essential to control the selection of sites at which the final assembly takes place, the direction of synthesis, and the integration of the process with the biosynthesis of other macromolecules and cell division. It should be pointed out again that peptidoglycan synthesis per se can be relatively easily uncoupled from the synthesis of other macromolecules. Continued RNA, DNA, and protein synthesis may occur in the absence of peptidoglycan synthesis. For example, growth in the presence of inhibitors of peptidoglycan synthesis results in the formation of spheroplasts or L-forms in osmotically protective environments. Conversely, there are many examples of continued cellular synthesis of peptidoglycan in the absence of synthesis of informational macromolecules. In this latter case, the overall rate of peptidoglycan synthesis is virtually unchanged, but the topology of the process is altered. One of these changes is the wall-thickening phenomenon mentioned in Section V, A. Recent studies revealed specific changes in the ultrastructure of S. faecalis, well within the first hour, following the relatively selective inhibition of DNA, RNA, and protein synthesis [206]. Selected concentrations of various antibiotics were added to balanced exponentially growing cultures. The antibiotics and their concentrations were selected on the basis of their relative specificity as inhibitors of the synthesis of one type of macromolecule. This was essential since high concentrations of most of the agents used were not as selective in their target as one might judge from the literature. Even at the antibiotic concentrations selected, specificity was only relative in degree of inhibition or in time. All of the inhibitors used permitted the continuation of peptidoglycan synthesis at or near the preexisting rate for periods of 10 min or more. After this time, incorporation of lysine into peptidoglycan still continued but at reduced rates. This occurred upon inhibition of protein synthesis with chloramphenicol (50 μg/ml) or 5-azacytidine (5 μg/ml); RNA synthesis with actinomycin D (0.25 μg/ml) or rifampicin (0.1 μg/ml); or DNA synthesis with mitomycin C (0.5 μg/ml). As expected from previous work on amino acid starvation [196], chloramphenicol treatment was accompanied by a rapid thickening of the wall over the entire surface of the cocci. In contrast, mitomycin-C treatment was not accompanied by wall thickening during the first half-hour of treatment. When this was analyzed in more detail, it was found that, in contrast to the chloramphenicol-inhibited cells, the mitomycin-treated cells almost doubled in cell number. Also, the mitomycin-C-treated cells differed in shape. In general they were longer and had shorter nascent cross walls, as if the rate of splitting (R_2 in Fig. 34) remained high. Quantitative measurements of electron micrographs of central longitudinal sections demonstrated that the ratio of length of peripheral wall to cross wall increased after mitomycin treatment.

Results obtained with these inhibitors have been interpreted in terms of the model for cell-wall growth discussed in Section V, C and shown in Fig. 34. While selective inhibition of either protein or DNA syntheses is accompanied by continued peptidoglycan synthesis, the ultrastructure, size, and shape of the resultant cell population is strikingly different. On the one hand, inhibition of protein synthesis appears to be accompanied by inhibition of both wall elongation (R1 in Fig. 34) and cross-wall separation (R2 in Fig. 34). Thus, the additional peptidoglycan (and other wall polymers) made result in about the same number of cells with thicker walls. On the other hand, inhibition of DNA synthesis appears to be accompanied by both continued wall elongation and, more strikingly, cross-wall separation (R2). Also a large fraction of the cell population continues to divide. Thus, wall thickening is not an early consequence of inhibition of DNA synthesis. Only later, when other cellular processes slow and/or stop, and further cell division ceases, is wall thickening seen. As yet we know little about the timing or regulation of these processes during a cell division cycle. However, it does seem likely that, at the correct time in a cell division cycle, centripetal closure of a cross-wall would be favored by an inhibition of cross-wall separation, so that more of the wall elongation product would end up as cross-wall and less as peripheral wall [206, 239a].

F. Of Autolytic Enzymes

Potential functions of autolytic enzymes are discussed in Section IV,D, some of which are related to the occurrence and direction of cellular sites of peptidoglycan synthesis. At the very least, the hydrolytic action of such enzymes must be limited so that in the growing cell the protective nature of the wall is maintained. This is particularly true for those species that show extensive peptidoglycan turnover, and those species that excrete autolysins into the growth medium. In the latter case, there must be some mechanism that protects the wall not only from extracellular enzymes but during release from the cells.

Mechanisms of control of autolytic enzymes may differ. Two examples of physiologically related, but morphologically different species S. faecalis and L. acidophilus will serve to illustrate this point. In S. faecalis (Fig. 35) it appears that the N-acetylmuramidase is made as a proenzyme, which is transported to the wall sites engaged in wall enlargement, where it is activated by a proteinase [240]. Both active and latent enzymes have an exceedingly strong affinity for the wall so that once wall-bound, they cannot easily be released [207, 241-243]. This affinity for the wall plus the occurrence of the cytoplasmic form in a latent state and the absence of peptidoglycan turnover, are all properties which made it possible to experimentally find an association of the enzyme with newly made wall at nascent cross walls. Lactobacillus acidophilus also has an N-acetylmuramidase [175]. Many properties of the autolysins of the two species differ [199]. Important for consideration of the L. acidophilus system is (1) the apparent absence of a latent form and the presence of non-wall-bound soluble active autolysin in the cell, (2) the occurrence of peptidoglycan turnover, and (3) an affinity of autolysin for the wall that appears to be not as strong as that of S. faecalis enzyme. Thus, the techniques that have been used to localize autolytic activity in S. faecalis have not been successful with L. acidophilus.

All of the steps in the S. faecalis system from synthesis of the latent form to the activity of the active form could be subject to regulation. Evidence has been obtained that suggests that the conversion of latent to active form is carried out by a process requiring protein synthesis [196a]. More interesting is the probability that cellular autolysis is controlled at the level of autolytic enzyme activity in S. faecalis and perhaps also in L. acidophilus. In both organisms the ability of cells to autolyze drops sharply when a culture enters the stationary phase. Walls from stationary-phase L. acidophilus cultures apparently still contain substantial amounts of autolysin [175]. A more detailed study has been carried out with S. faecalis. In this organism, inhibition of protein synthesis with chloramphenicol or threonine starvation results in a very rapid drop in the ability of cells to autolyze while the isolated walls from these cells retain

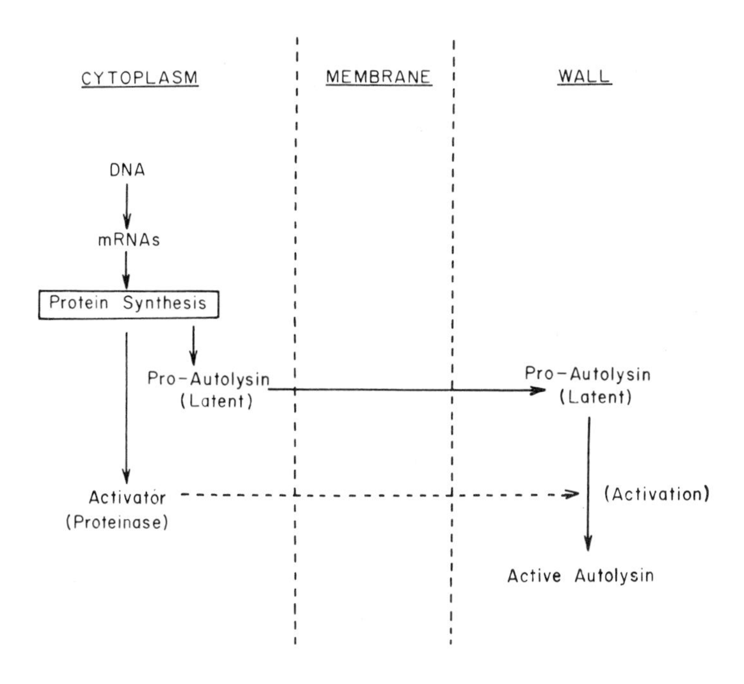

FIG. 35. Possible relationship between the latent and active forms of autolytic enzyme of Streptococcus faecalis 9790. The latent form is shown as a precursor (proenzyme) of the active form. Activation of the latent enzyme, by means of a proteinase, takes place after the enzyme has passed through the membrane and become bound to the cell wall. Reprinted from reference 240 by courtesy of American Society for Microbiology.

nearly the same amount of active and latent autolysin activity [196a]. The retention of autolysin activity long after inhibition of further protein synthesis suggests that the enzyme has a relatively long half-life, and that some other mechanism rapidly prevents its ability to lyse cells. Further studies, using selective inhibition of DNA, RNA, and protein synthesis showed that the rapid decrease in ability of cells to autolyze was much more closely correlated with inhibition of protein synthesis than with inhibition of RNA or DNA synthesis [244]. An as yet unidentified low molecular weight substance that inhibits autolytic activity can be extracted from cells of S. faecalis with hot water. Larger amounts of it can be extracted from chloramphenicol-inhibited than exponential-phase cells. Thus, availability of inhibitor could be inversely related to protein synthesis.

These observations on the effects of DNA, RNA and protein synthesis are consistent with a role for the autolysin in peeling apart the two layers of wall at nascent cross walls, as discussed in Section V, C. Inhibition of DNA synthesis is accompanied by a great deal of peeling apart so that short cross walls and long peripheral walls were seen. This correlates with the absence of an inhibition of autolytic activity. On the other hand, little peeling apart was observed after inhibition of protein synthesis, this correlating with the rapid drop in cellular autolysis seen under these conditions.

In order for an autolysin to carry out a morphogenetic function coupled to other cellular biosynthetic systems during the cell cycle, it would be essential to have a mechanism that would control autolytic activity in a rapid and sensitive way. A small molecule that can bind to an exocellular protein to modify its activity would be such a tool.

There is little evidence of cycles of autolytic activity during the cell division cycle. Groves and Clark have obtained evidence for an increase in autolytic activity before division in synchronized cultures of E. coli B/r [245]. A cyclic decrease and increase in ability of L. acidophilus to autolyze during recovery from valine starvation has been noted [199]. Such cyclic autolytic behavior has not been observed with S. faecalis. It is not yet known if the L. acidophilus cultures were dividing with some degree of synchrony, but recovery of growth after a period of amino acid starvation resulted in synchronized division in S. faecalis and other species.

ACKNOWLEDGMENTS

The work in the authors' laboratories was supported by research grants 515 and 1000 from the Fonds de la Recherche Fondamentale Collective, Brussels, and l'Institut pour l'Encouragement de la Recherche Scientifique dans l'Industrie et l'Agriculture, Brussels (1699) to J. M. Ghuysen, and by research grants from the National Science Foundation (GB 20813) and U. S. Public Health Service (AI 05044) from the National Institute of Allergy and Infectious Diseases) to G. D. Shockman. We wish to thank the following for providing some of the illustrations used in this chapter: Dr. D. J. Tipper (Fig. 18), Dr. J. Fleck (Fig. 26), Dr. D. J. Ellar (Fig. 28), Dr. R. G. E. Murray (Figs. 29 and 30), and Dr. M. L. Higgins (Figs. 32 and 33); and for their comments on this chapter, Dr. O. Kandler, Dr. H. R. Perkins, Dr. F. C. Neuhaus, Dr. H. Martin, Dr. P. Reynolds, Dr. R. Tinelli, Dr. M. Guinand, Dr. J. F. Petit, Dr. M. L. Higgins, and Dr. L. Daneo-Moore.

REFERENCES

1. M. R. J. Salton, The Bacterial Cell Wall, Elsevier Publishing Co., Amsterdam, 1964.

2. C. Weibull, J. Bacteriol., 66, 688 (1953).

2a. S. Brenner, F. A. Dark, P. Gerhardt, M. H. Jeynes, O. Kandler, E. Kellenberger, E. Klieneberger-Nobel, K. McQuillen, M. Rubio-Heurtos, M. R. J. Salton, R. E. Strange, J. Tomcsik, and C. Weibull, Nature (London), 181, 1713 (1958).

2b. L. B. Guze, Microbial Protoplasts, Spheroplasts and L-Forms, Williams & Wilkins Co., Baltimore, Maryland, 1968.

3. J. T. Park, J. Biol. Chem., 194, 897 (1952).

4. H. J. Rogers and H. R. Perkins, Cell Walls and Membranes, E. F. and N. Spon Ltd., London, 1968.

5. J. M. Ghuysen, Bacteriol. Rev., 33, 425 (1968).

6. A. M. Glauert and M. J. Thornley, Ann. Rev. Microbiol., 23, 1525 (1969).

7. M. J. Osborn, Ann. Rev. Biochem., 38, 701 (1969).

8. W. Weidel and H. Pelzer, Advan. Enzymol., 26, 193 (1964).

9. J. M. Ghuysen, J. L. Strominger, and D. J. Tipper, in Comprehensive Biochemistry (M. Florkin and E. H. Stotz, eds.), Vol. 26A, pp. 53-104, American Elsevier Publishing Co., New York, 1968.

10. D. Ellar, in Organization and Control in Prokaryotic and Eukaryotic Cells (H. P. Charles and B. C. J. Knight, eds.), p. 167, Cambridge University Press, 1970.

11. D. J. Tipper, J. Systematic Bacteriol., 20, 361 (1970).

12. J. L. Strominger, The Harvey Lectures, 64, 179 (1970).

13. H. H. Martin, Ann. Rev. Biochem., 35(2), 457 (1966).

14. H. J. Rogers, Bacteriol. Rev., 34, 194 (1970).

14a. J. S. Thompson, J. Theoret. Biol., 33, 63 (1971).

15. R. W. Jeanloz, N. Sharon, and H. M. Flowers, Biochem. Biophys. Res. Commun., 13, 20 (1963).

16. N. Sharon, T. Osawa, H. M. Flowers, and R. W. Jeanloz, J. Biol. Chem., 241, 223 (1966).

17. D. J. Tipper, Biochemistry, 7, 1441 (1968).

18. J. Fleck, M. Mock, R. Minck, and J. M. Ghuysen, Biochim. Biophys. Acta, 233, 489 (1971).

19. T. Y. Liu and E. C. Gotschlich, J. Biol. Chem., 242, 471 (1967).

20. J. M. Ghuysen, D. J. Tipper, and J. L. Strominger, Biochemistry, 4, 474 (1965).

21. A. R. Archibald, J. Baddiley, and N. L. Blumson, Advan. Enzymol., 30, 223 (1968).

22. I. Azuma, D. W. Thomas, A. Adam, J. M. Ghuysen, R. Bonaly, J. F. Petit, and E. Lederer, Biochim. Biophys. Acta, 208, 444 (1970).

23. A. D. Warth and J. L. Strominger, Proc. Nat. Acad. Sci. U.S., 64, 528 (1969).

24. O. Hoshino, U. Zehavi, P. Sinay, and R. W. Jeanloz, J. Biol. Chem., 247, 381 (1972).

25. R. W. Wheat and J. M. Ghuysen, J. Bacteriol., 105, 1219 (1971).

26. R. W. Wheat, S. Kulkarni, A. Cosmatos, E. Scheer, and R. Steele, J. Biol. Chem., 244, 4921 (1969).

26a. B. Cziharz, K. H. Schleifer, and O. Kandler, Biochemistry, 10, 3574 (1971).

26b. D. Bogdanovsky, E. Interschick-Niebler, K. H. Schleifer, F. Fiedler, and O. Kandler, Eur. J. Biochem., 22, 173 (1971).

27. O. Kandler, Intern. J. Systematic Bacteriol., 20, 491 (1970).

28. F. Fiedler, K. H. Schleifer, B. Cziharz, E. Interschick, and O. Kandler, Reports from the Conference on the Taxonomy of Bacteria held in the Czechoslovak Collection of Microorganisms, J. E. Purkyne University, Brno, September 1969, pp. 111-122.

29. O. Kandler, K. H. Schleifer, E. Niebler, M. Nakel, H. Zahradnik, and M. Ried, Reports from the Conference on the Taxonomy of Bacteria held in the Czechoslovak Collection of Microorganisms, J. E. Purkyne University, Brno, September 1969, pp. 143-156.

29a. K. Schleifer and O. Kandler, Bacteriol. Rev., 36, 407 (1972).

30. J. van Heijenoort, L. Elbaz, P. Dezelee, J. F. Petit, E. Bricas, and J. M. Ghuysen, Biochemistry, 8, 207 (1969).

31. M. Nakel, J. M. Ghuysen, and O. Kandler, Biochemistry, 10, 2170 (1971).

32. T. A. Krulwich, J. C. Ensign, D. J. Tipper, and J. L. Strominger, J. Bacteriol., 94, 734 (1967).

32a. T. A. Krulwich, J. C. Ensign, D. J. Tipper, and J. L. Strominger, J. Bacteriol., 94, 741 (1967).

33. D. J. Tipper and M. F. Berman, Biochemistry, 8, 2183 (1969).

34. D. J. Tipper, Biochemistry, 8, 2192 (1969).

35. J. M. Ghuysen, E. Bricas, M. Leyh-Bouille, M. Lache, and G. D. Shockman, Biochemistry, 6, 2607 (1967).

36. K. D. Hungerer, J. Fleck, and D. J. Tipper, Biochemistry, 8, 3567 (1969).

37. M. Leyh-Bouille, R. Bonaly, J. M. Ghuysen, R. Tinelli, and D. J. Tipper, Biochemistry, 9, 2944 (1970).

38. O. Kandler, R. Plapp, and W. Holzapfel, Biochim. Biophys. Acta, 147, 252 (1967).

39. J. M. Ghuysen, D. J. Tipper, C. H. Birge, and J. L. Strominger, Biochemistry, 4, 2244 (1965).

40. J. F. Petit, E. Munoz, and J. M. Ghuysen, Biochemistry, 5, 2764 (1966).

41. E. Munoz, J. M. Ghuysen, M. Leyh-Bouille, J. F. Petit, H. Heymann, E. Bricas, and P. Lefrancier, Biochemistry, 5, 3748 (1966).

42. K. H. Schleifer and O. Kandler, Biochem. Biophys. Res. Commun., 28, 965 (1967).

43. J. M. Ghuysen, E. Bricas, M. Lache, and M. Leyh-Bouille, Biochemistry, 7, 1450 (1968).

44. J. J. Campbell, M. Leyh-Bouille, and J. M. Ghuysen, Biochemistry, 8, 193 (1969).

45. H. R. Perkins, Biochem. J., 102, 29c (1967).

46. M. Guinand, J. M. Ghuysen, K. H. Schleifer, and O. Kandler, Biochemistry, 8, 200 (1969).

47. N. Sharon and S. Seifter, J. Biol. Chem., 239, PC2398 (1964).

48. D. M. Chipman, J. J. Pollock, and N. Sharon, J. Biol. Chem., 243, 487 (1968).

49. J. J. Pollock and N. Sharon, Biochemistry, 9, 3913 (1970).

50. M. Welsch, Rev. Belge Pathol. Méd. Exp., 28(2), 1 (1947).

51. J. M. Ghuysen, Arch. Intern. Physiol. Biochim., 65, 173 (1957).

52. J. C. Ensign and R. S. Wolfe, J. Bacteriol., 91, 524 (1966).

53. D. J. Tipper and J. L. Strominger, Biochem. Biophys. Res. Commun., 22, 48 (1966).

54. D. J. Tipper, M. Tomoeda and J. L. Strominger, Biochemistry, 10, 4683 (1971).

55. D. J. Tipper, J. L. Strominger, and J. C. Ensign, Biochemistry, 6, 906 (1967).

56. P. E. Kolenbrander and J. C. Ensign, J. Bacteriol., 95, 201 (1968).

57. D. J. Tipper, J. Bacteriol., 97, 837 (1969).

58. J. M. Ghuysen and J. L. Strominger, Biochemistry, 2, 1110 (1963).

59. J. M. Ghuysen and J. L. Strominger, Biochemistry, 2, 1119 (1963).

60. J. Coyette and J. M. Ghuysen, Biochemistry, 9, 2935 (1970).

61. D. Carlström, J. Biophys. Biochem. Cytol., 3, 669 (1957).

62. M. V. Kelemen and H. J. Rogers, Proc. Natl. Acad. Sci. U.S., 68, 992 (1971).

63. E. Oldmixon and M. L. Higgins, in preparation.

64. M. L. Higgins and G. D. Shockman, CRC Crit. Rev. Microbiol., 1, 29 (1971).

65. J. L. Strominger, in The Bacteria (I. C. Gunsalus and R. Y. Stanier, eds.), Vol. III, p. 413, Academic Press, New York, 1962.

66. K. C. Gunetileke and R. A. Anwar, J. Biol. Chem., 243, 5570 (1968).

67. D. Hendlin, E. O. Stapley, M. Jackson, H. Wallick, A. K. Miller, F. J. Wolf, T. W. Miller, L. Chaiet, F. M. Kahan, E. L. Foltz, H. B. Woodruff, J. M. Mata, S. Hernandez, and S. Moekales, Science, 166, 122 (1969).

68. B. G. Christensen, W. J. Leanza, T. R. Beattie, A. A. Patchett, B. H. Arison, R. E. Ormond, F. A. Kuehl, Jr., G. Albers-Schonberg, and O. Jardetzky, Science, 166, 123 (1969).

69. K. D. Hungerer and D. J. Tipper, Biochemistry, 8, 3577 (1969).

70. D. J. Tipper and I. Pratt, J. Bacteriol., 103, 305 (1970).

71. F. C. Neuhaus, C. V. Carpenter, M. P. Lambert and R. J. Wargel, Proc. Symp. Mol. Mechanisms of Antibiotic Action on Protein Biosynthesis and Membranes (E. Munoz, F. Ferrandiz and D. Vazquez, eds.), pp. 339-362, Elsevier Co., Amsterdam, 1972.

72. F. C. Neuhaus, Antimicrobial. Agents Chemother., 304 (1967).

73. U. Roze and J. L. Strominger, Mol. Pharmacol., 2, 92 (1966).

74. J. Lynch and F. C. Neuhaus, J. Bacteriol., 91, 449 (1966).

74a. W. A. Wood and I. C. Gunsalus, J. Biol. Chem., 190, 403 (1951).

75. R. B. Johnston, J. J. Scholz, W. F. Diven and S. Shepard, in Pyridoxal Catalysis: Enzymes and Model Systems (E. E. Snell, A. E. Braunstein, E. S. Severin, and Y. M. Torshinsky, eds.), p. 537, Interscience Publishers, New York, 1966.

75a. M. M. Johnston and W. F. Diven, J. Biol. Chem., 244, 5414 (1969).

76. F. C. Neuhaus and J. L. Lynch, Biochemistry, 3, 471 (1964).

77. F. C. Neuhaus, C. V. Carpenter, J. Lynch-Miller, N. M. Lee, M. Cragg, and R. A. Stickgold, Biochemistry, 8, 5119 (1969).

78. F. C. Neuhaus and W. G. Struve, Biochemistry, 4, 120 (1965).

79. Y. Higashi, J. L. Strominger, and C. C. Sweeley, Proc. Nat. Acad. Sci. (Wash.), 57, 1878 (1967).

80. Y. Higashi, J. L. Strominger, and C. C. Sweeley, J. Biol. Chem., 245, 3697 (1970).

81. D. Brooks and J. Baddiley, Biochem J., 115, 307 (1969).

82. M. Stow, B. Starkey, I. C. Hancock, and J. Baddiley, Nature New Biol., 229, 56 (1971).

83. A. Wright, M. Dankert, P. Fennessey, and P. W. Robbins, Proc. Nat. Acad. Sci., 57, 1798 (1967).

84. R. J. Watkinson, H. Hussey, and J. Baddiley, Nature (London), 229, 57 (1971).

84a. R. G. Anderson, H. Hussey, and J. Baddiley, Nature (London), 229, 57 (1971).

84b. L. Glaser, F. Fiedler, and J. Mauck, Ann. N. Y. Acad. Sci., in press.

85. F. A. Troy, F. E. Frerman, and E. C. Heath, J. Biol. Chem., 246, 118 (1971).

86. M. Scher, W. J. Lennarz, and C. C. Sweeley, Proc. Nat. Acad. Sci., 59, 1313 (1968).

87. W. J. Lennarz, Ann. Rev. Biochem., 39, 359 (1970).

88. F. N. Hemming, in Terpenoids in Plants (J. B. Pridham, ed.), p. 23, Academic Press, New York, 1967.

89. Y. Higashi, G. Siewert, and J. L. Strominger, J. Biol. Chem., 245, 3683 (1970).

90. Y. Higashi and J. L. Strominger, J. Biol. Chem., 245, 3691 (1970).

91. J. G. Christenson, S. K. Gross, and P. W. Robbins, J. Biol. Chem., 244, 5436 (1969).

92. L. Rothfield and D. Romeo, Bacteriol. Rev., 35, 14 (1971).

93. F. C. Neuhaus, Accounts of Chemical Research, 4, 297 (1972).

93a. R. A. Stickgold and F. C. Neuhaus, J. Biol. Chem., 242, 1331 (1967).

94. W. G. Struve, R. K. Sinka, and F. C. Neuhaus, Biochemistry, 5, 82 (1966).

95. M. G. Heydanck and F. C. Neuhaus, Biochemistry, 4, 1474 (1969).

96. A. N. Chatterjee and J. T. Park, Proc. Nat. Acad. Sci. U.S., 51, 9 (1964).

97. P. M. Meadow, J. S. Anderson, and J. L. Strominger, Biochem. Biophys. Res. Commun., 14, 382 (1964).

98. W. G. Struve and F. C. Neuhaus, Biochem. Biophys. Res. Commun., 18, 6 (1965).

99. P. W. Robbins, D. Bray, M. Dankert, and A. Wright, Science, 158, 1536 (1967).

100. G. Siewert and J. L. Strominger, Proc. Nat. Acad. Sc. U.S., 57, 766 (1967).

101. G. Siewert, in Inhibitors Tools in Cell Research, (Th. Bücher and H. Sies, eds.), pp. 210-216, Colloquium der Gesellschaft für Biologische Chemie, Springer-Verlag, Berlin, 1969.

102. M. Matsuhashi, C. P. Dietrich, and J. L. Strominger, J. Biol. Chem., 242, 3191 (1967).

103. T. Kamiryo and M. Matsuhashi, Biochem. Biophys. Res. Commun., 36, 215 (1969).

104. J. Thorndike and J. T. Park, Biochem. Biophys. Res. Commun., 35, 642 (1969).

105. R. M. Bumsted, J. L. Dahl, D. Söll, and J. L. Strominger, J. Biol. Chem., 243, 779 (1968).

106. T. S. Stewart, F. J. Roberts, and J. L. Strominger, Nature (London), 230, 36 (1971).

107. B. Niyomporn, J. L. Dahl, and J. L. Strominger, J. Biol. Chem., 243, 773 (1968).

108. W. S. L. Roberts, J. F. Petit, and J. L. Strominger, J. Biol. Chem., 243, 768 (1968).

109. J. F. Petit, J. L. Strominger, and D. Söll, J. Biol. Chem., 243, 757 (1968).

110. R. Plapp and J. L. Strominger, J. Biol. Chemistry, 245, 3675 (1970).

111. A. Kaji, H. Kaji, and G. D. Novelli, J. Biol. Chem., 240, 1185 (1965).

112. A. Kaji, H. Kaji, and G. D. Novelli, J. Biol. Chem., 240, 1192 (1965).

113. K. Momose and A. Kaji, J. Biol. Chem., 241, 3294 (1966).

114. W. J. Lennarz, J. A. Nesbitt, III, and J. Reiss, Proc. Nat. Acad. Sci. U. S., 55, 934 (1966).

115. J. L. Strominger, Y. Higashi, and W. Staudenbauer, Abstr. Amer. Chem. Soc., 158th Natl. Mtg, 1969 (Biol.), p. 221.

116. W. L. Staudenbauer and J. L. Strominger in Methods in Enzymology, Vol. 17, p. 718, Academic Press, 1971.

117. G. Siewert and J. L. Strominger, J. Biol. Chem., 243, 783 (1968).

118. W. Katz, M. Matsuhashi, C. P. Dietrich, and J. L. Strominger, J. Biol. Chem., 242, 3207 (1967).

119. W. Hammes, Lunteren Lectures on Molecular Genetics, Lunteren, The Netherlands, Abstracts: Section 1, p. 2 (1971).

120. O. Kandler, personal communication.

121. E. M. Wise, Jr., and J. T. Park, Proc. Nat. Acad. Sci., 54, 75 (1965).

122. D. J. Tipper, and J. L. Strominger, Proc. Nat. Acad. Sci. U.S., 54, 1133 (1965).

123. D. J. Tipper and J. L. Strominger, J. Biol. Chem., 243, 3169 (1968).

123a. F. Lipmann, in Essays in Biochemistry (P. N. Campbell and G. D. Greville, eds.), Academic Press, London, 4, 1 (1968).

123b. F. Lipmann, Science, 173, 875 (1971).

124. Y. Araki, R. Shimai, A. Shimada, N. Ishimoto, and E. Ito, Biochem. Biophys. Res. Commun., 23, 466 (1966).

125. V. Araki, A. T. Shimada, and E. Ito, Biochem. Biophys. Res. Commun., 23, 518 (1966).

126. K. Izaki, M. Matsuhashi, and J. L. Strominger, Proc. Nat. Acad. Sci. U. S., 55, 656 (1966).

127. K. Izaki and J. L. Strominger, J. Biol. Chem., 243, 3193 (1968).

128. K. Izaki, M. Matsuhashi, and J. L. Strominger, J. Biol. Chem., 243, 3180 (1968).

129. R. J. Wargel, C. A. Shadur, and F. C. Neuhaus, J. Bacteriol., 105, 1028 (1971).

130. J. M. Ghuysen, M. Leyh-Bouille, R. Bonaly, M. Nieto, H. R. Perkins, K. H. Schleifer, and O. Kandler, Biochemistry, 9, 2955 (1970).

131. M. Leyh-Bouille, J. M. Ghuysen, R. Bonaly, M. Nieto, H. R. Perkins, K. H. Schleifer, and O. Kandler, Biochemistry, 9, 2961 (1970).

132. M. Leyh-Bouille, J. M. Ghuysen, M. Nieto, H. R. Perkins, K. H. Schleifer, and O. Kandler, Biochemistry, 9, 2971 (1970).

133. M. Leyh-Bouille, J. Coyette, J. M. Ghuysen, J. Idczak, H. R. Perkins, and M. Nieto, Biochemistry, 10, 2163 (1971).

134. M. Leyh-Bouille, M. Nakel, J. M. Frère, K. Johnson, J. M. Ghuysen, M. Nieto and H. R. Perkins, Biochemistry, 11, 1290 (1972).

135. J. J. Pollock, J. M. Ghuysen, R. Linder, M. R. J. Salton, H. R. Perkins, M. Nieto, M. Leyh-Bouille, J. M. Frère, and K. Johnson, Proc. Nat. Acad. Sci. U. S., 69, 662 (1972).

135a. J. M. Ghuysen, M. Leyh-Bouille, J. M. Frère, J. Dusart, K. Johnson, M. Nakel, J. Coyette, H. R. Perkins, and M. Nieto, Proc. Symposium on Mol. Mechanisms of Antibiotic Action on Protein Biosynthesis and Membranes (E. Munoz, F. Ferrandiz, and D. Vazquez, eds.), p. 406, Elsevier, Amsterdam, 1972.

136. P. J. Lawrence and J. L. Strominger, J. Biol. Chem., 245, 3653 (1970).

137. P. J. Lawrence and J. L. Strominger, J. Biol. Chem., 245, 3660 (1970).

138. P. E. Reynolds, Biochim. Biophys. Acta, 237, 239 (1971).

139. P. E. Reynolds, Biochim. Biophys. Acta, 237, 255 (1971).

140. M. Matsuhashi, M. Furuyama and B. Maruo, Biochim. Biophys. Acta, 184, 670 (1969).

141. J. J. Pollock, M. Distèch, R. Linder, M. R. J. Salton, and J. M. Ghuysen, Abstr. Ann. Mtng, Am. Soc. Microbiol., p. 153 (1973).

141a. M. Nieto, H. R. Perkins, J. M. Frère, and J. M. Ghuysen, Biochem. J., in press.

142. P. E. Reynolds, in The Molecular Basis of Antibiotic Action, pp. 49-120, Wiley, 1972.

143. J. F. Collins and M. H. Richmond, Nature (London), 195, 142 (1962).

144. B. Lee, J. Mol. Biol., 61, 463 (1971).

144a. M. Nieto, H. R. Perkins, M. Leyh-Bouille, J. M. Frère, and J. M. Ghuysen, Biochem. J., 131, 163 (1973).

144b. E. H. Böhme, H. E. Applegate, B. Toeplitz, and J. E. Dolfini, J. Amer. Chem. Soc., 93, 4324 (1971).

145. H. J. Rogers, Biochem. J., 103, 90 (1967).

146. J. L. Strominger, P. Blumberg, H. Suginaka, J. Umbreit, and G. Wickus, Proc. Symp. Mol. Mechanisms of Antibiotic Action on Protein Biosynthesis and Membranes (E. Munoz, F. Ferrandiz, and D. Vazquez, eds.), pp. 388-405, Elsevier Publ. Co., Amsterdam, 1972.

147. U. Schwarz, A. Asmus, and H. Frank, J. Mol. Biol., 41, 419 (1969).

148. R. Hartmann, J. V. Höltze, and U. Schwarz, Nature (London), 235, 426 (1972).

148a. J. M. Ghuysen, M. Leyh-Bouille, J. M. Frère, J. Dusart, A. Marquet, H. R. Perkins, and M. Nieto, Ann. N. Y. Acad. Sci., in press.

148b. J. L. Strominger, Ann. N. Y. Acad. Sci., in press.

149. H. R. Perkins and M. Nieto, Proc. Symp. Mol. Mechanisms of Antibiotic Action on Protein Biosynthesis and Membranes (E. Munoz, F. Ferrandiz and D. Vazquez, eds.), pp. 363-387, Elsevier Publ. Co., Amsterdam, 1972.

150. D. C. Jordan, Can. J. Microbiol., 11, 390 (1965).

151. G. Best and W. Durham, Arch. Biochem. Biophys., 111, 685 (1965).

152. M. Nieto and H. R. Perkins, Biochem. J., 123, 789 (1971).

153. M. Nieto and H. R. Perkins, Biochem. J., 123, 773 (1971).

154. M. Nieto and H. R. Perkins, Biochem. J., 124, 845 (1971).

155. H. R. Perkins and M. Nieto, Biochem. J., 116, 83 (1970).

156. M. Nieto, H. R. Perkins, and P. E. Reynolds, Biochem. J., 126, 139 (1972).

157. S. De Petris, J. Ultrastruct. Res., 19, 45 (1967).

157a. R. G. E. Murray, P. Steed, and H. E. Elson, Can. J. Microbiol., 11, 547 (1965).

157b. H. Frank and D. Dekegel, Folia Microbiol. Praha, 12, 221 (1967).

158. J. H. Freer and M. R. J. Salton, Microbiol. Toxins, 4, 67 (1971).

159. V. Braun and V. Sieglin, Eur. J. Biochem., 13, 336 (1970).

160. M. E. Bayer, Virology, 2, 346 (1968).

161. T. Miura and S. Mizushima, Biochim. Biophys. Acta, 150, 159 (1969).

162. H. R. Kaback, in Methods in Enzymology, Vol. XXII, p. 99, Academic Press, New York, 1971.

163. D. J. Ellar, D. G. Lundgren, and R. A. Slepecky, J. Bacteriol., 94, 1189 (1967).

164. A. N. Chatterjee, D. Mirelman, H. J. Singer, and J. T. Park, J. Bacteriol., 100, 846 (1969).

165. J. Coyette and J. M. Ghuysen, Biochemistry, 7, 2385 (1968).

166. R. C. Hughes, P. J. Tanner, and E. Stokes, Biochem. J., 120, 159 (1970).

167. M. V. Nermut and R. G. E. Murray, J. Bacteriol., 93, 1949 (1967).

168. D. C. Ellwood and D. W. Tempest, Biochem. J., 111, 1 (1969).

169. A. H. Hughes, M. Stow, I. C. Hancock, and J. Baddiley, Nature New Biol., 229, 53 (1971).

170. R. G. E. Murray, M. M. Hall, and J. Marak, Can. J. Microbiol., 16, 883 (1970).

171. G. D. Shockman, Bacteriol. Rev., 29, 345 (1965).

172. G. D. Shockman, Proc. Soc. Exp. Biol. Med., 101, 693 (1959).

173. A. Tomasz, A. Albino, and E. Zanati, Nature (London), 227, 138 (1970).

174. G. D. Shockman, J. S. Thompson, and M. J. Conover, Biochemistry, 6, 1054 (1967).

175. J. Coyette and J. M. Ghuysen, Biochemistry, 9, 2952 (1970).

176. J. L. Mosser, and T. Tomasz, J. Biol. Chem., 245, 287 (1970).

177. W. C. Brown and F. E. Young, Biochem. Biophys. Res. Commun., 38, 564 (1970).

177a. K. Takumi, T. Kawata, K. Hisatsune, Japan J. Microbiol., 15, 131 (1971).

177b. R. Tinelli, C. R. Acad. Sci., 266, 1792 (1968).

177c. R. Tinelli, Bull. Soc. Chim. Biol., 51, 283 (1969).

178. T. Wadstrom and O. Vesterberg, Acta Pathol. Microbiol. Scand., Section B, 79, 248 (1971).

179. H. P. Browder, W. A. Zygmunt, J. R. Young, and P. A. Tavormina, Biochem. Biophys. Res. Commun., 19, 383 (1965).

180. E. Huff, C. S. Silverman, N. J. Adams, and W. S. Awkard, J. Bacteriol., 103, 761 (1970).

181. I. Takebe, H. J. Singer, E. M. Wise, Jr., and J. T. Park, J. Bacteriol., 102, 14 (1970).

182. P. Mitchell and J. Moyle, J. Gen. Microbiol., 16, 184 (1957).

183. W. Weidel, H. Frank, and H. H. Martin, J. Gen. Microbiol., 22, 158 (1960).

184. G. D. Shockman, J. J. Kolb, and G. Toennies, J. Biol. Chem., 230, 961 (1958).

185. G. Toennies and G. D. Shockman, Proc. Fourth Int. Congr. Biochem., 13, 365 (1958).

186. T. A. Krulwich and J. C. Ensign, J. Bacteriol., 96, 857 (1968).

187. S. L. Kingan and J. C. Ensign, J. Bacteriol., 96, 629 (1968).

187a. D. Mirelman, R. Bracha, and N. Sharon, Proc. Nat. Acad. Sci., U. S., 69, 3355 (1972).

188. D. White, M. Dworkin and D. J. Tipper, J. Bacteriol., 95, 2186 (1968).

189. C. Forsberg and H. J. Rogers, Nature (London), 229, 272 (1971).

190. D. P. Fan, J. Bacteriol., 103, 494 (1970).

191. D. P. Fan and M. M. Beckman, J. Bacteriol., 105, 629 (1971).

192. A. Tomasz, Proc. Nat. Acad. Sci., 59, 86 (1968).

193. H. M. Pooley, G. D. Shockman, M. L. Higgins, J. Porres-Juan, J. Bacteriol., 109, 423 (1972).

194. G. D. Shockman, J. Biol. Chem., 234, 2340 (1959).

195. G. Toennies, B. Bakay, and G. D. Shockman, J. Biol. Chem., 234, 3269 (1959).

196. M. L. Higgins and G. D. Shockman, J. Bacteriol., 103, 244 (1970).

196a. H. M. Pooley and G. D. Shockman, J. Bacteriol., 103, 457 (1970).

197. M. L. Higgins and G. D. Shockman, J. Bacteriol., 101, 643 (1970).

198. M. L. Higgins, H. M. Pooley, and G. D. Shockman, J. Bacteriol., 103, 504 (1970).

199. J. Coyette, D. Boothby, M. L. Higgins, and G. D. Shockman, Fed. Proc. Fed. Amer. Soc. Exp. Biol., 30, 1174 (1971).

200. C. Frehel, A. M. Beaufils, and A. Ryter, Ann. Inst. Pasteur, 121, 271 (1971).

201. D. W. Pitel and C. Gilvarg, J. Biol. Chem., 245, 6711 (1970).

201a. L. Rothfield and M. Pearlman-Kothencz, J. Mol. Biol., 44, 477 (1969).

202. D. Boothby, L. Daneo-Moore, M. L. Higgins, J. Coyette, and G. D. Shockman, J. Biol. Chem., 248, 2161 (1973).

203. J. Mauck, L. Chan, and L. Glaser, J. Biol. Chem., 246, 1820 (1971).

203a. J. Chaloupka and P. Krackova, Folia Microbiol. Praha, 16, 372 (1971).

204. D. Mirelman, D. R. D. Shaw, and J. T. Park, J. Bacteriol., 107, 239 (1971).

205. D. C. Barker and K. J. I. Thorne, J. Cell. Sci., 7, 755 (1970).

206. M. L. Higgins, L. Daneo-Moore, D. Boothby, and G. D. Shockman, in preparation.

207. G. D. Shockman, H. M. Pooley, and J. S. Thompson, J. Bacteriol., 94, 1525 (1967).

208. R. M. Cole and J. J. Hahn, Science, 135, 722 (1962).

209. R. M. Cole, Bacteriol. Rev., 29, 326 (1965).

210. K. L. Chung, R. Z. Hawirko, and P. K. Isaac, Can. J. Microbiol., 10, 473 (1964).

211. J. Swanson, K. C. Hsu, and E. C. Gotschlich, J. Exp. Med., 130, 1063 (1969).

212. M. Wagner, Zentralblatt Bakteriol. Parasit. Infekt. Hyg., 195, 87 (1964).

213. E. B. Briles and A. Tomasz, J. Cell Biol., 47, 786 (1970).

214. R. C. Hughes and E. Stokes, J. Bacteriol., 106, 694 (1971).

215. K. L. Chung, R. Z. Hawirko, and P. K. Isaac, Can. J. Microbiol., 10, 43 (1964).

216. E. H. Beachey and R. M. Cole, J. Bacteriol., 92, 1245 (1966).

217. R. M. Cole, Science, 143, 820 (1964).

218. J. W. May, Exper. Cell Res., 31, 218 (1963).

219. R. P. Van Tubergen and R. B. Setlow, Biophys. J., 1, 589 (1961).

219a. J. Mauck, L. Chan, L. Glaser, and J. Williamson, J. Bacteriol., 109, 373 (1972).

220. J. Lederberg, Proc. Nat. Acad. Sci. U.S., 42, 574 (1956).

221. M. E. Bayer, J. Gen. Microbiol., 46, 237 (1967).

222. W. D. Donachie and R. J. Begg, Nature (London), 227, 1220 (1970).

223. R. G. E. Murray, W. H. Francombe, and B. H. Mayall, Can. J. Microbiol., 5, 641 (1959).

224. H. J. Rogers, Nature (London), 213, 31 (1967).

225. S. Scott and G. D. Shockman, unpublished observations (1971).

226. E. H. Stonehill and D. J. Hutchison, J. Bacteriol., 92, 136 (1966).

227. M. L. Higgins, H. M. Pooley, and G. D. Shockman, J. Bacteriol., 105, 1175 (1971).

228. W. Katz, and H. H. Martin, Biochem. Biophys. Res. Commun., 39, 744 (1970).

229. E. P. Previc, J. Theoret. Biol., 27, 471 (1970).

230. H. Matsuzawa, M. Matsuhashi, A. Oka, and Y. Seguno, Biochem. Biophys. Res. Commun., 36, 682 (1969).

231. E. J. J. Lugtenberg, in "Escherichia coli mutants impaired in the synthesis of murein," Ph. D. Thesis, Rijksuniversiteit te Utrecht, Holland, 1971.

232. H. J. W. Wijsman, in "Een genetische studie over de celwand-synthese bij Escherichia coli," Ph. D. Thesis, Univ. of Amsterdam, Holland, 1970.

233. E. J. J. Lugtenberg, J. Bacteriol., 108, 233 (1971).

234. E. J. J. Lugtenberg, and P. G. de Haan, Autonie van Leeuwenhock, 37, 537 (1971).

235. H. J. W. Wijsman, Genetical Research, 20, 269 (1972).

236. V. M. Reusch and F. C. Neuhaus, Fed. Proc. Abstr., 29, 1341 (1970).

237. M. Martinez-Carrion and W. T. Jenkins, J. Biol. Chem., 240, 3538 (1965).

238. R. Berberich, M. Kaback, and E. Freese, J. Biol. Chem., 243, 1006 (1968).

239. V. P. Marshall and J. R. Sokatch, J. Bacteriol., 95, 1419 (1968).

239a. G. D. Shockman, L. Daneo-Moore, and M. L. Higgins, Ann. N.Y. Acad. Sci., in press.

240. H. M. Pooley and G. D. Shockman, J. Bacteriol., 100, 617 (1969).

241. G. D. Shockman, J. S. Thompson, and M. J. Conover, J. Bacteriol., 90, 575 (1965).

242. G. D. Shockman and M. C. Cheney, J. Bacteriol., 98, 1199 (1969).

243. H. M. Pooley, J. M. Porres-Juan, and G. D. Shockman, Biochem. Biophys. Res. Commun., 38, 1134 (1970).

244. M. Sayare, L. Daneo-Moore, and G. D. Shockman, J. Bacteriol., 112, 337 (1972).

245. D. Groves, Ph. D. Thesis, University of British Columbia, 1971.

Chapter 3

BIOSYNTHESIS AND ASSEMBLY OF LIPOPOLYSACCHARIDE AND THE OUTER MEMBRANE LAYER OF GRAM-NEGATIVE CELL WALL

Hiroshi Nikaido

Department of Bacteriology and Immunology
University of California
Berkeley, California

I. NATURE OF THE OUTER MEMBRANE LAYER

The cell wall of gram-negative bacteria appears to be very different from that of gram-positive bacteria. This is seen, for example, when thin sections are observed under the electron microscope. Cell walls of gram-positive bacteria consist mainly of a thick, structureless layer, whereas the gram-negative cell envelope region generally consists of (a) an outermost, unit membranelike, 75-85 Å thick layer (outer membrane), (b) a rather structureless layer of variable thickness, and (c) an innermost unit-membranelike layer, which is presumed to be the cytoplasmic membrane [1,2] (Fig. 1). Some gram-negative bacteria produce additional layers located outside the outer membrane [1].

The gram-negative cell wall is different from the gram-positive cell wall also in terms of chemical composition [2-4]. The latter consists mostly of peptidoglycan, with some teichoic acid or polysaccharide or both. From gram-negative bacteria it was not possible until quite recently to prepare cell wall fractions lacking cytoplasmic membranes; the earlier studies were done on "cell envelope" fractions, which contain both the cell wall and the cytoplasmic membrane. This situation poses a severe limitation on the interpretation of results, yet it has been agreed by most investigators that gram-negative cell walls usually contain only small amounts of peptidoglycan, and that most gram-negative cell walls are likely to contain lipopolysaccharides (LPS), lipids, and proteins.

It has been possible to pinpoint the location of some of the components of gram-negative cell wall, without attempting to separate the cell wall layer from the cytoplasmic membrane. For example, by using *Veillonella*, which produces a cell wall exceptionally rich in peptidoglycan among gram-negative bacteria, Bladen and Mergenhagen [5] showed that the extraction of LPS with hot 45% phenol resulted in the disappearance of the outer membrane layer. With *Salmonella* Shands [6, 7] could show, by the use of ferritin-labeled antibodies to LPS, that the carbohydrate chains of LPS extend outward up to 300 Å into the medium from the outer membrane layer of cell wall. Weidel's group [8] showed first that LPS and some cell wall proteins function as receptors for phages and must therefore be accessible from the outside, and then that the removal of these components from crude cell envelope preparations of *E. coli* revealed a thin underlying layer of peptidoglycan. All these studies indicated that the peptidoglycan layer of gram-negative cell wall is covered by LPS and proteins. Furthermore, lysozyme treatment of cells and spheroplasts was shown to cause the disappearance of the structureless layer beneath the outer membrane [5,9]; this layer was therefore identified as consisting of peptidoglycan (Fig. 1).

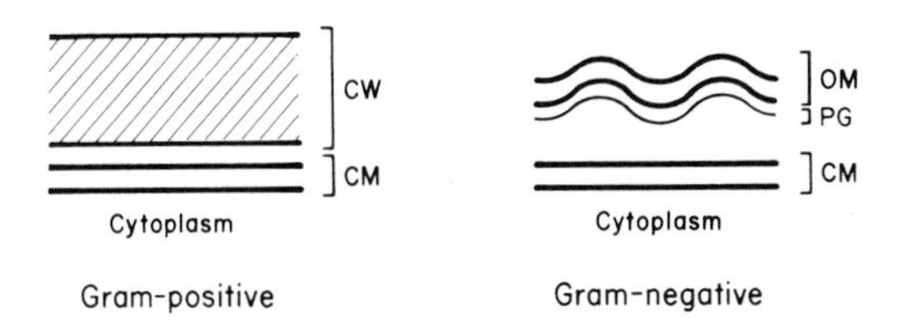

FIG. 1. Appearance of the surface layers of bacteria in thin sections. CW, cell wall; CM, cytoplasmic membrane; OM, "outer membrane" layer of cell wall; and PG, peptidoglycan layer of cell wall. Both gram-positive and negative cells may contain extra layers of cell wall in addition to those shown here.

It is obvious that more detailed studies on gram-negative cell walls require the isolation and purification of cell wall preparations uncontaminated by the cytoplasmic membrane. The first preparation of the "pure" outer membrane was obtained, rather unexpectedly, from a mutant defective in the biosynthesis of lysine [10]. This mutant lacks the enzyme meso-diaminopimelic acid decarboxylase, and excretes large amounts of diaminopimelic acid into the medium if grown under lysine-limiting conditions. Bishop and Work [10] found that a product containing both lipid and carbohydrate was excreted by this mutant under these conditions, and a major component of the excreted complex was later identified as LPS [11]. Electron microscopic studies revealed that the outer membrane formed protrusions ("blebs") and that these blebs became pinched off as vesicles during the excretion process [12]. Thus, the excretion appeared to be a consequence of unbalanced growth, in which protein synthesis is inhibited while the synthesis of outer membrane components and peptidoglycan continues. The composition of the excreted complex [60% LPS, 26% lipids (mostly phosphatidylethanolamine), 11% protein] may then be considered to reflect the composition of the outer membrane to some extent [13].

This system was further exploited by Rothfield and Pearlman-Kothencz [14], who found that the complex was excreted whenever protein synthesis was inhibited (by addition of chloramphenicol or by withdrawal of the required amino acids) in any strain of E. coli or Salmonella. The excreted complex contained 50% LPS, 35% phospholipids (two-thirds phosphatidylethanolamine, one-third phosphatidylglycerol), and 15% protein. Proteins were examined by acrylamide gel electrophoresis in sodium dodecyl sulfate (SDS) and in urea-acetic acid, and were found to contain one major species plus several minor ones.

Under these conditions, however, the synthesis of outer membrane proteins was also inhibited; there was no guarantee, therefore, that the composition of the fragments that were shed corresponded exactly to that of the normal outer membrane. Indeed, it was shown that the fragments excreted at the end of the starvation period contained much less protein than those shed at the beginning [14].

In view of this difficulty, it was most important that a method be found for the physical separation of outer membrane from the inner, cytoplasmic membrane. Such a method was first described by Miura and Mizushima [15], and included the osmotic lysis of EDTA-lysozyme spheroplasts of E. coli, followed by extensive dialysis of the cell envelope fraction against 3 mM ethylenediaminetetraacetic acid (EDTA) and equilibrium centrifugation of this fraction in a sucrose density gradient. The lighter fraction contained cytochromes, succinic dehydrogenase, and ATPase, and presumably corresponded to the cytoplasmic membrane. The heavier fraction was enriched for carbohydrate and presumably corresponded to the outer membrane. However, the carbohydrate content of this fraction was only about 5%, and the comparison with the composition of excreted outer membrane fractions mentioned in the preceding text suggests an extensive loss of LPS during purification, probably due to the use of EDTA (see Section IV). More recently, this method has been modified by Schnaitman [16] and by Osborn's group [17], and the latter group showed that the loss of LPS could be prevented by using an EDTA concentration just sufficient to prevent the aggregation of membranes. The outer membrane prepared from a UDP-glucose-4-epimerase mutant of S. typhimurium in this manner contained proteins, LPS, and phospholipids at a weight ratio of 1.0:0.3:0.3 [17]. The phospholipid composition of the outer membrane was qualitatively similar to that of the cytoplasmic membrane, but quantitatively the outer membrane was enriched for phosphatidylethanolamine and contained very little cardiolipin [17].

Another approach to the isolation of outer membrane was discovered when DePamphilis and Adler [18] and Schnaitman [19] found that outer membrane does not disintegrate in 1.5-2% Triton X-100 (a nonionic detergent) provided Mg^{2+} is present. Under these conditions cytoplasmic membrane becomes completely dissociated. Unfortunately about one-half of LPS and two-thirds of phospholipids of the outer membrane are lost during the Triton treatment [18, 19]. In contrast, there seems to be no significant loss of proteins [19]; this procedure is therefore convenient for the study of the protein composition of outer and cytoplasmic membranes [20]. Acrylamide gel electrophoresis in SDS has already shown the presence of about 30 protein bands in the cytoplasmic membrane and of about 10 bands in the cell wall [20] (Fig. 2). It is most interesting that among these 40 bands, none seems to be present in both the outer membrane and the cytoplasmic membrane, except the "major" protein band of the outer membrane [20].

The most recently developed method for the isolation of outer membrane involves the direct fractionation, by a particle electrophoresis

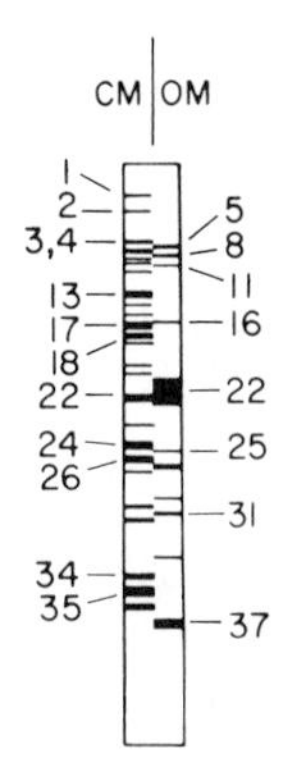

FIG. 2. Acrylamide electrophoresis of E. coli cell envelope proteins in sodium dodesylsulfate. Reprinted from Schnaitman's work [20] by courtesy of the American Society for Microbiology. CM, Cytoplasmic membrane; OM, outer membrane layer of cell wall.

apparatus, of the cell envelope prepared by mechanical disruption of cells [21]. This method seems advantageous in that it avoids lengthy incubation during spheroplast formation and the use of high concentrations of EDTA.

II. THE STRUCTURE OF LIPOPOLYSACCHARIDES

LPS is by far the most extensively studied component of gram-negative cell wall. Numerous reviews have appeared on this topic. These deal with the chemical structure of LPS [22-27], its physical structure [28], biosynthesis [26, 27, 29-32], and genetic control [27, 29, 33, 34]. The interest in LPS also owes very much to the fact that it is the major somatic antigen in enteric bacteria and is therefore important from the viewpoint of immunochemistry and taxonomy; these aspects have been covered by Lüderitz et al. [22, 24] and Nikaido [25]. LPS is also a powerful toxic agent, and its biological activity as "endotoxin" is dealt with in the reviews compiled by Kadis et al. [35].

Many methods have been described for the isolation of LPS. The phenol-water method [36] seems to be by far the best one for use with most organisms. Here the organisms are extracted with a single phase solvent (45% aqueous phenol at 65°-68°C), and the extract is then cooled down so that the substances extracted will be partitioned either into the phenol layer or into the aqueous layer. In most cases LPS is recovered in the aqueous phase, together with RNA and soluble polysaccharides. Since LPS exists as large aggregates in water, it can usually be separated from other substances by ultracentrifugation [36].

Some LPS preparations contain many hydrophobic sugar residues. LPS from the mutants that have lost most of the hydrophilic polysaccharide portion (see Section III, B, 2) is also rather hydrophobic as a whole. In these cases, LPS tends to partition in the phenol phase and its recovery from the aqueous phase will be poor. A method involving extraction with a phenol-petroleum ether-chloroform solvent has been devised for such strains and gives preparations that are free from contaminants [37].

The structure of LPS has been studied in most detail in S. typhimurium (Fig. 3) [24]. An important feature is the presence of three different regions, i.e., O side chain, core oligosaccharide, and lipid A. LPS from other bacteria are also assumed to consist of these three regions, although detailed studies have so far been performed only on a few Salmonella, E. coli, and Shigella flexneri strains (see below).

The outermost region, O side chain, consists of tetrasaccharide (or pentasaccharide, see Section III, C, 6) repeating units in S. typhimurium. The presence of repeating units seems to be a general feature; O side chains in many Salmonella serotypes, a few E. coli strains, and several Shigella flexneri strains have been found to contain repeating units (Table 1). This is the region which has undergone tremendous diversification during evolution in certain organisms [25], and in many strains the repeating unit contains rather rare sugars such as 6-deoxyhexoses and 3,6-dideoxyhexoses. Various portions of the repeating unit serve as determinant groups in the interaction of LPS with O-antibodies. Thus, it is possible to classify LPS preparations by the use of specific antibodies; this procedure establishes various O-serotypes of LPS and of the parent bacteria. The resolving power of this immunological technique is seen from the fact that Salmonella typhimurium LPS and Salmonella newport LPS do not show appreciable cross-reaction serologically, in spite of the identical sugar composition of their O repeating units (Table 1) [22].

Before we discuss the structure of the core oligosaccharide, we must emphasize the fact that the study of LPS structure was greatly accelerated, or even made possible, by the availability of different kinds of mutants defective at various stages of LPS biosynthesis [29, 33, 34]. The innermost region of LPS, i.e., the $(KDO)_3$-lipid A region, appears to be indispensable for the survival of bacteria, and mutants defective in its biosynthesis have never been isolated (see Section V). However, the remaining portion of LPS, i.e., the O side chain and most of the core oligosaccharide, is not necessary for the survival of these organisms under pure culture conditions; mutants can therefore be easily isolated. In the organisms of the enteric group, upon which most of the studies were done, the wild-type organisms produce LPS with the O side chain, and form smooth colonies on solid media, presumably owing to the water-retaining properties of the O side-chain polysaccharide which covers the surface of

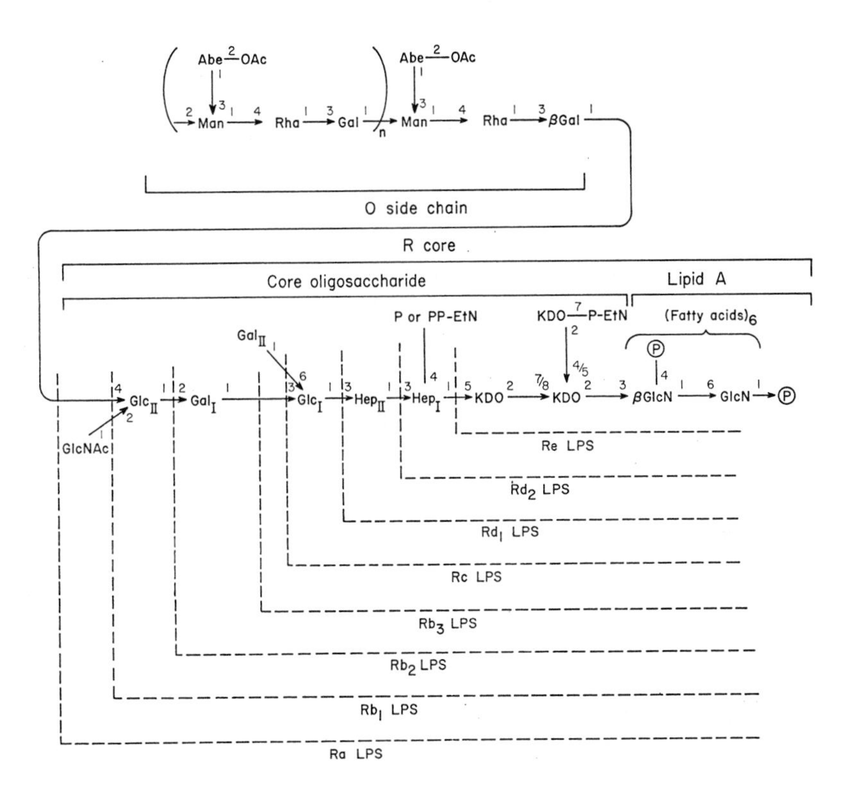

FIG. 3. Structure of S. typhimurium LPS. This structure shows a basic "monomer" unit, which is presumably cross-linked with other units. Studies that were performed in many laboratories and led to the elucidation of this structure are listed in references 22-24. Abbreviations: Abe, abequose; Man, D-mannose; Rha, L-rhamnose; Gal, D-galactose; GlcNAc, N-acetyl-D-glucosamine; Glc, D-glucose; Hep, L-glycero-D-manno-heptose; KDO, 3-deoxy-D-manno-octulosonic acid; EtN, ethanolamine; Ac, acetyl. To some sugar residues in the core oligosaccharide region roman numeral subscripts have been added in order to differentiate one from the other. All sugar residues are α-anomers except where β-conformation is specified. (The rhamnose residues were recently found to be α-L-anomers (37a).) O side chain galactose residues may be glucosylated at their C-4 positions in O-12_2 (+) forms (Section III, C, 6), or at the C-6 positions in P22-lysogenic strains (Section III, C, 8). The biosynthesis of LPS starts at the lipid A portion and the core oligosaccharide is elongated toward the "left" (i.e., nonreducing end) in this scheme. Thus, mutants defective at various stages of core oligosaccharide produce incomplete LPS; the structure of these LPS molecules and their "chemotype" are indicated by dotted lines (see also Table II).

TABLE 1

Some O Side-Chain Repeating Units of *Salmonella* [a]

Serotype	Serogroup	O-Antigen	O Repeat unit
S. paratyphi A	A	2, 12	Par (1→3) Man; OAc–Rha; Glc (1→4) Gal Man $\xrightarrow{1\ 4}$ Rha $\xrightarrow{1\ 3}$ Gal $\xrightarrow{1\ 2}$
S. typhi	D_1	9, 12	Tyv (1→3) Man; Glc-2-OAc (1→4) Gal Man $\xrightarrow{1\ 4}$ Rha $\xrightarrow{1\ 3}$ Gal $\xrightarrow{1\ 2}$
S. typhimurium	B	4, 5, 12	Abe-2-OAc (1→3) Man; Glc (1→4) Gal Man $\xrightarrow{1\ 4}$ Rha $\xrightarrow{1\ 3}$ Gal $\xrightarrow{1\ 2}$
S. newington	E_2	3, 15	βMan $\xrightarrow{1\ 4}$ Rha $\xrightarrow{1\ 3}$ βGal $\xrightarrow{1\ 6}$

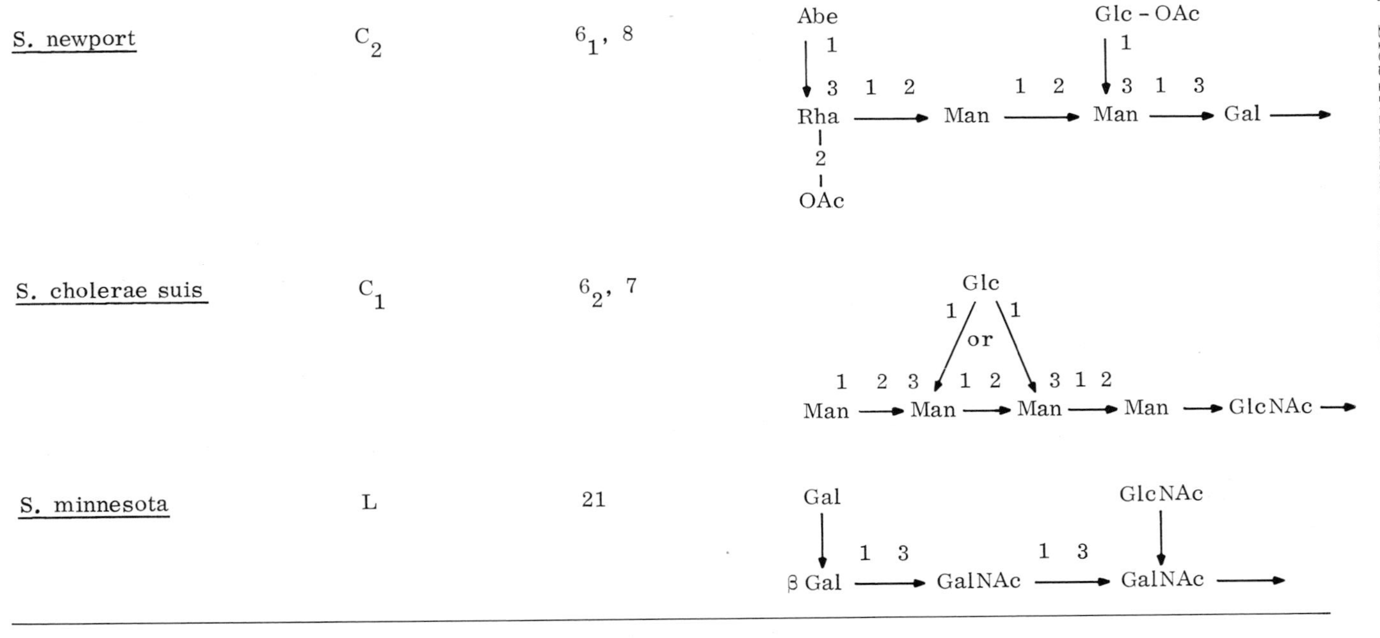

S. newport	C_2	6_1, 8	Abe (1→3) on Rha; OAc (2) on Rha; Rha 1→2 Man 1→2 Man 1→3 Gal →; Glc-OAc (1→3) on second Man
S. cholerae suis	C_1	6_2, 7	Man 1→2 Man 1→2 Man 1→2 Man → GlcNAc →; Glc (1→3) on second or third Man ("or")
S. minnesota	L	21	β Gal 1→3 GalNAc 1→3 GalNAc →; Gal → β Gal; GlcNAc → second GalNAc

[a] For references and the structure in other serotypes, see ref. 24. Abbreviations: Par, paratose (3,6-dideoxy-D-glucose); Tyv, tyvelose (3,6-dideoxy-D-mannose). For other abbreviations see the legend to Fig. 3. Anomeric configurations of the sugar residue is α except as specified. (In a few cases it is unknown.) The repeating units in Groups A, D_1, B, and E are related to each other, but others seem very different.

cells. In contrast, mutants producing LPS devoid of O side chains generally produce rough-surfaced colonies on solid media. It was soon recognized, however, that the appearance of the colonies is not always a reliable criterion of LPS structure. Thus, it was decided to call S (smooth) any organism that fully reacts with anti-O-antibodies and thus is presumed to produce complete LPS, and R (rough) any organism that fails to do the same.

The analysis of LPS from various mutants showed that when a mutant fails to synthesize O side chains, its LPS does not contain any part of O side chain, and is of the R type (in Fig. 3). (However, there are a small group of exceptional mutants, called SR, which are in a sense intermediate between S and R [Section III, C, 5, b].) LPS from such O side-chain-defective mutants is called R core, since in S-form LPS it corresponds to the central portion underlying the O side chains. The R core itself is divided further into the "core oligosaccharide" and "Lipid A."

LPS from other kinds of R mutants contains _incomplete_ R core, i.e., proximal parts of the core oligosaccharide, attached to the lipid A. These mutants are defective at various steps of the synthesis of the core oligosaccharide, and thus produce incomplete core structures which cannot accept the transfer of O side chains.

One classification used for these defective LPS (and for the mutants that produce the LPS) is based on the sugar composition of LPS. Each class may be further subdivided on the basis of detailed structural studies. The structure and composition of these mutant LPS are summarized in Fig. 3 and Table 2. Those mutants that lack most of the core oligosaccharide (and all of O side chain), for example, Rd_1 and Re, show readily recognizable phenotypes (Section V,B), and are often called "deep roughs."

The structure of the core oligosaccharide seems to vary little among _Salmonella_ [24], in contrast to the extreme diversity found in the structure of O side chains. _Escherichia coli_ and _Shigella_, on the other hand, seem to produce core oligosaccharides with structures slightly different from that of _Salmonella_ (reviewed in references 24 and 25). In bacteria not closely related to the _E. coli-Salmonella-Shigella_ group, more extensive deviations in the structure of core are likely to be encountered. It is noteworthy, however, that a characteristic component of this portion of LPS, 3-deoxy-D-_manno_-octulosonic acid (2-keto-3-deoxyoctonic acid or KDO), is present in most gram-negative bacteria [38], and that another component, L-_glycero_-D-_manno_-heptose, is found in LPS from organisms such as _Pasteurella pseudotuberculosis_ and _Proteus vulgaris_, although absent in LPS from _Xanthomonas_ or _Anacystis_ [24]. Thus, the structure of LPS seems to be most diverse at its most peripheral portion (O side chains), less diverse in the outer portion of the core oligosaccharide, and presumably rather similar in its KDO-containing, innermost region. The

TABLE 2

Classification of _Salmonella_ R Mutants

Chemotype	Further subdivision of chemotype	Molar ratio of sugars in core oligosaccharide					Corresponding genotype[a]	Probable defect[a]
		KDO	Hep	Glc	Gal	GlcNAc		
Ra							_rfb_	Synthesis of O repeating unit
		3	2	2	2	1	_rfbT_	Transfer of O side chains to R core
							rfaL	Transfer of O side chains to R core
Rb	Rb_1	3	2	2	2	0	_rfaK_	GlcNAc transferase
	Rb_2	3	2	1	2	0	_rfaJ_	Glc_{II} transferase
	Rb_3	3	2	1	1	0	_rfaH_	Gal_I transferase
Rc		3	2	1	0	0	_galE_	UDP-glucose 4-epimerase
Rd							_rfaG_	Glc_I transferase
	Rd_1	3	2	0	0	0	_galU_	UDP-glucose pyrophosphorylase
							pgi	phosphoglucoisomerase
	Rd_2	3	1	0	0	0	_rfaF_	Hep_{II} transferase ?
Re		3	0	0	0	0	_rfaE_	Hep_I transferase or Heptose synthesis ?

[a] See Section III, B, 2.

implications of this finding in connection with the possible protective function of LPS have been discussed [25, 39] (see also Section V).

In addition to heptose and KDO, phosphate residues are present in the inner region of the core oligosaccharide. Furthermore, ethanolamine, which Tauber and Russell [40] first recognized as a component of LPS, was found to occur as O-phosphorylethanolamine [41] and O-pyrophosphorylethanolamine [26] also in this region of the core oligosaccharide. Studies from Lüderitz's laboratory showed that the O-phosphorylethanolamine group is linked, via phosphodiester linkage, to the 7-position of one of the KDO residues [42] (see Fig. 3) in the LPS of a S. minnesota mutant. In another mutant of S. minnesota defective in UDP-glucose pyrophosphorylase (see Section III, B, 1), some of the first heptose residues (Hep_I of Fig. 3) are substituted with O-pyrophosphorylethanolamine at C-4 [43].

Before the discovery of phosphorylated ethanolamine groups in LPS, the phosphate residues linked to heptose residues were generally thought to be involved in the cross-linking of neighboring core oligosaccharide chains via phosphodiester bridges [44]. In view of the more recent results mentioned, a careful reexamination of this hypothesis was thought to be necessary. Such a study using the UDP-glucose pyrophosphorylaseless mutant of S. minnesota showed that a fraction of the Hep_I residue (Fig. 3) carried phosphomonoester residue at C-4 [43], but there was no evidence for the presence of phosphodiester cross-linkages between oligosaccharide chains. However, it may be difficult for this particular mutant to synthesize such cross-links, since Mühlradt's results [45] suggest that the phosphorylation of the heptose residue is normally done after the transfer of Glc_I residue (Fig. 3), which does not occur in this mutant because of the lack of UDP-glucose.

Thus at present uncertainties still exist concerning the presence of phosphodiester cross-links between core oligosaccharides. Core oligosaccharide preparations obtained from R mutant LPS by the removal of lipid A appeared to have much higher molecular weights than expected for monomers [44, 46], and these preparations could be degraded into oligosaccharide monomers by hydrofluoric acid at 0° C [47], a procedure known to specifically cleave phosphodiester linkages. However, in at least one case the apparent high molecular weight has been shown to be a result of aggregation [47a].

The molecular weight of an Rc mutant LPS, determined by ultracentrifugation in benzene after acetylation, was 10,300 [48]. When a S. london wild-type LPS was deacylated by treatment with Dictyostelium discoideum, it had a molecular weight of 15,400 [49]. Calculation indicates that in both cases an LPS molecule consists of an average of three "monomer" units, each of which contains one core oligosaccharide and one acylated glucosamine dissaccharide (i.e., "lipid A") as shown in Fig. 3. Although these results are compatible with the presence of cross-links between core oligosaccharides, it is equally possible that they are due to the presence of cross-links at the level of lipid A. In fact, Lüderitz's group [49a] has recently shown that lipid A molecules are cross-linked through pyrophosphate bridges, as follows.

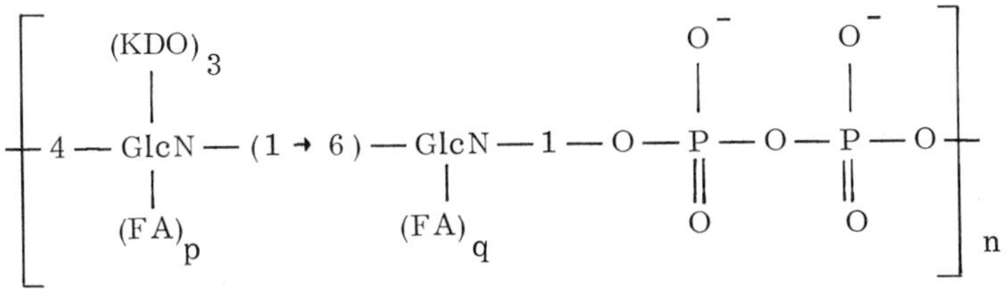

$$\left[-4 - \underset{\substack{| \\ (FA)_p}}{\overset{\substack{(KDO)_3 \\ |}}{GlcN}} - (1 \rightarrow 6) - \underset{\substack{| \\ (FA)_q}}{GlcN} - 1 - O - \underset{\substack{\| \\ O}}{\overset{\substack{O^- \\ |}}{P}} - O - \underset{\substack{\| \\ O}}{\overset{\substack{O^- \\ |}}{P}} - O - \right]_n$$

At present the detailed structure of the inner core region has been studied only in _Salmonella_. It seems probable that the LPS of other organisms have somewhat different structures in this region. For example, LPS preparations from pseudomonads contain much higher proportions of phosphate than _Salmonella_ LPS [50, 51], and most of these phosphate residues are likely to be located in the inner core region.

The hydrophobic region of LPS corresponds to "lipid A." This is an unusual phospholipid, which contains no glycerol and instead utilizes D-glucosamine as a skeleton. Mild acid hydrolysis (e.g., 1% acetic acid for 30-90 min at 100° C) of LPS splits the acid-labile glycosidic linkages of KDO and results in the production of free lipid A and the polysaccharide [44]. Structural studies on such preparations of lipid A indicated the presence of glucosamine oligosaccharides as the backbone [52, 53], but suffered from the difficulty that some degradation of lipid A was inevitable during the mild acid hydrolysis. Gmeiner et al. [54] circumvented this difficulty by using _Salmonella_ mutants that produce LPS of extremely reduced structure. These mutants, called Re, have lost the ability to synthesize or transfer L-_glycero_-D-_manno_-heptose residues, and thus their "LPS" (or "glycolipid") contains only the lipid A and KDO residues (Fig. 3 and Table II). Hydrazinolysis of this preparation gave a pentasaccharide, in which the KDO trisaccharide (see Fig. 3) is linked ketosidically to C-3 of D-glucosamine, which in turn is linked to the reducing end D-glucosamine residue through a β(1 → 6) linkage [54, 55]. The same β-D-glucosaminyl-(1 → 6)-D-glucosamine backbone has been found in _Serratia marcescens_ LPS [56], whereas β(1 → 4)-linked glucosamine oligosaccharides have been found in _Shigella flexneri_ and in one strain of _E. coli_ [57].

Acid hydrolysis of LPS in most cases produces small amounts of amino acids. This observation led to the suggestion that LPS is covalently linked to the protein(s) in the cell. Although the presence of such a covalent linkage cannot be excluded, the evidence for it is not conclusive even in the most rigorous work so far published [57a]. The successful and efficient extraction of protein-free R-type LPS by phenol-chloroform-petroleum ether solvent in the cold [37] also seems to speak against the presence of covalent LPS-protein linkages.

More information has recently been obtained on the structure of lipid A, at least in Salmonella. The glucosamine disaccharide is substituted with a maximum of two phosphate groups, one at C-1 of the reducing sugar residue, the other at C-4 of the nonreducing sugar [54, 55]. This glucosamine disaccharide diphosphate constitutes the repeating unit of the oligomer cross-linked through pyrophosphate bridges [49a] (see p. 143), and probably is a product of the hydrolytic degradation of the oligomer. The amino groups of both glucosamine residues are substituted with 3-hydroxytetradecanoic acid (β-hydroxymyristic acid) residues via amide linkages [58], and all the available hydroxy groups of the disaccharide are also substituted with ester-linked fatty acids. 3-Hydroxytetradecanoic acid, which was shown to have the D-configuration [58a], is never found in the usual glycerophospholipids and thus is a characteristic component of lipid A. (Lipid A preparations from Pseudomonas aeruginosa [59] and Neisseria catarrhalis [60] contain mostly 3-hydroxydodecanoate rather than 3-hydroxytetradecanoate, and that from Veillonella contains 3-hydroxytridecanoate [61].)

The results of fatty acid analysis on Salmonella LPS (Table 3) indicate that lipid A also contains various medium-chain saturated fatty acids in addition to 3-hydroxytetradecanoate, and that the fatty acid composition of lipid A is strikingly different from that of the glycerophosphatides in the same cell. Furthermore, the results of Rietschel [58] show clearly that a total of six fatty acid residues, including three 3-hydroxytetradecanoic acid residues, are present in a unit containing a glucosamine disaccharide. Since only C-3 and C-4 of the reducing end sugar, C-6 of the nonreducing end sugar, and the two amino groups are available for substitution, it follows that all of the three OH groups must be esterified and furthermore one of the six fatty acid residues must be linked to a group other than the OH and NH_2 groups of the disaccharide. Rietschel found that this particular fatty acid is released through β-elimination reaction in alkali, and concluded that this acid (which happens to be mostly tetradecanoic acid) is ester-linked to the hydroxy group of the 3-hydroxytetradecanoic acid, which in turn is linked to one of the hydroxyl groups of the disaccharide. The hydroxyl groups of amide-linked 3-hydroxytetradecanoic acid residues appear to be free. The overall structure of LPS in Salmonella can thus be schematically expressed as seen in Fig. 4.

The presence of microheterogeneity is a pronounced characteristic of LPS structure. First, if LPS from "wild-type" laboratory strains of Salmonella is analyzed, the length of O side chains shows a remarkable heterogeneity. This was shown, for example, by gel filtration of polysaccharides obtained after the treatment with 1% acetic acid to split off lipid A

TABLE 3

Fatty Acid Composition of Lipid A and Glycerophospholipids from Salmonella [a]

Fatty acid	Lipid A[b]			Glycerophospholipids [62]
	Rietschel [58]		Romeo et al. [48]	
12:0	0.9[c]	1.3[d]	1.2	-
14:0	1.0	0.9	1.2	0.2
14:1	-	-	0.4[f]	-
14:0(3-OH)[e]	3.2	2.9	2.6	-
16:0	0.8	0.9	0.6	2.2
16:1	-	-	-	1.1
17:cy	-	-	-	1.0
18:0	-	-	-	0.1
18:1	-	-	-	1.2
19:cy	-	-	-	0.2
Total:	5.9	6.0	(6.0)	(6.0)

[a] Values are expressed as molar ratios. In Rietschel's analysis, they correspond to moles per two moles of glucosamine. For others, the data were recalculated so that the total will make 6.0 mole.

[b] Rietschel used Re glycolipid (LPS) from S. minnesota, Romeo et al. Rc LPS from S. typhimurium.

[c] After alkaline hydrolysis (4 N NaOH, 5 h, at 100°C).

[d] After acid hydrolysis (4 N HCl, 5 h, at 100°C).

[e] The values for 3-hydroxytetradecanoic acid and Δ^2-tetradecenoic acid (which is generated from the former compound during hydrolysis) were pooled. Romeo et al. [48] report the presence of a peak which they tentatively identify as Δ^3-tetradecenoic acid; this, however, is likely to be Δ^2-tetradecenoic acid in view of the β-elimination mechanism that is responsible for the generation of this acid [58].

[f] This is different from the Δ^2- (or Δ^3)-tetradecenoic acid. It is likely to be Δ^7- tetradecenoic acid.

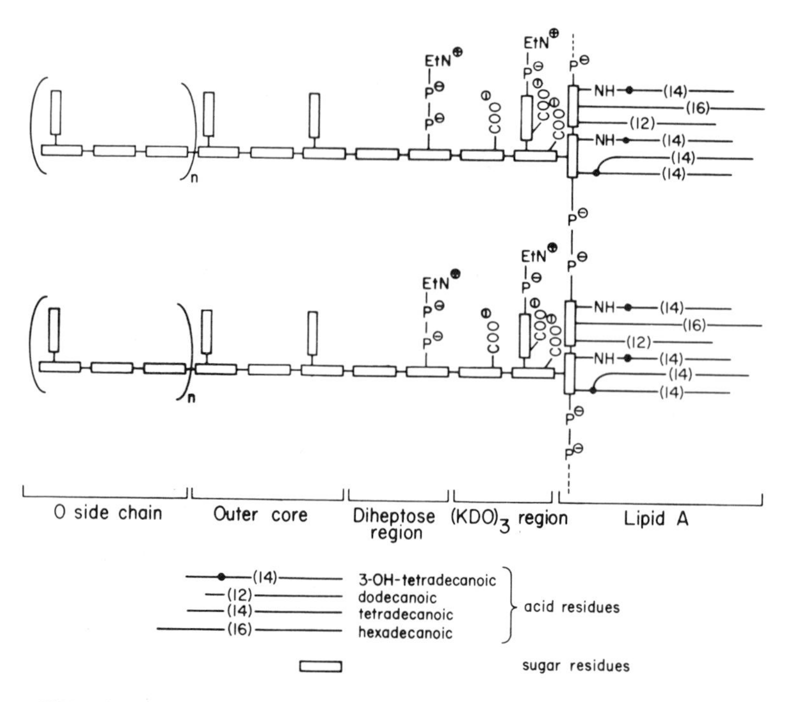

FIG. 4. A schematic representation of a LPS molecule. This scheme is based on the structure shown in Fig. 3 and on the work of Rietschel [58]. A molecule containing only two subunits is shown, although on an average three sub-units are believed to be cross-linked together [48]. Pyrophosphate bridges between lipid A molecules were recently reported by Lüderitz [49a]. In addition, there might be cross-links at the heptose region (see text for further explanation). The figure is not drawn to scale; also arbitrarily drawn are directions of protruding ionized groups and the location of the three hydroxyl-linked fatty acids in Lipid A. EtN = ethanolamine.

[63, 64]; the presence of a very similar kind of heterogeneity was first demonstrated with Aerobacter aerogenes LPS by Koeltzow and Conrad [65]. It is interesting that the distribution of the size (which reflects the length of O side chain) is not smooth. Rather, there are polysaccharides with very long O side chains and those with very short side chains, but nothing in between [64, 65]. A similar result was obtained by acrylamide gel electrophoresis of S. typhimurium LPS in the presence of SDS [66], and by gel filtration of E. coli O111 LPS [66a].

Conrad's group have found that, in both A. aerogenes and Salmonella LPS, a class of exceptionally long O side chains exist in a form linked to lipid A, rather than to the core oligosaccharide, through an acid-labile linkage [63, 65]. The detailed structural features of this linkage are still unknown.

Variable results have been reported on the possible presence of unsubstituted core stubs in the LPS from smooth, wild-type strains.

1. If unsubstituted stubs exist in *S. typhimurium*, the average O side-chain length computed from the ratio of mannose in the terminal repeating unit to that in the internal units should be different from the chain length computed from the ratio of mannose (or rhamnose) to one of the core sugars. These ratios can be determined from the results of methylation studies by Hellerqvist and co-workers [67] on LPS from *S. typhimurium* LT2; in contrast to the result predicted above, very similar ratios 8 and 10 were found. Thus, most core oligosaccharides seem to carry O side chains.

2. Many LPS preparations from smooth *Salmonella* strains, however, have been found to cross-react with anti-R-core sera [68]. Furthermore, acetic acid treatment of *S. anatum* LPS gave rise to unsubstituted core oligosaccharides [63]. The extent of the covering of core stubs, therefore, seems to be very much strain-dependent.

3. Some methylation studies suggest that the extent may also vary even in a single strain, presumably owing to differences in the conditions of growth [69].

These observations reveal the difficulties encountered when we try to determine the length of O side chains. We see that average values mean very little in view of the presence of a discontinuous populations of very short chains, long chains, and probably Conrad's exceptionally long chains. [An exception is the LPS from *rfc* mutants, where each O side chain consists uniformly of a single repeating unit [70] (see Section III, C, 5, b).] Obviously, it will be important to find out the mode of association, by cross-linking, of LPS polysaccharide chains, e.g., whether a polysaccharide containing a long O side chain is cross-linked with another such polysaccharide, with a polysaccharide containing a short O side chain, or with an unsubstituted core oligosaccharide.

Only a few points will be mentioned about the physicochemical behavior of LPS in solution. The reader is referred to a recent review [28] for details. LPS is a typical amphipathic molecule and contains both the hydrophilic (polysaccharide) and the hydrophobic (lipid A) ends. Consequently, LPS forms molecular aggregates in aqueous solution. The types of aggregates seen depend on the organism LPS was isolated from, on the presence or absence of divalent cations, and on the method of preparation of LPS [28]. It seems, however, that with LPS from wild type *Salmonella* or

E. coli, containing long O side chains, ribbonlike structures are most common, and that with LPS from "deep rough" mutants, lacking all of O side chains and most of the core oligosaccharide, vesicular structures are frequently observed. It must be emphasized that in both of these structures the hydrophobic interaction between the lipid A portions of LPS is assumed to play a central role. Evidence favoring this interpretation include the trilaminar, unit-membranelike profiles frequently seen, and the presence of carbohydrate chains on the surface of these structures, as revealed by the binding of ferritin-labeled antibody [28].

An important feature of the LPS structure is the very large number of ionizable groups, especially anionic groups, which are clustered together in the heptose-KDO region of the molecule (Fig. 4). DePamphilis [71] pointed out after a survey of literature that LPS always has a tendency to assume vesicular structures whenever Mg^{2+} was present during its preparation. Since different investigators have used different organisms, it is difficult to conclude that the availability of Mg^{2+} was the sole determining factor in these experiments. However, it is tempting to speculate that in the presence of enough Mg^{2+}, electrostatic repulsion between neighboring LPS molecules is neutralized, thus enabling LPS to assume a bimolecular leaflet structure, which could form closed vesicles. In contrast, in the absence of Mg^{2+}, LPS aggregate will assume a form which would minimize the repulsion between anion-rich regions. As a purely speculative possibility, it may be thought that LPS molecules aggregate in a radial pattern with lipid A portions in the center, and that the piling up of such a two-dimensional, disclike structure forms a cylindrical structure, which might be recognized as ribbonlike (cf. "hexagonal I" phase of Luzzatti [71a]).

Various reagents can disaggregate the LPS aggregate in water. Olins and Warner [72], for example, showed that EDTA dissociates the 70-100 S aggregate of Azotobacter vinelandii LPS into 5.7 S (1.34×10^5 daltons) units. DePamphilis [71] found that vesicular E. coli LPS aggregate (diameter 100-500 nm) is converted into discs (about 20 nm in diameter) and rods upon EDTA treatment. EDTA presumably acts by increasing the electrostatic repulsion as was mentioned previously. In addition, detergents that presumably bind to the hydrophobic portions of LPS are also capable of dissociating LPS aggregates [65, 66, 71, 72, 73]. An intriguing observation is that several proteins can cause disaggregation of LPS [74-76]. In many cases what first appeared to be an enzymatic degradation of LPS turned out to be a case of disaggregation due to complex formation. Possibly proteins undergo a hydrophobic interaction with the lipid A portion. Yet some proteins that are effective in disaggregation are basic proteins, and the possibility of electrostatic interactions with lipid A phosphate groups or the anion-rich inner core region, or both, cannot be dismissed lightly. In this connection, the observation [77] that triethylamine brings about the disaggregation of LPS also seems interesting.

LPS from wild-type Salmonella or E. coli is insoluble in organic solvents, undoubtedly due to the presence of the large polysaccharide portion. LPS from "deep rough" mutants, however, are soluble in solvents such as chloroform-methanol [78, 79].

III. BIOSYNTHESIS OF LPS

A. Biosynthesis of Lipid A

Very little is known about the biosynthesis of lipid A. Its fatty acid composition is very different from that of the glycerophospholipids (Table 3), and seems to be rather constant. According to Rietschel [58], dodecanoic acid, tetradecanoic acid, hexadecanoic acid, and 3-hydroxytetradecanoic acid occur in a molar ratio of 1:1:1:3. The results of Romeo et al. [48] also fit this pattern, if we pool together tetradecanoic and hexadecanoic acids. In view of these results it seems possible that specific transferases transfer each fatty acid to a specific location on the glucosamine disaccharide skeleton. Nesbitt and Lennarz [80], however, showed that an L-form of Proteus produces an LPS of fatty acid composition rather different from that of the LPS of the parent bacillary form; the presumed specificity of the transferases, therefore, may not entirely explain the control over the fatty acid composition of lipid A.

Since 3-hydroxytetradecanoic acid is a unique component of lipid A, attempts have been made to demonstrate the enzymatic transfer of this fatty acid onto partially degraded LPS or lipid A preparations. The results, however, have so far been negative, and the only transfer reaction demonstrated has been that involving the transfer of this acid from its ACP (acyl carrier protein) derivative to lysophosphatidylethanolamine [81].

It is conceivable that the product of this reaction, i.e., phosphatidylethanolamine containing 3-hydroxytetradecanoic acid, then transfers the fatty acid to lipid A. If this is the normal pathway for lipid A synthesis, phosphatidylethanolamine containing this acid can be expected to accumulate whenever lipid A biosynthesis is shut off by the lack of glucosamine disaccharide. This hypothesis was tested by using mutants defective in glucosamine biosynthesis [82]; no accumulation of the expected phospholipid was found [83].

Recently Humphreys et al. [83a] showed that cell envelopes of a Pseudomonas strain incorporate 3-hydroxydodecanoyl residues into an endogenous acceptor, if incubated with dodecanoyl CoA. The incorporation product was shown to be the lipid A portion of the endogenous LPS.

B. Biosynthesis of Core Oligosaccharide

1. Reaction Sequence

In contrast to O side chains, which are independently synthesized and then transferred to the R core (Section III, C), the core oligosaccharides are made by the stepwise extension of the preexisting lipid A structure. The first step may be the transfer of a KDO residue onto lipid A, as is clear from Fig. 3. Heath and co-workers [52] have demonstrated, in a cell-free system, a reaction that presumably corresponds to this first step. In this system KDO is transferred from its activated derivative, CMP-KDO, to an acceptor in the presence of a soluble enzyme from *E. coli*. LPS or alkali-deacylated LPS is inactive as acceptor presumably because the site of attachment for KDO is already blocked. Lipid A prepared by mild acid hydrolysis of LPS is active, but a far more active acceptor can be prepared by the controlled hydrolysis of lipid A with acid and alkali. This acceptor appears to correspond to a glucosamine oligosaccharide in which the amino groups are substituted by 3-hydroxytetradecanoic acid residues. These results could mean that the transfer of KDO normally occurs before the O-acylation of the glucosamine skeleton of lipid A. It is equally possible, however, that the increased activity of the degraded lipid A preparations simply is the result of their increased solubility.

Nothing is known about the addition of the two remaining KDO residues and of heptose moieties. Nucleoside diphosphate derivative of L-*glycero*-D-*manno*-heptose has not yet been isolated. It has been shown, however, that transketolase is needed for the synthesis of this heptose [84]; sedoheptulose-7-phosphate, therefore, appears to be its biosynthetic precursor. If the mutant is grown in the presence of ^{14}C-sedoheptulose-7-phosphate, the LPS becomes labeled specifically in its heptose residues [84]. In LPS heptose residues are phosphorylated, and are believed by some investigators to be cross-linked via phosphodiester linkages (Section II). Some of the steps in the phosphorylation of LPS have been clarified by Mühlradt [45, 85]. A "soluble" enzyme, obtained by repeated washing of EDTA-lysozyme-treated *Salmonella* cells, catalyzes the transfer of the γ-phosphate from ATP to LPS in the heated cell-envelope fraction [45]. As the acceptor, cell envelopes from mutants (P^-), which cannot phosphorylate heptose residues in LPS, were used. The phosphate groups are transferred to the C-4 position of Hep_I (Fig. 3) [43]. Although LPS lacking Glc_I (chemotype Rd_1, Fig. 3) can be phosphorylated, LPS containing this glucose residue is a much better acceptor [45]. On the basis of these findings Mühlradt proposed that the transfer of glucose (Glc_I of Fig. 3) normally precedes the phosphorylation of the Hep_I residue [45].

If the core oligosaccharide chains are to become cross-linked through phosphodiester bridges between heptose residues, the phosphomonoester groups transferred to the C-4 of Hep_I residues might eventually become linked to a heptose residue of another chain. In this connection it should be noted that some Hep_I residues in an Rc LPS carry pyrophosphorylethanolamine at the C-4 position [43]. It is thus likely that the phosphorylation of Hep_I described above is followed by a transfer of phosphorylethanolamine group onto the phosphate residue just added. It is also conceivable that this is a step necessary for the subsequent formation of phosphodiester cross-bridges, if such cross-bridges indeed exist (Section II).

The reactions involved in the transfer of Glc_I and Gal_I residues were the first steps in LPS biosynthesis to be studied in cell-free systems, because of the availability of UDP-glucose- and UDP-galactose-deficient mutants. It was found that mutants unable to synthesize UDP-galactose, the donor of galactosyl residues including Gal_I, produce a very incomplete LPS containing only heptose, glucose and KDO as sugar components of its polysaccharide portion [86-88]. It could also be shown that the alteration in the composition of LPS was the result of the defective synthesis of UDP-galactose (due to the loss of UDP-glucose-4-epimerase activity; 1, of Fig. 5), since these mutants could synthesize UDP-galactose from exogenous galactose (Fig. 5) and could thus produce an LPS of "normal" composition and structure if grown in media containing galactose [86, 87]. Very similar results were obtained with mutants defective in phosphoglucoisomerase [89] (2, of Fig. 5) or UDP-glucose pyrophosphorylase (3, of Fig. 3) [90-92]. Both of these mutants produce an Rd_1 LPS (Fig. 3), because of the defective synthesis of UDP-glucose; LPS synthesis proceeds only up to the transfer of the Hep_{II} residue. In the phosphoglucoisomerase mutant, the defect in LPS synthesis can be "cured" by growing the cells in media containing glucose, which is converted into UDP-glucose through the pathway outlined in Fig. 5.

These mutants were useful in studying the chemical structure of the heptose-KDO backbone region of LPS, and more importantly, in establishing cell-free systems of LPS biosynthesis. It is difficult to demonstrate the *in vitro* transfer of core sugars in wild-type cells, which contain most of their LPS in its completed form (Fig. 3), and thus have few available empty sites for the sugar to be transferred. In contrast, the cell envelope fraction from these mutants contains all of its LPS in an incomplete form (Rc or Rd_1, Fig. 3), which is ready to accept galactose or glucose residues. Incubation of this fraction with labeled UDP-galactose or UDP-glucose thus resulted in a rather efficient addition of these sugars onto the endogenous incomplete LPS [93-95].

Use of this unfractionated system already gave us the following important conclusions:

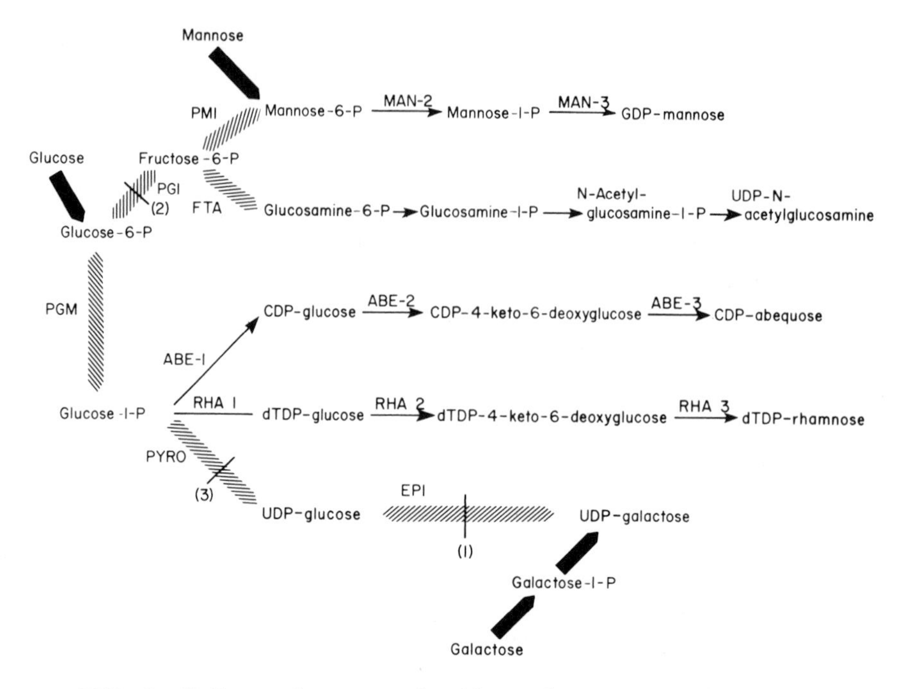

FIG. 5. Pathway of sugar nucleotide synthesis in S. typhimurium. → and ↔ denote anabolic reactions exclusively involved in LPS synthesis; ▱ and ⬡ , amphibolic reactions; and ➡ , catabolic reactions which nevertheless can be utilized, as seen, for the conversion of exogenous sugars to sugar nucleotides. The enzymes catalyzing these reactions are shown in shorthand notations. PGI, phosphoglucoisomerase; PGM, phosphoglucomutase; PMI, phosphomannoisomerase; PYRO, UDP-glucose pyrophosphorylase; EPI, UDP-glucose-4-epimerase; FTA, fructose-6-P-transaminase. The following notations are also used in the genetic map of Fig. 7: ABE-1, CDP-glucose-pyrophosphorylase; ABE-2, CDP-glucose-oxidoreductase; RHA-1, dTDP-glucose-pyrophosphorylase; RHA-2, dTDP-glucose-oxidoreductase; MAN-2, phosphomannomutase; MAN-3, GDP-mannose-pyrophosphorylase. Reactions in the final steps of CDP-abequose and dTDP-rhamnose synthesis need more than one enzyme in each case; for the sake of brevity, however, ABE-3 and RHA-3 are used as symbols.

1. The incorporation product is LPS, rather than the free core oligosaccharide devoid of lipid A. Together with the observation that these mutants accumulate incomplete LPS rather than incomplete oligosaccharides, this result suggests that core oligosaccharide chains remain linked to lipid A while it is being extended.

2. The incorporated glucose is linked to a heptose residue in the endogenous, acceptor LPS [96], and the galactose becomes linked to a glucose residue, presumably identical with the residue incorporated by envelopes of UDP-glucose-deficient mutants, via the $\alpha(1 \rightarrow 3)$ linkage [97].

3. Osborn and co-workers found that one can get further extension of the oligosaccharide chain in this system [98, 99]. Thus, the envelope fraction from a UDP-galactose-deficient mutant incorporates ^{14}C-glucose from ^{14}C-UDP-glucose if nonradioactive UDP-galactose is added together. It also incorporates ^{14}C-N-acetylglucosamine from its UDP-derivative in the presence of nonradioactive UDP-galactose and UDP-glucose. These observations suggest the sequential addition of galactose, glucose, and N-acetylglucosamine in this order (D-F, Fig. 6). Since the LPS from O-side-chain-defective mutants (R core, or Ra LPS of Fig. 3) has the sequence GlcNAc-Glc-Gal-(Gal-)-Glc-$(Hep)_2$-$(KDO)_3$-, it appears that the complete R core has been synthesized in a cell-free system starting from the Rc-type LPS. This conclusion was confirmed by partial hydrolysis of incorporation products and of Ra-LPS [31, 100]. The only difference was the absence, from the *in vitro* product, of one of the galactose residues (Gal_{II} of Fig. 3), which occurs as a branch. The transfer of this residue was more recently demonstrated in a cell-free system from *S. minnesota* [45, 85].

The most important finding obtained by the use of the unfractionated cell envelope fraction was the fact that core oligosaccharide is made by the *sequential* addition of monosaccharide units. Thus, in the example discussed above, the enzymatic transfer of Glc_{II} residue in the UDP-galactose-deficient mutant does not require the *simultaneous* presence of UDP-galactose; cell envelopes first incubated with UDP-galactose and then washed free of remaining UDP-galactose can incorporate Glc_{II} residues very well [98, 99]. Each reaction step, therefore, only requires the presence of the proper transferase, the sugar nucleotide, and the acceptor molecule.

In *E. coli*, Edstrom and Heath [101, 102] have demonstrated that the core oligosaccharides are synthesized by a sequential transfer mechanism very similar to that found in *Salmonella*.

2. Genetics of Core Oligosaccharide Synthesis

The concept of the sequential transfer of monosaccharide units is also supported by the studies on the structure of LPS produced by various core-defective mutants. Mutants deficient in UDP-glucose or UDP-galactose synthesize the core LPS normally up to the point at which these sugar nucleotides are needed (see Section III, B, 1). These mutations (*pgi*, *galU*, and *galE* for phosphoglucoisomerase, UDP-glucose pyrophosphorylase, and

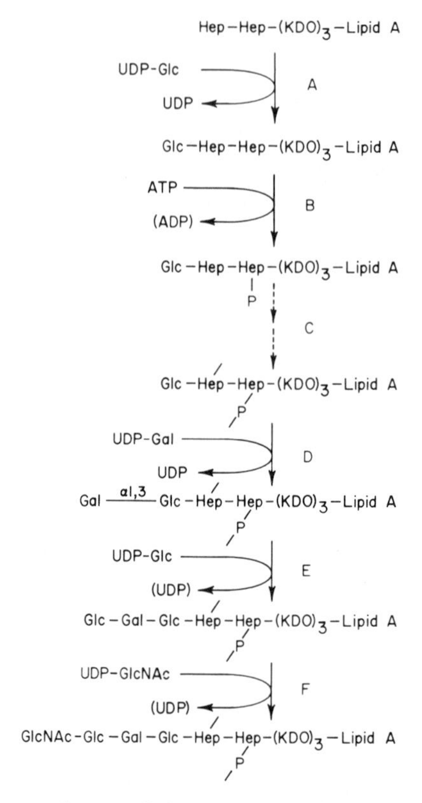

FIG. 6. Biosynthesis of the R core. The phosphorylation of one of the heptose residues is assumed to occur before the transfer of the first galactose residue, according to the results of Mühlradt [44]. In step C, chains might become cross-linked via phosphodiester linkages; this reaction, however, has not yet been demonstrated in cell-free systems.

UDP-glucose 4-epimerase) map at various places on the chromosome (Fig. 7).

In addition to these sugar-nucleotide-deficient mutants, there should exist mutants defective in many glycosyl transferases, which are presumed to be necessary for the assembly of the core LPS. Such mutants were indeed isolated as a group of R mutants, and have been classified by Stocker's group on the basis of various phenotypic criteria ([103] for review see Mäkelä and Stocker [33]; Stocker and Mäkelä [34]). One criterion that was extremely useful is based on the finding of Beckmann et al. [104] that the aqueous phase after phenol-water extraction of R-core-defective mutants contained haptenic O side chains, which are not linked to the R core, whereas mutants defective in O side chain assembly did not produce such haptenic polysaccharides. Another was the use of various phages which required LPS at various intermediate stages of biosynthesis as receptors (for review see Stocker and Mäkelä [34] and Rapin and Kalckar [105]). Genetic mapping showed that all mutations involved exclusively in core biosynthesis were located in the xyl-ilv region of the chromosome [103]. These mutants are called "rough A" or rfa. They were subdivided on the basis of their phenotypes and also by genetic mapping, and at present seven classes are known (Fig. 7). It is interesting that genes located outside of the xyl-ilv region, such as galE or pgi, have important functions other than core biosynthesis. Among the rfa genes, five (rfaF, rfaG, rfaJ, rfaK, and rfaL) map in a very small region of the chromosome between cysE and pyrE, but at least two (rfaE and rfaH) map outside this region. It is not known whether the genes in the former group form a single unit of transcription.

The analysis of mutant LPS showed that each rfa gene probably determines one of the glycosyl transferases responsible for core biosynthesis. This correspondence, shown in Table 2, is impressive in that mutants are known for almost every step of the biosynthetic scheme; these results confirm the sequential mechanism of core assembly. The loss of enzyme activity, however, has been ascertained in only two classes of mutants, rfaH and rfaG [106]. The rfaH mutants still attach the Gal_{II} residues to their LPS; thus the presence of Gal_{I} residue is apparently not necessary for this transfer reaction. One class of mutants, rfaP, known only in S. minnesota, lacks the ability to phosphorylate the heptose residue of LPS. This gene is closely linked to other rfa loci [107]. Another class of rfa mutants which deserves special mention is rfaL. Unlike other rfa mutants, the rfaL mutants produce complete R core; thus they are not core-defective. Their extracts were shown to be defective in the "ligase" reaction which links together the R core with the independently synthesized O side chain [26] (Section III, D).

Series of core-defective mutants have been characterized in Shigella flexneri [108] as well as E. coli O111 [109], although the mutants do not seem to have been analyzed genetically.

Escherichia coli K12 is a rough strain. However, one can isolate from K12 mutants that have lost further parts of the core oligosaccharide. Thus, several "deep rough" mutants were isolated from this strain [110] by taking advantage of their increased sensitivity to certain antibiotics (see Section V, B). These mutants contain no heptose or only reduced amounts of heptose in their LPS. The mutations have been mapped in two regions of the chromosome; one group ("lpcA") is located between leu and lac, and the other ("lpcB") close to thyA [110]. None of the rfa mutants of Salmonella has so far been mapped in these regions of the chromosome.

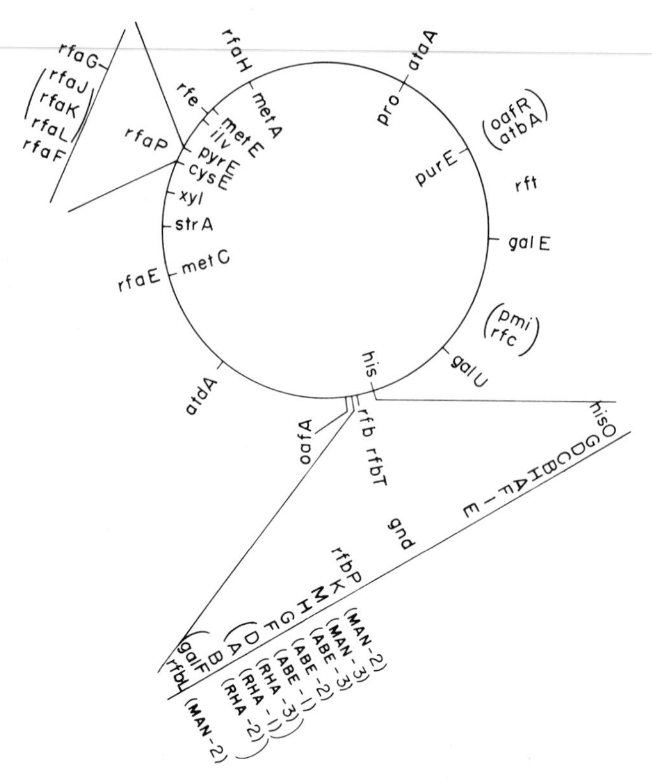

FIG. 7. Chromosomal location of genes involved in LPS synthesis. In this map, genes involved in LPS synthesis are shown by gene symbols drawn outside the circle depicting the chromosome; the gene symbols inside denote other genes, shown for reference. A gene symbol (e.g., *rfaG*), according to the standard convention in bacterial genetics, consists of a three-letter lower case symbol (*rfa*) denoting a specific function of a group of genes, and a capital letter (*G*) denoting individual cistrons involved in the expression of such a function. The functions coded for by the genes are *rfa*, biosynthesis of the core oligosaccharide; *rfb*, biosynthesis of O side chain; *rfc*, polymerization of O repeating units; *rfe*, biosynthesis of O side chains in group C_1 *Salmonella* (?); *rft*, biosynthesis of T1 side chain; *oafA*, acetylation of abequose residues in O side chain to produce antigen O-5; *oafR*, glucosylation of galactose residues in O side chain to produce antigen O-12_2; *galU*, UDPG pyrophosphorylase; *galE*, UDPG-4-epimerase; *pmi*, phosphomannoisomerase; *pro*, *pur*, *his*, *met*, *cys*, *pyr*, and *ilv*, biosynthesis of proline, purine, histidine, methionine, cyteine, pyrimidine, and isoleucine-valine, respectively; *xyl*, D-xylose fermentation; *gnd*, 6-phosphogluconate-dehydrogenase; and *str*, sensitivity to streptomycin. The symbol *atx* was recently proposed for phage attachment sites [102a] : *ata*, *atb*, and *atd* thus correspond to the attachment sites for converting phages P22, ϕ27, and $\phi 6_1$, respectively. The enzymes coded for by *rfb* genes are indicated by the shorthand notations (Fig. 5) of the reactions they catalyze. The *rfa* and *his-rfb* regions are enlarged for convenience. The genes *rfaP* and *rfbT* have not been precisely mapped relative to other *rfa* and *rfb* genes. The order of genes in brackets is not known.

3. Role of Phospholipids in Glycosyl Transferase Reactions

In the earlier studies of core LPS synthesis discussed in Section III, B,1, the cell envelope fractions used as the enzyme also contained the acceptor LPS. For the detailed studies of glycosyl transferase reactions involved in core synthesis, one obviously had to separate enzyme from the acceptor. This important step was achieved by Rothfield et al. [95] when they found that much transferase activity could be released into the soluble fraction by prolonged sonication. The reaction mixture containing such a soluble fraction now needs the acceptor which preferably does not contain active transferase, and this can be supplied in the form of a heat-inactivated cell envelope fraction. In this manner, the acceptor specificity of transferases was demonstrated. For example, Gal_I transferase acts only on Rc LPS. Rd_1 LPS is presumably inactive because of the lack of the glucose residue (Glc_I of Fig. 3) that comprises the attachment site for the Gal_I residue; Ra and S LPS are inactive because they already contain the Gal_I residues [95].

In this system, the purified Rc LPS was completely inactive as an acceptor. The reason for this inactivity was elucidated by Rothfield and Horecker [111], who found that extraction of phospholipids from the heated cell envelope preparation made it inactive as an acceptor, and that the activity could be restored by adding back the phospholipids. Furthermore, if phospholipids were added to purified LPS and the mixture heated and slowly cooled, a very active acceptor preparation was obtained. These results suggested that LPS-phospholipid complex, rather than LPS, is the acceptor for the galactosyl transferase reaction.

The formation of LPS-phospholipid complex was studied by Rothfield and co-workers with various techniques. (1) This complex can be isolated by isopycnic banding in a sucrose gradient, as it has a buoyant density intermediate between LPS and phospholipids [112]. (2) Addition of LPS into the aqueous phase increases the surface pressure of phospholipid monolayers at the air-water interphase [113, 114]. This is at least consistent with the idea that LPS molecules become inserted among phospholipid molecules which are aligned perpendicular to the surface. (3) In aqueous suspensions, phospholipids usually exist as a bilayer structure. The results of monolayer studies just described lead us to expect a similar insertion of LPS molecules to occur, whenever LPS is added to such a suspension of phospholipids. Electron micrographs that appear to indicate such a process have been obtained [113, 114].

The LPS-phospholipid complex, but not LPS alone, can then bind the enzyme in the presence of Mg^{2+}. Rothfield and Takeshita showed this reaction by adding crude extracts to a suspension containing the LPS-phospholipid complex [114, 115]. When the complex was centrifuged down, it was found that the transferase that uses the particular LPS in the complex as acceptor had specifically become bound to the complex and thus had become depleted from the crude supernatant. The ternary complex was also isolated by isopycnic banding [112], and was shown to interact with UDP-hexose, which then transfers the hexose residue. Further addition of Mg^{2+} is not necessary for this last step of the reaction.

The galactosyl transferase reaction, for example, thus seems to involve the following steps.

(galactose-deficient, i.e., Rc) LPS + phospholipids

$\longrightarrow$ LPS-phospholipid complex (1)

LPS-phospholipid complex + (Gal_I transferase) enzyme

$\xrightarrow{Mg^{2+}}$ LPS-phospholipid-enzyme complex (2)

LPS-phospholipid-enzyme complex + UDPGal

$\longrightarrow$ galactosyl LPS + ? (3)

These steps have now been studied in detail by Rothfield and his co-workers. First, the $galactose_I$ transferase [116] and the $glucose_I$ transferase [117] have been purified to homogeneity. Both enzymes were soluble and their molecular weights were close to 20,000. The amino acid analysis of the galactosyl transferase revealed nothing unusual except for the absence of cysteine residues [116], in marked contrast to another purified enzyme from bacterial membrane, polyisoprenol phosphokinase [118], which is extremely rich in hydrophobic amino acids. It is interesting, however, that both glucosyl and galactosyl transferases tended to aggregate in aqueous environment, and that the latter enzyme contained a lipid of unknown structure that was responsible for causing the enzyme molecules to aggregate [116]. It is of interest also, in view of the extremely rare occurrence of glycoproteins among procaryotes, that a fraction of the galactosyl transferase molecules appeared to contain glucosamine and probably other neutral sugar(s) [116].

A great deal of work was done on the mechanism of formation of LPS-phospholipid-enzyme ternary complex. It was first found that the phospholipids must have certain structural features in order to produce a complex active in the transferase reaction [119]. Phosphatidylethanolamine, phosphatidylglycerol, and phosphatidic acid were all active, but phosphatidylcholine (which rarely occurs in bacteria) was completely inactive. Fatty acid composition also had a strong influence. Lipids containing two unsaturated fatty acid residues were most active, followed by those containing one unsaturated (or cyclopropane) fatty acid and one saturated fatty acid.

Those containing two saturated, long-chain fatty acids were inactive. More recent studies with the monolayer technique produced very similar results [48]. LPS penetrated rapidly and extensively into films consisting of phosphatidylethanolamine containing unsaturated or cyclopropane fatty acids, but little penetration was observed when phosphatidylethanolamine with saturated fatty acids or phosphatidylcholine was used for manufacturing the monolayer. In the presence of excess LPS, the mixed monolayer contained one molecule of LPS per 5.6 molecules of phosphatidylethanolamine. The molecular model of LPS, the construction of which was based on the assumption that cross-linking exists at the level of heptose residues but not at that of lipid A (see Section II), suggested that phosphatidylethanolamine molecules can be inserted into the cleft that might exist between the three lipid A moieties of the same LPS molecule [48]. The significance of this model, however, is somewhat doubtful now that the cross-linking at the level of lipid A has been demonstrated [49a].

If, after the formation of a film containing phospholipids and the Rc LPS, the purified galactosyl transferase is injected into the aqueous subphase, the enzyme becomes bound to the monolayer [120], resulting in a large increase in surface pressure. The enzyme is also able to interact with monolayers containing phospholipids alone. The latter kind of interaction could actually be of primary importance in holding the enzymes on the surface of the cytoplasmic membrane (see below). Yet the significance of the enzyme-phospholipid interaction was somewhat underestimated at first, because of the following observations [120]. (1) The enzyme interacts with monolayers consisting of any phospholipid, including phosphatidylcholine, which is not capable of forming an active LPS-phospholipid-enzyme complex, and (2) the phospholipid monolayer saturated with the enzyme cannot interact with LPS. The latter situation is avoided in intact cells, presumably because lipid A is first synthesized within the membrane and then the core oligosaccharide is built on the membrane-linked lipid A. Thus, LPS never leaves the membrane during its synthesis and therefore does not have to be inserted into the membrane from the aqueous phase as in the monolayer model experiments.

The LPS-phospholipid-enzyme ternary complex is enzymatically active in the form of the film [120]. The amount of galactose transferred, however, was proportional to the amount of enzyme in the film, and one enzyme was seen to catalyze maximally the transfer of two sugar residues. This is in marked contrast to the system where an aqueous suspension of LPS-phospholipid complex was used [116, 121]; in this case an enzyme molecule could catalyze the transfer of many galactose residues. It appears that both LPS and enzyme molecules are relatively immobile in the monolayer, whereas in the aqueous suspension, enzyme or LPS or both can carry out translational movement along the plane of a bilayer structure, or "leap" from one bilayer to another possibly as a result of collision. Furthermore, the monolayer experiment was done at 25°C, in contrast to the aqueous suspension experiment done at 37 °C; thus, the difference in the fluidity of the hydrocarbon portions of the mono- and bilayer might also be responsible for this difference.

The kinetic study of the galactosyl transferase reaction in the aqueous suspension system produced additional information on the interaction of various components of this system [121]. The complex formation between acceptor LPS and phosphatidylethanolamine increases the yield of the reaction product, galactosyl-LPS, but at the same time V_{max} of the reaction (at infinite concentration of LPS) is increased 13-fold. If the complex formation merely serves to introduce space between LPS molecules and to make LPS more accessible to the enzyme, it should be possible to saturate the enzyme with acceptor LPS even in the absence of phospholipid and therefore phospholipid should not alter the V_{max}. The observation just described clearly shows that this is not the case, and that phospholipid should play a more positive role in the formation of the ternary complex, perhaps by activating the enzyme through direct interaction with sites on the enzyme molecule. Thus, it is likely that in the ternary complex the components interact in a way schematically illustrated as follows.

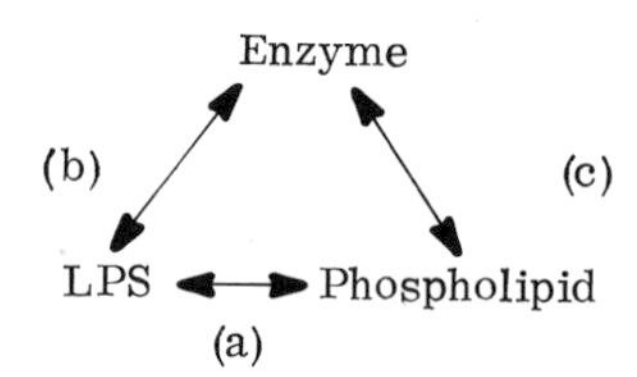

We have already discussed the nature of the LPS-phospholipid interaction (a); here the hydrophobic portions of the molecules are mainly responsible, but a tight "fit" between hydrophilic head groups might also be involved. Interaction (b), between the enzyme and LPS, probably involves the core oligosaccharide. This is seen, for example, in the observation that LPS containing the heptose-KDO region of the core, or even lipid A-free, core oligosaccharide can inhibit the reaction competitively [116]. The least clear part in this scheme is interaction (c), between enzyme and phospholipid. The kinetic data suggest the existence of this interaction, as described above; it is not clear, however, whether the enzyme interacts with only the hydrophilic head groups of phospholipids or also with the hydrophobic portion. A very large increase in surface pressure is produced when the enzyme interacts with phospholipid or LPS-phospholipid monolayers; this observation is consistent with, but does not prove, the idea that at least a portion of the enzyme penetrates into the hydrophobic portion of the monolayer. Phosphatidylethanolamines containing short-chain saturated fatty acids, e.g., decanoic acid did not form complexes with the enzyme in aqueous suspension [121]; this observation also suggests the interaction of the hydrophobic portion of the phospholipid with the enzyme. In monolayer studies, however, the enzyme interacted very well with the layer of dihexanoyl phosphatidylethanolamine [120]; these results are rather difficult to interpret at present.

Recent studies with the aqueous suspension system showed that one can isolate, by isopycnic density centrifugation, a complex which can catalyze the sequential transfer of glucose$_I$ and galactose$_I$ [122]. This complex is made by adding an Rd$_1$-type LPS, phosphatidylethanolamine, and purified glucose$_I$ and galactose$_I$ transferases. Although it was not rigorously proved that both enzymes are located in the same complex, this seems very likely in view of the high initial rates of galactose transfer after the preincubation with UDP-glucose. What emerges from these results is that the transferase enzyme (in this case galactose$_I$ transferase) can enter the complex in the absence of its proper substrate LPS (in contrast to the results obtained with crude extracts — see reference 115). Perhaps the "nonspecific" interaction of the enzymes with phospholipids — interaction (c) in the above scheme — is primarily responsible for retaining these enzymes within the membranelike structure of the complex. This hypothesis then explains why the enzyme never leaves the complex after the completion of the reaction [123].

4. A Tentative Model

On the basis of results already described, Rothfield, Romeo, and Hinckley [123] have proposed a model for the organization of various components involved in the assembly of core LPS (Fig. 8). In this model, the cytoplasmic membrane is assumed to have the basic structure of a phospholipid bilayer. Lipid A is synthesized within this basic structure, and sugar residues are sequentially added while the incomplete LPS molecules remain anchored to the phospholipid bilayer structure through the insertion of lipid A fatty acid chains among the fatty acid chains of phospholipids. Glycosyl transferases are also nonspecifically bound to the bilayer portion of the membrane. When the incomplete LPS encounters the proper transferase, a specific interaction develops, followed by the transfer of the sugar residue. If the donor sugar nucleotide is absent as in the UDP-glucose 4-epimerase mutants, this tight and specific complex persists and can be released as such from the cells with EDTA treatment [124]. But when the sugar nucleotide is available, the tight complex dissociates after the transfer reaction and the LPS then binds with the transferase catalyzing the next step.

This scheme adequately explains most of the results so far obtained. Furthermore, it clarifies the nature of the remaining problems that should be attacked in the near future. For example, is it LPS or enzymes that

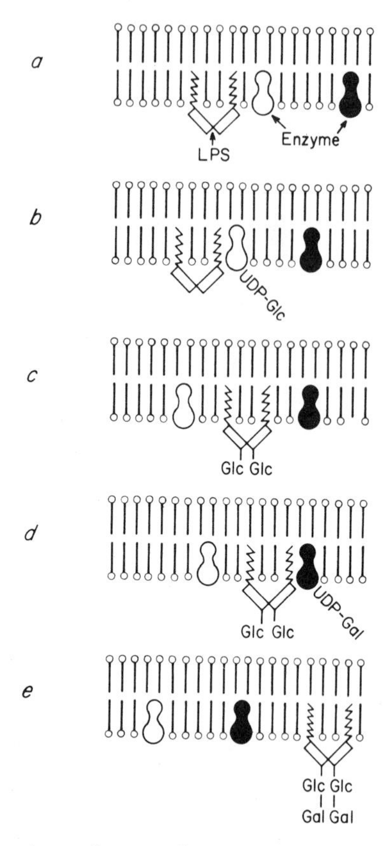

FIG. 8. Mechanism of core oligosaccharide extension on the surface of the membrane. The sequence of events beginning at (a) and ending at (e) are shown, which results in the successive transfer of glucose and galactose residues. It is based on the proposal of Rothfield and Romeo [30]. In essence, incomplete LPS molecules are supposed to move along the phospholipid bilayer membrane, and to accept sugar residues when they encounter transferase enzymes, which are half buried in the bilayer. Possibly transferase enzymes are arranged in the correct order, through the formation of a protein aggregate within the membrane; this possibility is not shown here.

move along the membrane during the sequential transfer reaction? It seems certainly possible that LPS moves, since a rapid translational movement (or cis migration according to the terminology of Rothfield et al. [123]) of phospholipid molecules in bilayer models has been demonstrated [125]. Kulpa and Leive [126] have recently found that newly made LPS enters the outer membrane via discrete, specialized regions; it may be then expected to diffuse along the surface of the outer membrane. No evidence is so far available on the mobility of transferases in the membrane.

Another question that comes to mind is that of protein-to-protein interaction in the membrane. If the incomplete LPS molecules move from one enzyme to the next, the synthesis may be facilitated greatly by placing various enzymes in the correct sequence. This might be achieved by having transferase molecules specifically interacting with each other, or by having a central "structural," protein to which transferase molecules become attached.

The third question concerns the flip-flop transition, or trans-migration of the completed R core. It is likely, as we shall see later, that O antigen polysaccharide is polymerized on the outside surface of the cytoplasmic membrane. The completed R core, then, must be transported across the membrane in order to become linked with the O side chain. Flip-flop transitions are known to be extremely slow with phospholipid molecules [127] ; perhaps the situation is different with LPS, but no experimental evidence is available at the moment.

C. Biosynthesis of O Side Chains

1. Earlier Studies

Mutants were very useful in the study of core oligosaccharide synthesis, and it was natural that mutants defective in the synthesis of sugar nucleotides were first used also in the study of O side-chain synthesis in cell-free systems. Thus, in Salmonella typhimurium mutants defective in the biosynthesis of dTDP-rhamnose [103, 128] or of GDP-mannose [99] were utilized. It was readily found that cell envelope fractions from these mutants incorporated various component sugars of the O side chain (i.e., abequose, mannose, rhamnose, and galactose in this serotype) into LPS if all of the four sugars were supplied as proper derivatives of nucleotides (cf. Fig. 5) [129, 130]. The sugar residues were linked together in a proper sequence, and these in vitro-synthesized O side chains were linked to the endogenous R core.

In contrast to the R core, O side chains are made up of oligosaccharide repeating units. But this conclusion was based on the observation that, for example in S. typhimurium, galactosyl-mannosyl-rhamnose is the main oligosaccharide product obtained by mild acid hydrolysis. (Abequosyl residues are extremely acid-labile and are completely split off during the partial acid hydrolysis, but the results of periodate oxidation studies indicate that they are attached to mannosyl residues.) This trisaccharide, however, is obtained because rhamnosyl linkages just happen to be more acid-labile than galactosyl or mannosyl linkages; it does not necessarily follow that rhamnose comprises the reducing end of O side chain. In other words, the "biological" repeating unit may be different from the "chemical" repeating unit, although polysaccharides produced by polymerization of these units will have an identical sequence except at the reducing and nonreducing ends. In S. typhimurium, examination of cosubstrate requirements in the cell-free incorporation system established the structure of the biological repeating unit, and showed that galactose comprises its reducing terminal sugar [130]. Thus, incorporation of rhamnose requires the presence of UDP-galactose but not of other sugar nucleotides; that of mannose, the presence of dTDP-rhamnose and UDP-galactose; and finally, that of abequose requires the presence of GDP-mannose in addition to these two sugar nucleotides [130].

It was thought possible that the repeating units are first assembled on a carrier molecule, and then transferred to the R core either before or after polymerization into O side chains. If this is true, repeating units are not expected to be transferred to the R core until they are completed. Although an incomplete repeating unit, rhamnosyl-galactose, can readily be transferred to the R core in a cell-free system [131], such a transfer does not seem to occur in intact cells in view of the observation that the Ra LPS without any part of the O side chain is produced by mutants lacking dTDP-rhamnose or GDP-mannose. That the repeating unit oligosaccharide is indeed assembled on a carrier lipid was discovered in the laboratories of Robbins [132] and Osborn [133], soon after the discovery of similar lipid intermediates involved in the peptidoglycan synthesis [134]. The details of these reactions are described in the next section.

2. Biosynthesis of Repeating Unit Oligosaccharide

Instead of mutants of S. typhimurium used by other workers, Robbins' group used wild-type cells of group E Salmonella such as Salmonella newington. These cells produce O side chains that have a structure very similar to those of S. typhimurium, except for the absence of abequosyl branches (Table I). These workers found a good incorporation of sugars into LPS when GDP-mannose, dTDP-rhamnose, and UDP-galactose were added together to the cell-envelope fraction [135]. In contrast to the cell-free synthesis of core oligosaccharide, one does not need defective mutants to accomplish the in vitro assembly of O side chains. This may be related to the fact that a large fraction of LPS from a wild type S. newington strain is in the form of unsubstituted R core [63], whereas LPS corresponding to the intermediate stages of core synthesis are not found in noticeable amounts, at least in S. typhimurium [136].

When only dTDP-rhamnose and UDP-galactose were added, the incorporation product was not LPS and was released from the envelope fraction as rhamnosylgalactose by phenol-water extraction [132]. Wright et al. [132] then found that the galactose and rhamnose residues were transferred to a "lipid" carrier, and that the true incorporation product, rhamnosyl-galactosyl-"lipid," could be extracted from the reaction mixture with organic solvents. The phenol-water extraction at 68°C was assumed to result in the hydrolysis of the bond between the "lipid" and rhamnosyl-galactose, which then appears in the aqueous phase. It was also found, by the use of ^{32}P-labeled UDP-galactose that the entire galactose-1-phosphate moiety of this sugar nucleotide is transferred to the lipid acceptor [137]. The phosphate group of the transferred galactose-1-phosphate becomes linked to a phosphate group of the acceptor "lipid," to form a pyrophosphate linkage, since the kinetics of hydrolysis of this linkage in acid was exactly the same as that of the pyrophosphate linkage in UDP-glucose [137]. Another evidence consistent with this structure is the observation that the lipid intermediate containing rhamnosyl-galactose is labile to alkali, and produces rhamnosyl-galactosyl-1-phosphate presumably via the 1,2-cyclic phosphate intermediate [132]. These results then indicate the occurrence of the following series of reactions. The acceptor lipid is represented here as P-ACL, or antigen carrier lipid phosphate.

UDP-galactose + P-ACL → Gal-1-PP-ACL + UMP

dTDP-rhamnose + Gal-1-PP-ACL → Rha-Gal-1-PP-ACL + (dTDP)

GDP-mannose + Rha-Gal-1-PP-ACL → (Man-Rha-Gal-1-PP-ACL) + (GDP)

↓
↓
↓

(Man-Rha-Gal-)$_n$-(R core)

Although the presumed product of the third step, trisaccharide-PP-ACL, could not be isolated, the occurrence of this step has been suggested by the observation that the addition of GDP-mannose to a reaction mixture preincubated with ^{14}C-dTDP-rhamnose and UDP-galactose rapidly converted the organic-solvent-extractable radioactivity into a form linked to LPS [132]. Thus in this system, the transfer to the R core (catalyzed by "O-antigen: LPS ligase") does not take place at least until the repeating unit is completed, in contrast to *S. typhimurium* where incomplete repeating units are readily transferred to R core in a cell-free system [131].

The reactions described by Robbins and coworkers are formally analogous to reactions of the "lipid cycle" in the biosynthesis of peptidoglycan [134]. Here again the final product, peptidoglycan, consists of oligosaccharide (here disaccharide) repeating units. The repeating unit is synthesized on a lipid carrier. The first sugar, N-acetylmuramic acid, is transferred also as sugar-1-phosphate (actually with covalently linked pentapeptide) to the carrier, whereas the second sugar, N-acetylglucosamine, is transferred by the usual mechanism.

Studies with the S. typhimurium system, carried out in Osborn's laboratory, also established the presence of a lipid cycle similar to that operating in S. newington [133, 138, 139]. These workers utilized a UDP-galactose-deficient mutant of S. typhimurium, which, owing to the incompleteness of the core oligosaccharide, cannot transfer the O side chains to the core and thus accumulates intermediates of O side chain synthesis in a cell-free system. The reactions of S. typhimurium lipid cycle are shown in Fig. 9, with the topological features that will be described later. The first reaction is the transfer of galactose-1-phosphate to P-ACL [reaction a in Fig. 9], and the product, galactose-1-PP-ACL, has been isolated from the reaction mixture [140]. The reaction is freely reversible, and the apparent equilibrium constant is about 0.5 [140]. This step is then followed by the sequential transfer of rhamnose and mannose. The trisaccharide intermediate, mannosyl-rhamnosyl-galactosyl-1-PP-ACL, accumulated to a certain extent when a low incubation temperature was used [133], and could be isolated from such a reaction mixture. As soon as the temperature was raised to 37°C, the trisaccharide-PP-ACL molecules interacted with each other, and polymerized into a polysaccharide, presumably still linked to PP-ACL.

3. The Polymerization of Repeating Units

Osborn's group studied the polymerization of repeating units (a reaction catalyzed by "polymerase") by using endogenously generated oligosaccharide-PP-ACL intermediates. Since abequose occurs as branches of O side chain (Fig. 3), it was conceivable that it could be added later after the polymerization of the trisaccharide (Man-Rha-Gal) to form the main chain sequence. However, Osborn and Weiner [141] have shown, again by using incubation at low temperature, that abequose residues can be added to the trisaccharide-PP-ACL, and that the tetrasaccharide-PP-ACL thus generated becomes polymerized more rapidly than the trisaccharide-PP-ACL. These results obviously suggest that abequose is normally transferred before the polymerization of the repeating unit. Later studies on a mutant defective in CDP-abequose synthesis conclusively showed that in intact cells the presence of the abequosyl group in the repeating unit

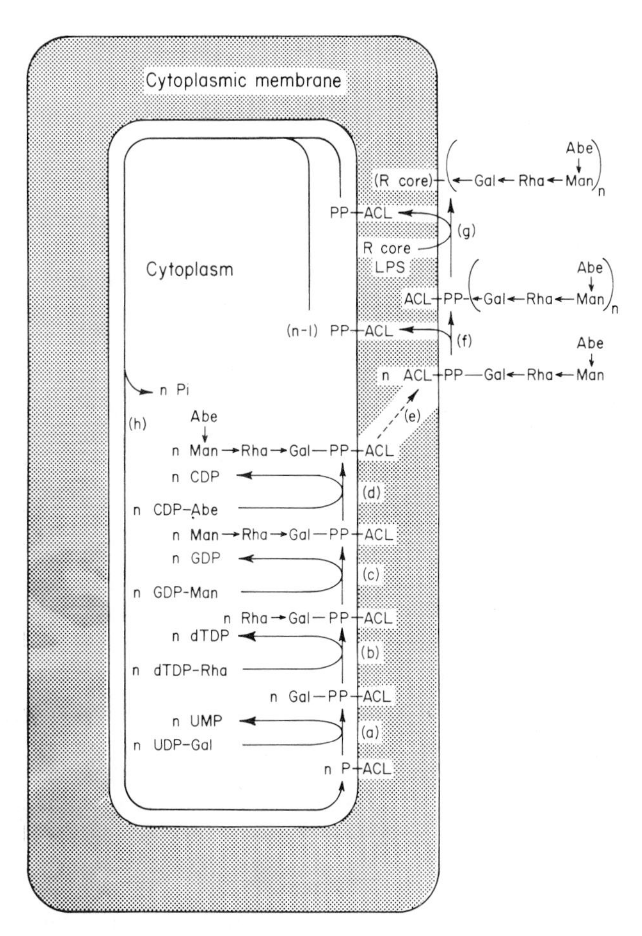

FIG. 9. Lipid cycle in the biosynthesis of O side chains in S. typhimurium. The tetrasaccharide repeating unit is synthesized on a carrier lipid (P-ACL) (reactions a through d), then the units are polymerized (reaction f), and finally the polymerized O side chain is transferred to the R core (reaction g). PP-ACL, a product of reactions f and g, is hydrolyzed to regenerate the carrier, P-ACL (reaction h). This figure also incorporates the hypothetical transport of (repeating unit)-PP-ACL across the membrane (reaction e) and assumes that reactions a through d take place at the inner surface of the cytoplasmic membrane, whereas reactions f and g take place at the outer surface. These hypothetical features of the topology are discussed fully in Section III, C, 9.

oligosaccharide is obligatory for the polymerization to occur [142] (see Fig. 9). The ready polymerization of the trisaccharide repeating units by the cell envelope fraction is thus an example of the loss of specificity in cell-free systems; another example of this phenomenon is the premature transfer of incomplete repeating unit to R core in a cell-free system [131] (see also pp. 164 and 183).

The polymerization reaction has been studied by Kanegasaki and Wright in S. anatum system [143]. The trisaccharide-PP-ACL can be added to the cell envelope fraction from this organism, and the polymerization reaction can be demonstrated after repeated freezing-thawing of the mixture. The freeze-thawing is assumed to cause the lipid intermediate to penetrate into the membrane structure [143].

It seems possible that the enzymes catalyzing the successive steps of repeating unit synthesis and the polymerase form a tightly organized complex in which P-ACL intermediates are sequentially transferred from one enzyme to the next. This hypothesis then predicts that only the endogenously generated trisaccharide-PP-ACL will interact with the polymerase molecule. In contrast, if lipid intermediates are mobile within the hydrophobic interior of the membrane, polymerase molecules will be able to interact with exogenous or endogenous trisaccharide-PP-ACL molecules associated with the membrane. The results of Kanegasaki and Wright [143] clearly favor this latter possibility.

Although results obtained in cell-free systems are not absolutely reliable, as seen in the cases of the specificity of polymerase (this section) and O antigen: LPS ligase (Section III, C, 2), there is at least one suggestive evidence for the mobility of ACL-linked intermediates within the membrane. The biosynthesis of peptidoglycan appears to require a lipid carrier of the same structure as ACL [144]. Thus, it is conceivable that an accumulation of oligosaccharide-PP-ACL in mutants defective in the O repeating unit assembly may tie up even that portion of ACL normally used for peptidoglycan synthesis and thus kill the cell. Indeed it is difficult to isolate point mutants defective in the later stages of repeating unit assembly and such mutants, when isolated, tend to be extremely unstable [142].

A very important observation was made by Bray and Robbins [145, 146] concerning the direction of chain elongation by polymerase. The polymerase catalyzes the joining of a repeating unit monomer to a growing chain containing more than one repeating unit. Here two mechanisms are possible (Fig. 10). Either the growing chain is transferred onto the nonreducing end sugar (mannose) of the repeating unit monomer, or the repeating unit monomer is transferred to the nonreducing end of the growing chain. It was conclusively shown, by pulse labeling experiments both in intact cells and in a cell-free system, that the former mechanism is operating in S. anatum [145, 146]. Thus, the extension of O side chain is formally analogous to protein synthesis and fatty acid synthesis, where the chain extension is accomplished also by the transfer of growing chains onto monomers.

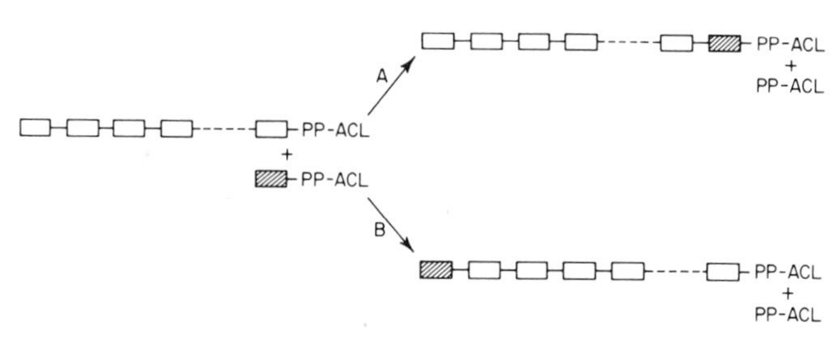

FIG. 10. Two theoretically possible mechanisms of chain elongation in O side chain synthesis, as considered by Bray and Robbins [145, 146]. In the polymerase-catalyzed elongation reaction, there are two possibilities. (A) A growing chain (already containing several repeating units) is transferred onto a repeating unit monomer, linked to PP-ACL. (B) A monomer is added on to the non-reducing end of the growing chain. Pulse-labeling experiments showed that the reaction takes place by the mechanism A.

In mutants defective in R core synthesis (rfa mutants), phenol extraction produces free, haptenic O side chains as well as the R-type LPS [140] (Section III, B, 2). When envelope fractions from such mutants are used for the *in vitro* biosynthesis of O side chains, the polymerized O side chains remain attached to the envelope fraction, and are released as free O side chains upon phenol extraction [147]. This *in vitro* synthesized product and the O hapten accumulated by rfa cells are very similar in structure [148]. Both contain about 30 repeating units, and the reducing end corresponds to galactose-1-phosphate. These results, the actual isolation of the (repeating unit)$_2$-PP-ACL from a *in vitro* reaction mixture [141], and the mechanism of chain elongation discussed above all suggest that the polymerized O side chains remain attached to PP-ACL until their transfer to the R core, which cannot take place in rfa mutants. Phenol extraction or treatment with cold trichloroacetic acid is assumed to cleave the pyrophosphate bond, releasing the O-polysaccharide-phosphate [148].

4. Structure of Antigen Carrier Lipid

P-ACL comprises only a very small portion (about 0.1%) of the phospholipids of bacteria. It has been purified as rhamnosyl-galactosyl-PP-ACL, and its structure was determined by mass spectrometry [149]. ACL is a C_{55}-polyisoprenoid alcohol ("undecaprenol"), which becomes P-ACL when phosphorylated at its primary hydroxyl group, as shown below.

$$CH_3\text{-}\underset{}{\overset{CH_3}{\overset{|}{C}}}\text{=CH-}CH_2\text{-(}CH_2\text{-}\overset{CH_3}{\overset{|}{C}}\text{=CH-}CH_2)_{10}\text{-O-}\underset{\underset{O}{\|}}{\overset{\overset{OH}{|}}{P}}\text{-OH}$$

The carrier lipid which functions in the biosynthesis of peptidoglycan was also found to be a C_{55}-polyisoprenoid phosphate [144] with two internal trans double bonds [150], and this lipid from Staphylococcus and Micrococcus functions well in a cell-free system for Salmonella O side-chain synthesis [26]. Thus, the two lipids are likely to be identical, although we must keep in mind that the activity of a lipid in a cell-free system is not a proof that it is identical to the natural compound. Indeed, it is known that ficaprenol, which contains three internal trans double bonds, can substitute for the natural C_{55}-lipid with two internal trans double bonds [151] in another cell-free system that requires such a carrier lipid, i.e., the biosynthesis of mannan by Micrococcus lysodeikticus [152].

The enzymatic synthesis of PP-ACL has been described in group E Salmonella [153]. The reaction utilizes Δ^3-isopentenyl pyrophosphate (presumably 8 molecules) and farnesyl pyrophosphate (97% of which was apparently the trans, trans-isomer [154]), and is catalyzed by the cell envelope fraction.

$$CH_3-\overset{CH_3}{\overset{|}{C}}=CH-CH_2-(CH_2-\overset{CH_3}{\overset{|}{C}}=CH-CH_2)_2\text{-pyrophosphate} + 8\ CH_2=\overset{CH_3}{\overset{|}{C}}-CH_2-CH_2\text{-pyrophosphate} \longrightarrow CH_3-\overset{CH_3}{\overset{|}{C}}=CH-CH_2-(CH_2-\overset{CH_3}{\overset{|}{C}}=CH-CH_2)_{10}\text{-pyrophosphate}$$

Since lipid carriers similar to ACL are known to have two internal trans double bonds [150, 151], the utilization of trans, trans-farnesyl pyrophosphate suggests that the reaction proceeds by the sequential addition of cis isopentenyl units at the polar end of the molecule, and that the trans double bonds in the final product will be located next to the nonpolar end of the molecule.

In any event, the direct product of the reaction described above must be a polyisoprenol pyrophosphate. Since the membrane-bound product functions as an acceptor of galactose-1-phosphate group, it must have been first hydrolyzed to a monophosphate ester. This hydrolytic cleavage of PP-ACL into P-ACL must also be a part of the lipid cycle (Fig. 9); this reaction is known to be inhibited by bacitracin [155].

5. Genetics of O Side-Chain Synthesis

a. The rfb Gene Cluster. We have already seen that several of the enzymes exclusively involved in the core oligosaccharide synthesis are coded for by a cluster of genes (rfa, G, J, K, and L) located between pyrE and cysE (Section III, B, 2). Similarly, most genes exclusively involved in O side chain synthesis seem to occur in a cluster located between metG locus and the his operon (Fig. 7). It was first found by Subbaiah and Stocker [103] that a number of rough S. typhimurium mutants mapped near his. These mutants were assumed to be defective in O side-chain synthesis, since (1) all of them produced Ra LPS which corresponds to the complete R core [104], (2) none of them (except rfbT mutants, which will be discussed later) produced O hapten [104], and (3) one of the mutants was shown to be incapable of O side-chain synthesis owing to its defect in the biosynthesis of dTDP-rhamnose [128].

Results of intergroup crosses of Salmonella by Mäkelä [156] also indicated the clustering of O side-chain genes near his, and most importantly showed that nearly all of the specific information for O side-chain synthesis is contained in this rfb cluster. Salmonella montevideo (group C_1), for instance, synthesizes O side chains containing $(Man)_4$-GlcNAc repeating units with glucose branches [157] (see S. cholerae suis, Table I). Thus, S. typhimurium (group B), which synthesizes the Abe-Man-Rha-Gal repeating unit, should contain little genetic information useful for the synthesis of C_1-type repeating unit except perhaps those needed for the synthesis of GDP-mannose. Yet incorporation of the rfb region of S. montevideo into S. typhimurium enables the recombinants to synthesize the montevideo-type O side chains [156]. Clearly, all specific information on O side chain synthesis must have been contained in the rfb cluster in the S. montevideo donor. Analysis of levels of enzymes involved in sugar nucleotide synthesis further showed that the rfb cluster of S. montevideo contains genes responsible for GDP-mannose synthesis, and that of S. typhimurium contains those determining the biosynthesis of CDP-abequose, dTDP-rhamnose, and GDP-mannose [158].

Mapping of genes in S. typhimurium rfb cluster was made easier by the availability of a number of chromosomal deletion mutants. The analysis of enzyme levels in these mutants, obtained from Philip Hartman's laboratory, indicated the order of rfb genes shown in Fig. 7 [159]. It is seen that there is a further clustering within this region: genes involved in GDP-mannose synthesis, those involved in dTDP-rhamnose synthesis, and those involved in CDP-abequose synthesis each form subclusters. Although UDP-galactose is needed for the assembly of O repeating units, the gene responsible for its synthesis (galE) is responsible also for the catabolism of galactose, and is located outside the rfb region (Fig. 7). Similarly, the gene for phosphomannoisomerase (pmi) is involved in the catabolic breakdown of mannose and is located outside the rfb cluster. An interesting situation was found in the case of UDP-glucose pyrophosphorylase, which is required for both the catabolism of galactose and the biosynthesis of UDP-glucose. Its structural gene (galU) is located outside the rfb cluster, but the protein specified by this gene seems to be "modified" by the product of a gene (galF) located within the rfb cluster [160-162].

Among the glycosyl transferases, only the gal-1-P transferase has been mapped, close to the right end of the rfb cluster ([163]; see Fig. 7). Mapping of the point mutants is hampered by the fact that an overwhelming majority of the point rfb mutants are defective in the first enzyme of repeating unit synthesis, UDP-galactose: P-ACL galactose-1-P-transferase [163]. One possible interpretation of this finding (trapping of P-ACL by oligosaccharide-PP-ACL intermediates in mutants defective in later steps of repeating unit biosynthesis, thus causing a lethal block) has been discussed (see Section III, C, 3).

Enzyme levels in polar mutants indicate that at least a major portion of the rfb cluster (from rfbD to rfbM) constitutes a single unit of transcription, read from "left to right," or counterclockwise in the usual representation of the Salmonella chromosome (Fig. 7) [164]. Since the his operon is read in the clockwise direction, deletion of parts of the his and rfb clusters results in the joining of the two transcriptional units "tail to tail." As expected, rapid transcription in one direction interferes with the transcription in the opposite direction under such circumstances [164]. Although this is an artificial situation created by the chromosomal deletion, it is possible that this is a natural control mechanism regulating other transcriptional units. Examples of such "bipolar" transcription of both strands of DNA have recently been found in phage λ [165, 166].

One group of rfb point mutants are unusual in that they produce O haptens [167]. Because of this finding, these mutants (provisionally called rfbT) were assumed to be defective in the joining of the completed O side chain to the core LPS; no biochemical studies, however, have been done on these mutants. The mutations appear to map near his [167], but the precise location is still unknown.

Few studies have been done on the rfb cluster in organisms other than group B and group C_1 Salmonella. Group D_1 Salmonella also has the rfb cluster near his, and the cluster seems to contain all the genetic information necessary for the synthesis of O repeating units, which are very similar in structure to those of the group B organisms [168, 169]. Results of crosses involving other serogroups also suggest the presence of his-linked gene cluster involved in O repeating unit synthesis [170]. Kochibe [171], however, obtained evidence that individual recombinants of interspecific crosses involving groups U, R, or G sometimes produce both the donor- and recipient-type O side chains, the former linked to R core, the latter occurring as O-haptens. In E. coli, Ørskov and Ørskov [172] found, by 1962, that loci linked to his determine the immunological specificity, and by inference, the structure of O side chains. This was confirmed recently in three O-serotypes of E. coli: O-8 [173], O-100, and O-111 [174]. In Shigella flexneri, the structure of the main chain of the O repeating unit seems to be determined by the his-linked gene cluster [175, 176]; the O chain is then modified by the addition of glucosyl residues at various places [177]. This modification, which results in the production of about a dozen different serotypes of Sh. flexneri, may be in many cases directed by the genomes of converting phages [178-180] (see also Section III, C, 8).

b. The rfc Gene(s). Although the transfer of rfb gene cluster from group C Salmonella into group B recipient ensures the production of normal O side chains in recombinants, the transfer in the opposite direction produces a different result. Most of the group C recipients that have received the rfb cluster of the group B donor produce an LPS with donor-type O repeating units, but the amount of the O side-chain material is very small, resulting in a phenotype intermediate between smooth and rough [156]. These semirough (SR) recombinants are assumed to be defective in the polymerization of O repeating units, since their LPS have very short O side chains each of which corresponds to the unpolymerized tetrasaccharide unit [70, 181, 182]. The polymerase defect can be cured by the introduction of the chromosomal region between the gal operon (not the galU or galF gene) and the trp operon [156]. Thus, a gene (or genes), called rfc, is needed for the polymerization of group B repeating units. Mutants of this gene were also isolated [70, 183] and were shown to have the SR-phenotype.

The simplest interpretation of the results is that rfc is the structural gene for polymerase. There is some evidence, however, indicating that the "polymerase" reaction may be more complex than it seemed at first. Some temperate phage genomes often contain genes which alter the structure of host O side chains. This phenomenon ("phage conversion," see Section III, C, 8) sometimes involves the alteration of linkages between repeating units, i.e., the linkages presumably determined by polymerase. In most Salmonella serotypes of group B, this linkage is Gal-(1 → 2)-Man (Fig. 3), but when the bacteria become lysogenic for a phage ϕ 27, the linkage now becomes Gal-(1 → 6)-Man [184]. In an analogous system of conversion in group E Salmonella, Robbins' group has established that the preexisting polymerase is inhibited [185], while a new polymerase coded for by a phage gene [186] is synthesized. This led to the prediction that the rfc mutants of group B Salmonella will be "cured" by lysogenization with ϕ 27 owing to the production of a new polymerase, but this was not borne out by the experimental results [187]. It could be that ϕ 27 simply modifies the preexisting polymerase, but it is also possible that polymerization is a rather complex reaction and rfc is involved in the "nonspecific" part of the reaction sequence. Similar experiments with a temperature-sensitive polymerase mutant of group E Salmonella, however, gave predicted results, i.e., the converting phage "cured" the host defect [188]. Unfortunately the host mutation has not been mapped and it is not clear as to whether it is comparable to rfc.

Long O side chains are apparently synthesized by group B Salmonella recipients that received only the rfb region from group C_1 donors [156].

This could mean, among other possibilities, that (1) the polymerase gene is a part of the rfb gene cluster in group C_1, or (2) C_1 side chains are random polymers. The latter seems unlikely in view of the results of structural studies so far performed [157].

c. rfe Gene(s). Some mutants of S. montevideo (group C_1) and S. minnesota (group L) do not seem to synthesize O side chains, yet the responsible mutations are not linked to his, and are between ilv and metE (Fig. 7) [107]. The gene(s), called rfe, is not required for the synthesis of group-B-type O side chains [189], and does not seem to act as a regulator of rfb genes either [189]. Thus, the rfe^+ allele seems likely to cause a special modification of P-ACL or to supply some (perhaps yet unknown) components of the C_1 repeating units. Yet the puzzling thing is that S. typhimurium can supply the function performed by the rfe^+ allele, as S. typhimurium strains whose rfb cluster was replaced by that of S. montevideo can readily produce S. montevideo-type O side chains [156].

6. Modification of O Side Chains

The structure of O side chains shows extreme variation from one serotype to the other [25]. Even within a single serotype, there are "optional" features (e.g., presence of glucose branches, O-acetyl residues, etc.), which may or may not be present in the O side chains of a particular strain. These features can usually be acquired and lost even in a single strain by mutation or mutation-like processes, thus giving the strain an additional possibility of altering its surface structure [25]. It follows also that the presence or absence of such features should not affect the synthesis of the basic O side-chain structure. Indeed, at least in one case these "optional" features seem to be added to the growing O side chains that are linked to PP-ACL. These reactions may be grouped together as modification reactions.

a. Glucosylation Reaction. Some Salmonella strains produce LPS that contain glucosyl residues as branches of the O side chain (Table I). In group B Salmonella, the glucosylation at C-4 of the main chain galactose residue is determined by a gene (or genes) called oafR, located close to the pro locus (Fig. 7) [190]. This gene apparently can exist in two interconvertible "states." When it is in the positive state, glucosylation occurs and this is recognized as the production of O-12_2 antigen; no glucosylation occurs when the gene is in the negative state. Furthermore, the gene undergoes conversion from one state to the other at frequencies much higher than normally encountered in spontaneous mutations [190]; this phenomenon has been called "form variation" by Kauffmann [191]. A population of any group B Salmonella strain, then, is likely to contain both $12_2(+)$ cells with glucosylated LPS and $12_2(-)$ cells with nonglucosylated LPS, and it is clear that the O repeating units are polymerized irrespective of the presence or absence of glucosyl branches. This can be contrasted with the addition of abequosyl branches to the repeating unit (see preceding discussion), which is absolutely necessary for polymerization of the units in intact cells [142].

Some group E Salmonella also carry glucosyl branches on their O side chains. In this case, the glucosylation is a result of lysogenization by a converting phage, ϵ^{34} [192]. In this system, it was demonstrated by Uchida and others [193] that glucose is transferred from UDP-glucose to the O side chains of the endogenous LPS, if cell envelope fractions of ϵ^{34}-lysogenic strains were used. The glucosylation occurred in the absence of other sugar nucleotide precursors of O side chain, and it was clear that the already polymerized O side chains were being "modified" by the transfer of glucose.

The mechanisms underlying these glucosylation reactions have recently been studied both in group E Salmonella [194, 195] and in group B Salmonella [196-198]. In both systems, it was found that UDP-glucose was not the direct donor of glucose, and that glucosyl-P-ACL acted as the intermediate carrier, as follows.

$$\text{UDP-glucose} + \text{P-ACL} \rightarrow \text{glucosyl-P-ACL} + \text{UDP}$$

$$\text{Glucosyl-P-ACL} + \text{Acceptor} \rightarrow \text{glucosyl-Acceptor} + \text{P-ACL}$$

It should be emphasized that, in contrast to the first reaction in repeating unit biosynthesis (Fig. 9), glucose, rather than glucose-1-P, is transferred to P-ACL. It is also known that the glucose moiety undergoes an inversion, from UDP-α-glucose to β-glucosyl-P-ACL. Since in the final product (i.e., O side chains) the glucose is linked as an α-glucoside, the second step of the reaction sequence also involves the inversion of anomeric configuration. Although the final product of the incorporation in vitro was LPS, in intact cells O side chains seem to become glucosylated before their transfer to the R core [195, 198]. O side chains, which accumulate in R core-defective mutants presumably in a form still linked to PP-ACL, are fully glucosylated except for the reducing terminal galactose residue [198]. In view of our knowledge on the direction of chain growth in O side-chain synthesis (Section III, C, 3), the most likely mechanism of glucosylation seems to be as shown in Fig. 11. Thus, the glucosylation is assumed to take place as soon as a new repeating unit monomer is added at the reducing end of the growing chain. In this mechanism, the polymerase always catalyzes the reaction between nonglucosylated repeating unit monomer and the repeating unit oligomer with the nonglucosylated reducing end, and the glucosylation thus can be accomplished without interfering with the polymerase reaction.

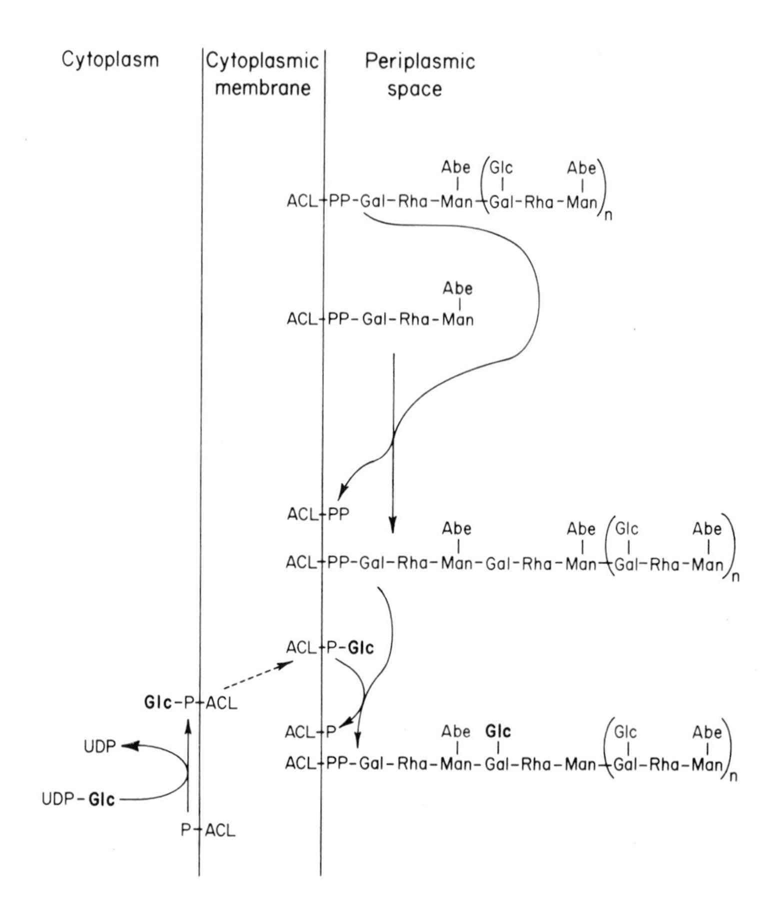

FIG. 11. The probable mechanism of O side chain glucosylation in S. typhimurium glucosyl-P-ACL, synthesized at the inside surface of the cytoplasmic membrane (a), is transported to the outside surface (b). Here, O side chains are being elongated by the transfer of polymerized chain onto a monomer repeating unit (c) (see Fig. 10). The galactose residue of the second repeating unit from the reducing end is then glucosylated by reaction with glucosyl-P-ACL (d).

The generation of many O antigenic specificities in Shigella flexneri appears to be done almost exclusively by the glucosylation of O side chains [177]. Thus, the main linear polysaccharide of the O-chain portion consists of repeat units containing N-acetylglucosamine and rhamnose. The O-chain portion of "variant y" serotype corresponds to such an unsubstituted main chain, but in all other serotypes the main chain carry glucosyl side branches at various positions. In most cases, it was shown that the glucosylation is determined by gene(s) located in the pro-lac region of the chromosome [199, 200]; it may be recalled that this is where the oafR gene, which also determines glucosylation, is located in group B Salmonella [190]. Elimination of pro-lac region from the chromosomes of most Shigella flexneri serotypes results in the production of "variant y" LPS [201]. Simmons' results, however, seemed to suggest that LPS of certain serotype (e.g., "variant x" and 5a) had a main chain ("y_2-type") in which the N-acetylglucosamine and rhamnose residues are linked together in a way different from what was found in the variant y LPS [177]. This introduced considerable difficulty in the interpretation of genetic results, since the elimination of pro-lac region from serotype 5a or variant x also produced variant y, rather than the expected strain producing LPS with the y_2-type O chains [201]. However, methylation studies carried out in Lindberg's laboratory [201a] have now shown that there is only one main chain common to all the S. flexneri strains examined; thus the "variant y" O chain can now be regarded as the unglucosylated prototype, from which all other S. flexneri O chains (and hence serotypes) can be produced by the attachment of glucose branches at different places. The methylation study [201a] also revealed other serious errors in the structure previously proposed [177]; obviously a careful reexamination of LPS structure is needed for S. flexneri.

b. Acetylation. Many O side chains carry O-acetyl groups [25] (see Table I). Examples include the O-acetyl group on abequose residues in group B Salmonella LPS (responsible for antigen 5) and that on galactose residues in group E_1 Salmonella LPS (responsible for antigen 10). The acetylation reaction involved in antigen 10 synthesis has been examined in a cell-free system [202, 203]. The enzyme(s) responsible are found in the cell envelope fraction, and acetyl-CoA acts as substrate. It is not known at which stage of O side chain synthesis the acetylation takes place. In the cell-free system, endogenous LPS, endogenous O side chains linked presumably to PP-ACL, and oligosaccharide fragments of O side chain were all active as acceptors.

In S. typhimurium, the R-core-defective rfa mutants accumulate O chains that are acetylated [104]. The acetylation is therefore likely to take place on intermediates linked to PP-ACL. Some rfc mutants produce LPS with antigen 5 [156]. Thus, the acetylation may take place on repeating unit monomers, in contrast to the glucosylation reaction. In this serotype the gene responsible for acetylation (oafA) has been mapped at a position outside the rfb cluster (Fig. 7) [204].

7. Synthesis of T1 Side Chains

Each *Salmonella* serotype produces O side chains of a characteristic structure. However, a special kind of side chain, called T1 [205], is known to occur in strains which would normally belong to various unrelated O-serotypes. T1 side chains consist of ribofuranose and galactofuranose residues [206], and they apparently have identical structures in various strains as judged from their immunological cross-reactions. Genetic analysis showed that in Tl strains there is a gene (or genes) (*rft*) responsible for T1 side-chain production, and that the gene is located near the *gal* operon [207]. Furthermore, the usual T1 strains additionally have mutations in the *rfb* cluster, and this defect enables them to synthesize LPS exclusively containing T1 side chains [207].

The structure of the Tl side chain is not entirely clear, although at present structures containing separate polyribose and polygalactose chains appear most likely [208]. The latter may consist of oligogalactose repeating units [208].

Studies with various mutants led to the conclusion that the galactofuranose residues are synthesized from galactopyranose residues most probably by isomerization at the level of UDP-galactose [209]. It may be noted that UDP-galactofuranose was recently isolated from a reaction mixture containing extracts of a fungus, *Penicillium charlesii* [210]. Envelope fractions from *Salmonella* Tl strains were also shown to synthesize Tl side chains in a cell-free system [211]. UDP-Galactopyranose was the only substrate required, and a large portion of the product was linked to the endogenous R core [211].

8. Phage Conversion of O-Antigens

We have mentioned at several places that temperate phages may bring about alterations in the structure of host O side chains. Examples of this phenomenon are summarized in Table 4.

In the cases of conversion performed by P22 or ϵ^{34}, the phage genome should contain only genes for the enzymes of the transglucosylation reactions, the mechanism of which has been discussed in Section III, C, 6. In cases where the conversion results in the shift of the glucose branch, as in the conversion performed by ϕ14, the phage genome presumably contains an additional genetic information for the inhibition or repression of the preexisting transglucosylase enzymes.

Phage ϵ^{15} represses the production of host transacetylase. It seems likely that converting phages that induce, rather than repress, acetylation of O side chains will be discovered.

The mechanism of alteration of the linkage determined by polymerase has been discussed (Section III, C, 5, b).

Phage conversion thus seems to involve either the alteration of the polymerase-determined linkages or the introduction or elimination of the "modification" reaction. This is expected because alterations within the repeating unit might make it difficult for the polymerases to function. However, it should be noted that the polymerases of group B and D_1 Salmonella do not seem to be affected by the configuration of 3,6-dideoxyhexose (tyvelose in D_1, abequose in B) in the repeating unit, since rfb clusters can be freely exchanged between these two groups by conjugation [169] or transduction [214] without apparently affecting the degree of polymerization of the repeating units.

9. A Tentative Model

The enzymes directly involved in the biosynthesis of O side chains all appear to be membrane-bound enzymes. The intermediate carrier, P-ACL is also membrane-bound. Thus, the synthesis of O side chain must occur on or in the membrane, and the question to be considered becomes that of the detailed topology of the synthetic reactions.

There are several lines of evidence indicating that the cytoplasmic membrane, rather than the outer membrane, is the site of synthesis of O side chains. (1) In rfa mutants, the synthesized O side chains remain on the outside surface of the cytoplasmic membrane. This is indicated by the observation that neither anti-O antibodies [103] nor a phage specific for O side chain [148] will interact with the outer membranes of the intact cells, but the phage at least can be adsorbed to the spheroplasts [148], presumably to the cytoplasmic membrane. (2) Some of the enzymes of the lipid cycle (Fig. 9) have actually been located in the cytoplasmic membrane [17].

If the synthesis is done on the surface of the cytoplasmic membrane, the repeating unit seems likely to be synthesized at the inner surface of the membrane, since the reaction uses sugar nucleotides which are typically water-soluble components of the cytoplasm. Once the repeating unit is completed, it is assumed that the intermediate lipid now moves across the membrane and that the oligosaccharide portion become exposed on the outside surface of the membrane (reaction e, Fig. 9). The reasons for this assumption are as follows. (1) The most convincing evidence is the experiment of Shands [7], in which he located O-side-chain-containing material in lysed Salmonella spheroplasts by the use of ferritin-labeled antibodies. The picture (Fig. 2 of reference 7) clearly indicates that such material is located on both sides of the outer membrane, and on the outer side only of

TABLE 4

Examples of Phage Conversion of O Side Chains

Nature of alterations	Phage	Host organism	Repeating unit[a] Before conversion	After conversion	Reference
Glucosylation	P22	Group A, B, and D_1 *Salmonella*	DDH[b] ↓ Man → Rha → Gal →	α-Glc (1→6 to Gal) DDH ↓ Man → Rha → Gal →	212
	ϵ^{34}	Group E *Salmonella*	Man → Rha → Gal →	α-Glc (1→4 to Gal) Man → Rha → Gal →	192–195
	ϕ14	Group C *Salmonella*	α-Glc (1→3 to second Man) Man → Man → Man → Man → GlcNAc →	α-Glc (1→3 to fourth Man) Man → Man → Man → Man → GlcNAc →	157
	Several phages	*Shigella flexneri*	See text		

Acetylation	ϵ^{15}	Group E *Salmonella*	Ac (on Gal) Man → Rha → Gal →	Man → Rha → Gal →	202, 203, 213
Alteration in polymerase specificity	ϵ^{15}	Group E *Salmonella*	Ac (on Gal) 6 → Man → Rha → αGal → 1	6 → Man → Rha → βGal → 1	185, 186
	ϕ27	Group A, B, and D_1 *Salmonella*	DDH ↓ (on Man) 2 → Man → Rha → Gal → 1	DDH ↓ (on Man) 6 → Man → Rha → Gal → 1	184

[a] Alterations caused by the phage are printed in bold face.

[b] 3, 6-Dideoxyhexose (i.e., paratose, abequose, or tyvelose).

the inner, cytoplasmic membrane. (2) The adsorption of O-specific phage to the lysozyme-EDTA spheroplasts of rfa mutants [148] also indicates that at least some O side chains, presumably still linked to PP-ACL, are exposed on the outer surface of the cytoplasmic membrane. (3) In Section III, C, 6 we have seen that the glucosylation of O side chains is done concomitantly with the chain elongation on PP-ACL. If the chain elongation or polymerization were done at the inner surface of the membrane, there should be no difficulty in utilizing UDP-glucose as the glucosyl donor. But the cells first synthesize glucosyl-P-ACL, and this in turn acts as the true glucosyl donor. The simplest explanation is that the polymerization is done at or near the outer surface of the cytoplasmic membrane, and that glucosyl-P-ACL is synthesized so that glucose can be transported across the membrane.

These considerations suggest that the major function of the polyisoprenoid lipid carrier P-ACL is to transport both monosaccharides (e.g., glucose) and oligosaccharides (e.g., O repeating unit) across the membrane. Although flip-flop transitions of phospholipids are extremely slow [127], ACL-linked intermediates might undergo much faster transitions, perhaps owing to the unusual length of their hydrocarbon chains (55 carbon atoms).

Polyisoprenol phosphates have been shown to act as carriers in other systems. Examples include the biosynthesis of peptidoglycan [136, 150] capsular polysaccharide in *Aerobacter* [215], mannan in *Micrococcus luteus* [151, 152], and a glycoprotein in mammalian liver [216, 217]. Furthermore, similar lipid carriers are known to be involved in the synthesis of plant polysaccharides [218, 219] and some "teichoic acids" [220, 221], although the structure of the lipids is less well defined in these cases. The polymers synthesized in most cases are located outside the cytoplasmic membrane, and it is possible that the major function of the polyisoprenol phosphate carriers always is the outward transport of activated carbohydrate residues across the membrane.

D. Joining the O Side Chains with the Core Lipopolysaccharide

We have seen so far that the O side chains and the core LPS are synthesized completely independently, by enzymes associated with the cytoplasmic membrane. The synthesized O side chain must now be transferred to the R core. This reaction, called translocase by Robbins' group and later O-antigen: LPS ligase by Osborn, has been demonstrated in a cell-free system by Cynkin and Osborn [222]. O side chains, presumably still linked to PP-ACL, were generated by incubation of a cell envelope fraction from a rfa mutant with nucleotide sugars. Addition of the complete core LPS resulted in the transfer of the side chain to the core.

Mutants defective in rfaL gene produce an LPS that is active as the acceptor in this reaction, but its envelope fraction cannot carry out the ligase reaction [26]. It has been suggested that another gene (rfbT), which is located near his, is also needed for the ligase activity (Section III, C, 5, a). Since the "ligase" catalyzes a rather unusual reaction, i.e., the joining together of two macromolecules, a series of reactions might possibly be involved.

Theoretically, it is possible that repeating units are added one at a time to the R core, rather than being added after polymerization into a polysaccharide. The efficient transfer of unpolymerized repeating units in SR (or rfc) strains indeed suggests that the ligase is capable of transferring single units (Section III, C, 5, b). However, the sequential transfer mechanism seems rather unlikely in view of Kent and Osborn's pulse-labeling experiment [223], which showed that the O-hapten appears to be a precursor of O side chains in LPS. Furthermore, the elongation of O side chains at the reducing ends (Fig. 10) is also contrary to such a mechanism.

It is not known exactly where ligase and other enzymes of LPS synthesis are located in the membrane. Sonicated cell envelope fraction from S. typhimurium can transfer even incomplete repeating units to the core LPS [131], whereas this never occurs in intact cells. It could be that in intact membranes, the synthesis of core LPS and that of repeating unit-PP-ACL are carried out at different locations so that premature transfer is avoided.

IV. ASSEMBLY OF THE OUTER MEMBRANE

Before we consider the assembly of outer membrane components, we should know the molecular arrangement of these components in the outer membrane. We know (see Section I) that the outer membrane contains specific proteins, phospholipids, and LPS. Furthermore, the outer membrane looks like a typical unit membrane under the electron microscope (for example, see reference 6), and we know from the work of Rothfield's group (Section III, B, 3) that LPS molecules readily penetrate the phospholipid monolayer structure. The simplest interpretation of these results would be that the outer membrane has a structure typical of biological membranes, i.e., it is basically a phospholipid bilayer structure with proteins penetrating the structure at scattered places. LPS, as a typical amphipathic molecule, would behave in a manner very similar to phospholipids and would constitute a part of the bilayer structure. It is not clear, however, whether the LPS is distributed symmetrically, i.e., evenly between the outer and inner leaflets of the bilayer structure.

Shands [7] has shown, by the use of lysed spheroplasts, that LPS carbohydrate chains exist on both sides of the outer membrane. Yet this result is difficult to reconcile with his own finding [6] that the O side chains extend up to 300 Å into the medium, because the very close juxtaposition of peptidoglycan layer and the inner dense line of the outer membrane image in the sections of normal cells contradicts with the presence of such long carbohydrate chains on the inner side of the membrane. Possibly only the LPS molecules with very short O side chains (see the discussion on microheterogeneity of LPS structure, p. 144 ff.) become located in the inner leaflet of the bilayer structure. Or it seems possible that the distribution of LPS is completely asymmetric in normal cells, but becomes symmetrical owing to the rapid flip-flop transition once the peptidoglycan layer is removed during the preparation of spheroplasts (such as asymmetric distribution has been assumed in Fig. 12).

The model described above, however, is clearly oversimplified. The major difficulty lies in the fact that LPS can penetrate phospholipid monolayers (and by inference bilayers) in the absence of divalent cations [48] (Section III, B, 3), yet such cations are known to play an extremely important role in maintaining the integrity of the outer membrane. First, outer membrane structure withstands the treatment with nonionic detergents such as Triton X-100, if enough Mg^{2+} is present [18, 19]. Second, treatment of intact cells of various gram-negative organisms with EDTA releases a large fraction of LPS from the outer membrane [224-226]. Detailed studies on this most interesting phenomenon were carried out by Leive's group. It appears that about one-half of the total LPS is released from E. coli by this treatment [224, 227], and that the released substance has an overall composition (by weight) of 85-90% LPS, 5-10% proteins, and 5% phospholipids [227]. Comparison with the reported composition of the outer membrane layer (see Section I) makes it clear that the release does not correspond to the simple peeling off of the outer membrane. The next important question is whether only a special kind of LPS is released by EDTA. Utilizing a UDP-glucose-4-epimerase mutant, Levy and Leive produced a striking answer to this question [228]. Newly synthesized LPS first enters the nonreleasable fraction, but eventually becomes evenly distributed between the releasable and nonreleasable fractions. Furthermore, these two fractions are in constant equilibrium, since even the LPS that remained with the cells after EDTA treatment eventually becomes evenly redistributed into releasable and nonreleasable fractions [228]. A similar result was also reported by Robbins [229].

Where, then, are these two equilibrating fractions of LPS located in the outer membrane structure? It is unlikely that the releasable fraction of LPS is attached to the surface of the bilayer-like structure through Mg^{2+}-bridges, since LPS is an amphipathic molecule and such an interaction will presumably lead to the formation of an additional LPS bilayer structure, which nobody has seen yet. The removal of the outer half of the bilayer structure is obviously impossible, and was ruled out directly by a recent study by Leive and Lawrence [230], who showed that the rate of adsorption of an LPS-specific phage, ϵ^{15}, to EDTA-treated S. anatum is identical to the rate seen with untreated bacteria. Thus, the most likely possibility is that the outer membrane, like any other biological membrane [231], is a mosaic structure and that EDTA releases LPS only from sections of a particular kind. This is strongly suggested by Leive and Lawrence [230] who further showed that EDTA-treated bacteria could adsorb, at saturation, only half as many phages as the untreated bacteria; both the releasable and unreleasable fractions of LPS, therefore, must have been equally exposed on the outside surface of the outer membrane.

We can speculate on the nature of these different sections that may make up the outer membrane mosaic (Fig. 12). We have very little concrete data, and the hypothesis described below is highly speculative and may not be correct. Yet one feels that the nonreleasable LPS is probably located in sections that are essentially phospholipid-LPS bilayer in structure, because phospholipid-LPS monolayers are stable in the presence of EDTA [48]. Sections of the second kind might be rich in proteins; calculations (Table 5) show that proteins may contribute up to one-half the surface area in the outer membranes. The interaction of some cytoplasmic membrane proteins — $galactose_I$ and $glucose_{II}$ transferases [120, 123] — with the phospholipid-LPS monolayer requires Mg^{2+} (Section III, B, 3). One is thus tempted to assume that the interaction of outer membrane proteins with phospholipid-LPS bilayer structure may also need Mg^{2+} (region C, Fig. 12). Treatment with EDTA might then produce the ejection of LPS from such Mg^{2+}-stabilized sections. It may be noted also that Mg^{2+} is not required for the interaction of $galactose_I$ transferase with monolayers containing only phospholipids [123]. If an analogous situation exists also in the outer membrane, the removal of Mg^{2+} from the protein-phospholipid-LPS-containing section could result in the formation of a complex that is stable without Mg^{2+}, i.e., either phospholipid-LPS or phospholipid-protein, by ejecting protein or LPS, respectively. Why only LPS is ejected is not clear. It may be due to protein-to-protein interactions which could form a backbone of the membrane structure in these sections, or simply to the degree of hydrophobicity of these proteins.

The protein-rich sections, however, may not be the only regions where EDTA-releasable LPS is located. Perhaps some sections of the outer membrane are composed almost exclusively of LPS molecules, and Mg^{2+} is needed to reduce the electrostatic repulsion between them (region B, Fig. 12). We may recall that the bilayer structures seem to be formed from LPS alone, provided Mg^{2+} is present (Section II). EDTA will obviously release LPS from this Mg^{2+}-stabilized region of the outer membrane.

TABLE 5

Calculated Contributions of Outer Membrane Components to the Surface Area

Component	Composition (wt %)[a]	Average molecular weight	Molar ratio	Surface area per molecule Å^2	Fraction of surface area in the outer leaflet of the outer membrane	
					Symmetrical[e]	Asymmetrical[f]
Phospholipids	19	700	19.3	54	31	20
LPS	19	10,300[b]	1.3	250[b]	11	22
Protein	62	44,000[c]	1.0	1960[d]	58	58

[a] From the composition of the isolated outer membranes from a UDP-glucose-4-epimeraseless strain of *S. typhimurium* [17].

[b] Value for the LPS of a UDP-glucose-4-epimeraseless mutant of *S. typhimurium* [48].

[c] Molecular weight of the "major protein," which comprises about 70% of outer membrane proteins [16].

[d] Rough approximation of the equatorial cross-sectional area based on the assumption that the protein is globular and is 50 Å in diameter.

[e] "Molar ratio" was multiplied by "surface area per molecule" for each component, and these values were adjusted by dividing with a factor, so that the sum of the values would make 100%. The figures in this column thus correspond to the fraction of outer membrane area that is occupied by individual components, assuming a *symmetrical* structure for the membrane.

[f] If one assumes that the outer membrane is asymmetric in the sense that LPS is located exclusively in the outer leaflet of the lipid bilayer structure, the following calculations may be made. If we assume that most of the protein molecules penetrate both leaflets of the membrane, 58% of the outer leaflet surface should still be occupied by protein. Since LPS (11%) is assumed to be concentrated in the outer leaflet which constitutes 50% of the total membrane surface, it should carry (11/50) x 100 = 22% of the outer surface of the membrane. Thus protein, LPS, and phospholipids should contribute 58%, 22%, and (100-58-22) = 20%, respectively, to the area of the outer leaflet.

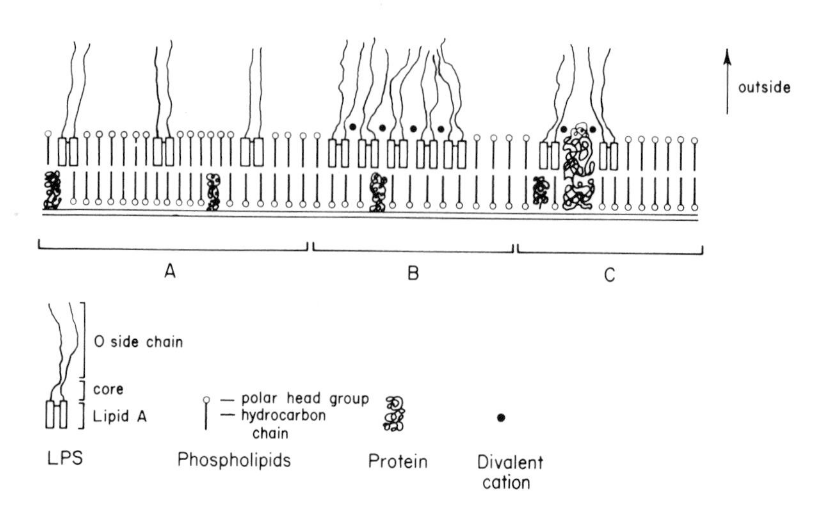

FIG. 12. A possible arrangement of molecules in the outer membrane layer. In region A, LPS molecules are interspersed among phospholipid molecules; electrostatic repulsion between the KDO-heptose regions of the core oligosaccharides does not create difficulty here, since the neighboring LPS molecules are placed apart by the phospholipid molecules inserted in between. In region B, divalent cations reduce this electrostatic repulsion, so that at least the outer leaflet of the bilayer is composed predominantly of LPS molecules. In region C, the protein molecules are thought to interact with phospholipids as well as LPS. LPS-protein interaction is here shown as though it involves divalent cations; this feature, however, is mostly speculative (see text for details). The outer membrane is assumed to be held in place by the penetration of the small protein molecules, which are placed 100 Å apart and are covalently linked to the peptidoglycan [233-236]. Although LPS is assumed to be present only in the outer leaflet portion of the membrane in this figure, other arrangements appear equally possible (see text for further explanation).

The model also explains the behavior of the outer membrane layer in Triton-Mg^{2+}. Although the morphological features are retained after such a treatment [18, 19], one analysis shows that one-half of LPS and one-third of phospholipids are removed from the membrane [19]. Triton may remove LPS from the phospholipid-LPS bilayer section of the outer membrane, but may extract neither LPS nor protein from the Mg^{2+}-stabilized sections. Thus, Triton-Mg^{2+} and EDTA extractions might attack complementary sections of the membrane mosaic.

After its synthesis on the cytoplasmic membrane, LPS must be transported across the peptidoglycan layer and inserted into the outer membrane.

If the model described above is to be taken seriously, the transported LPS is apparently inserted into the phospholipid-LPS bilayer region (since it is not extractable with EDTA [228, 229]) and then diffuses into the Mg^{2+}-stabilized sections. The constant equilibrium of LPS between these two kinds of sections appears to be a striking demonstration of the translational mobility of the membrane components, or the "fluidity" of the membrane mosaic, a phenomenon that is being demonstrated in increasing frequency (for example, see reference 232).

In addition to LPS, the proteins and phospholipids must also be transported and finally assembled as the outer membrane. This transport (and/or assembly) has to be a specific process, since O side chain-PP-ACL and cytoplasmic membrane proteins are left behind and never appear in the outer membrane. The transport also appears to be an irreversible process, since studies with a phosphomannoisomeraseless mutant showed that O side chains newly synthesized after the addition of exogenous galactose always become linked to newly synthesized R core, rather than to the preexisting incomplete core which is abundantly present in the outer membrane [283].

For the transport of outer membrane components, structural connection between cell wall and cell membrane may be necessary. Indeed electron microscopic studies have revealed that the peptidoglycan layer is attached to the inner surface of the outer membrane layer in *E. coli* or *Salmonella* (for review see references 1 and 2). A lipoprotein, covalently linked to the peptidoglycan layer, has been discovered in *E. coli* and related bacteria [234-237]. It is possible that noncovalent interaction of this lipoprotein with the outer membrane leads to the attachment of the peptidoglycan layer to the membrane. As regards the attachment of cytoplasmic membrane to cell wall, Cota-Robles [237a] and Bayer [238] obtained pictures of plasmolyzed *E. coli* cells, which show that such attachment does occur only in small areas of cell surface. Mühlradt [238a] has observed that the emergence of a new LPS on cell surface first occurs at a few, localized spots on the *Salmonella* cell surface; possibly LPS is transported into the outer membrane through such attachment sites.

V. FUNCTIONS OF THE OUTER MEMBRANE AND LIPOPOLYSACCHARIDE

A. Functions of O Side Chains

It is very rare to find mutants without O side chains in nature. However, if *E. coli* or *Salmonella* strains are kept as pure cultures in the laboratory, such mutants occur with a fairly high frequency and in most cases overgrow the parent organism. This is illustrated, for example, by the fact that the strains of *E. coli* most frequently used by biochemists and molecular biologists — K12, B, and C — are all rough strains.

These observations suggest that O side chains are needed only for something these organism encounter in their natural habitat. The situation is very clear for pathogenic organisms. O side chains enable these organisms to escape phagocytosis (and possibly killing by the antibody-complement system) in a host animal that does not have a proper antibody. There is a selective advantage for these pathogens if they could produce novel types of O side chain, because the level of antibody for such new antigen will be generally low among the population of host animals. This selective pressure presumably led to the tremendous diversification in O side chain structure, and to the abundance of mechanisms by which the organisms can modify the structure of their O side chains. Thus, it is possible to see why such a high proportion of temperate phages active on Salmonella, Shigella, etc., are converting phages; they change the bacterial antigen, and the lysogenic bacterium — and thus the temperate phage in it — survives. This aspect of the LPS function has been extensively discussed in recent reviews [25, 39].

O side chains and the more peripheral portions of the core act as receptors for many phages (reviewed in reference 105). This is difficult to consider as a "function" of LPS, since phages are more likely to have evolved to use LPS as receptors, rather than vice versa.

B. Functions Requiring the Outer Portions of the Core LPS

As we have seen in Section III, B, 2, a series of rough mutants with increasingly extensive deletion of the core oligosaccharide are available. Using these mutants of S. typhimurium, Roantree and co-workers [239, 240] showed that mutants producing Rd_1 or Rd_2 LPS are much more sensitive to bacitracin, vancomycin, erythromycin, polymyxin, and novobiocin, than the "superficial" rough mutants or the smooth wild-type strain. Rc mutants were sensitive to polymyxin and novobiocin, but their sensitivity to bacitracin, vancomycin, and erythromycin was almost normal.

The simplest interpretation of these results is that the outer membrane, like any biological membrane, acts as a barrier for diffusion, and that extensive changes in LPS structures result in the alteration of the properties of the membrane as a barrier, thus enabling these antibiotics to pass through the outer membrane and to reach their sites of action (cytoplasmic membrane or cytoplasm) more easily. We cannot exclude, however, alternative hypotheses. For example, some of these drugs may have detergent-like action on the outer membrane, and the efficacy of this action may be influenced by the structure of LPS.

This difficulty does not exist in the study involving various semisynthetic penicillins [239-241], for which oxacillin and nafcillin (but not penicillin G, ampicillin, or cephalosporin) were found to be much more active on Rd_1 and Rd_2 mutants than on the smooth parent strain. Since all penicillins may be presumed to act on a common target, i.e., a transpeptidase located in the cytoplasmic membrane[242], the different behavior of various penicillins must reflect the effect of LPS structure on the permeability of various penicillins through the outer membrane.

Schlecht and co-workers independently described similar correlation between the LPS structure and antibiotic sensitivity [243-245], and also sensitivity to certain dyes [244, 246]. A similar observation has been reported for the deep rough mutants of E. coli K12 [110]. The sensitivity to bile salts is also a characteristic of deep rough mutants, especially of Re and Rd_1 class [183, 246].

C. Functions of KDO-Lipid A Region and of the Outer Membrane

Mutants with more extensive defects than the Re (heptoseless) class have not yet been isolated. It seems likely that the presence of at least the Re-type glycolipid is essential for the survival of the cell, but the nature of this presumed indispensable function is not clear.

Perhaps LPS is needed for the proper assembly of the outer membrane. The outer membrane is very rich in phosphatidylethanolamine and rather deficient in acidic glycerophospholipids [17]. Interestingly, phospholipid mixtures of such a composition are known to have difficulty in forming closed vesicles in aqueous environment [246a, 246b]; the presence of a significant amount of acidic phospholipids is probably needed for the formation of vesicles. Conceivably the role of the negatively charged $(KDO)_3$-lipid A region in the assembly of outer membrane may be similar to the role of acidic glycerophospholipids in the other types of membranes.

It is also possible that LPS is needed for some functions of this outer membrane. For this discussion we must have a precise knowledge on these functions. An attempt will therefore be made below to describe the presumed functions of the outer membrane, although the available information is very limited at present. It is noteworthy here that a marine pseudomonad is known to lose its outer membrane on EDTA treatment or on simple washing with NaCl [247]. The resultant "mureinoplasts" may be useful for studies of outer membrane function in the future. One notes also that no enzyme activity has been demonstrated in the outer membrane except a Ca^{2+}-activated phospholipase A [248].

1. The Barrier Function

We have seen that the outer membrane functions as a barrier against penetration of penicillins and probably of other antibiotics, dyes, bile salts, etc. The barrier function against penicillin is consistent with the observation that gram-negative organisms are much more resistant to penicillin G than gram-positive organisms, in spite of the fact that the target enzyme, D-alanine transpeptidase, is inhibited by approximately the same levels of penicillin G in both kinds of organisms [242].

The barrier function of the outer membrane and the probable role of LPS in this function are also seen from the works on EDTA-induced permeability changes. EDTA releases about 50% of LPS from E. coli, and at the same time the cells lose their permeability barrier against actinomycin D [249, 250], penicillin G [251], etc. Salmonella typhi is made sensitive to polymyxin, novobiocin, bacitracin, actinomycin D, and penicillin G by treatment with EDTA [252]. Caution is needed in the interpretation of these results, since EDTA apparently affects also cytoplasmic membrane and even cytoplasmic components as judged by the leaking-in of β-galactosides and the degradation of ribosomes accompanying this treatment [224, 253]. However, the relationship between LPS release and the breakdown of permeability barrier is strongly supported by the work of Voll and Leive [254] on EDTA-resistant mutants. These workers isolated mutants that do not become permeable to actinomycin D upon EDTA treatment. When the LPS released by EDTA treatment was centrifuged through a sucrose density gradient, two bands were observed. The amount of the slowly sedimenting LPS fraction (presumably containing little phospholipid) was drastically reduced in the mutant, whereas the release of the faster sedimenting LPS (presumably complexed with phospholipid) was unaltered.

The outer membrane also acts as a barrier for penetration of macromolecules. Thus the enzymes that exist typically as exoenzymes in gram-positive organisms are trapped in between the outer membrane and the inner, cytoplasmic membrane, and become "periplasmic enzymes" in gram-negative bacteria [255]. The retention of these enzymes, however, may involve a more complex mechanism, since Lopes, Gottfried, and Rothfield have shown that some mutants become "leaky" (i.e., are able to degrade macromolecular substrates in the medium by the use of an enzyme normally periplasmic in location) only for one or two specific periplasmic enzymes [256]. At least one of the mutants had an altered outer membrane protein [256], and the study of such mutants appears to be a promising approach to the functions of such proteins.

2. Diffusion of Certain Compounds

The outer membrane is a barrier as we have seen in the preceding section; yet many substances (nutrients, excretion products) must pass through the outer membrane. No active transport mechanism was shown to exist in the outer membrane, and available data indicate that active transport systems are exclusively localized in the cytoplasmic membrane

[257, 258]. Thus, it seems likely that the outer membrane will have a specific mechanism to allow the penetration, probably by passive diffusion, of many substances. Whether such a mechanism exists and, if so, how it functions, will be important problems for future study. Some results exist, which indicate that the outer membrane is permeable to sucrose only to a certain extent — the sucrose-plasmolyzed, lysozyme-treated E. coli cells, when diluted in water, behave as though the main thrust of the osmotic pressure (owing to the presence of sucrose in the extended periplasmic space) is exerted against the outer membrane [259]. Similarly, the outer membrane can act as a partial diffusion barrier for β-galactosides under certain conditions [260], although the presence of such a barrier is not noticed in most experiments on the active transport of these compounds.

If diffusion pores exist, they are not likely to be located in the LPS-phospholipid bilayer sections of the outer membrane, since "liposomes" (i.e., closed vesicles bounded by bilayers) made up with bilayers containing LPS and phospholipid (cf. Kataoka et al. [261]) did not show very high diffusion rates for sugars and monovalent cations than the liposomes made with phospholipids alone [262].

The efficient inward transport of some substances across the outer membrane might not necessarily require the presence of specially porous regions. The periplasmic space seems to contain many "binding proteins," each of which is specific for a sugar, an amino acid, an inorganic ion, etc. [258]. The fact that most of these proteins were isolated from gram-negative bacteria is probably due to the availability of the "osmotic shock" technique [255]. Yet if these proteins indeed are present only in gram-negatives, one may speculate that they were evolved in order to overcome the problem of limited permeability of substrates through the outer membrane. It is thus conceivable that the outer membrane is poorly permeable to most substrates, yet the presence of binding proteins with high affinity for some substrates tends to "pull" these substrates into the periplasmic space.

3. Contribution to the Rigidity of Cell Wall

It is generally believed that the peptidoglycan layer is the sole component responsible for the rigidity of cell wall. But the peptidoglycan layer in many gram-negative organisms is extremely thin (about 20 Å and frequently very difficult to demonstrate, see reference 1), and it is rather difficult to imagine that this is the only rigid layer in the cell wall. There are indeed pieces of evidence suggesting the contribution of the outer membrane layer to the rigidity and to the shape of the cell wall. Birdsell and Cota-Robles [259] showed that treatment of E. coli cells with lysozyme alone in 0.5 M sucrose does not alter their morphology, yet peptidoglycan is likely to be degraded since dilution in water produces spherical forms.

That lysozyme degrades peptidoglycan in the absence of EDTA was convincingly shown by Feingold et al. [263] by the use of E. coli cells labeled specifically in the meso-diaminopimelic acid residues of the peptidoglycan. Here again the digestion of peptidoglycan did not produce spherical forms.

An alternative approach to this problem is the degradation of the outer membrane layer. In some pseudomonads, the partial degradation of the outer membrane by EDTA treatment alone is known to cause some lysis of the cells [264, 265].

VI. CONCLUSION

About 10 years ago, when the biochemistry of LPS began to become a very active field, the main emphasis was on the elucidation of the mechanism by which the cells assemble this extremely complex macromolecule, containing up to seven or eight different monosaccharide components in addition to phosphate, ethanolamine, O-acetyl groups, and fatty acids. As a result of collaboration as well as competition between many laboratories, the past decade has seen a tremendous progress in this area, and the main sequence of events in LPS biosynthesis has been elucidated. Some of the conclusions obtained are important in that they are, and will be, applicable to other fields as well. Among them are (1) the principle that the sequence of sugars in heteropolysaccharide is determined solely by the specificities of glycosyl transferases that act in succession (later found to be valid in mammalian glycolipid and glycoprotein biosynthesis [266, 267]), (2) the independent assembly of major parts (O side chains and R core) before they are joined together, (3) the prior assembly of oligosaccharide repeating units before their polymerization, and (4) the utilization of polyisoprenoid carrier lipid for the outward transport of O side-chain building blocks across the membrane.

During these several years, however, the major area of interest in this field appears to have shifted toward what can be called membrane biology. This shift has begun with the brilliant work by Rothfield's group on the mechanism of participation of phospholipids in R core biosynthesis. There are also several steps of LPS biosynthesis, which should undoubtedly be studied as membrane-associated phenomena; the examples are the polymerization of O repeating units, the "O-antigen: LPS ligase" reaction, the transport of LPS from the cytoplasmic membrane to the outer membrane, etc.

But in this field LPS biosynthesis is not the only phenomenon which requires study from the viewpoint of membrane biology. Studies by many electron microscopists, Miura and Mizushima's work on the isolation of the outer membrane, the work of Roantree's group and by Schlecht's group on the sensitivity of LPS mutants toward antibiotics and dyes, and the work of Leive's group on the release of LPS by EDTA all led to the realization that the LPS-containing "outer membrane" is a real membrane with properties of a permeability barrier. Since it contains only one major protein plus a few minor ones, this may be an ideal system for studying the interaction and function of various membrane components. The elegant reconstitution of the outer membrane (although depleted in phospholipids) by DePamphilis [71] may be a beginning in the right direction. It must be emphasized that, in contrast to some simple "membranes" (such as the glycoprotein envelope of viruses), the outer membrane is a dynamic system. In fact, the constant interchange between the EDTA-extractable and non-extractable fractions of LPS makes the outer membrane one of the most beautiful examples of what Singer and Nicolson [268] call the "fluid mosaic" structure of the biological membrane. Given the advantage of the possibility of isolation of mutants (the beginning has already been made — see references 239, 243, 254, and 256), it would not be surprising even if the outer membrane became one of the best studied biological membranes in the decade to come.

REFERENCES

1. A. M. Glauert and M. J. Thornley, Annu. Rev. Microbiol., 23, 159 (1969).

2. J. H. Freer and M. R. J. Salton, in Microbial Toxins (G. Weinbaum, S. Kadis, and S. J. Ajl, eds.), Vol. 4, pp. 67-126, Academic Press, New York, 1970.

3. M. J. Salton, The Bacterial Cell Wall, Elsevier, Amsterdam, 1964.

4. H. J. Rogers and H. R. Perkins, Cell Walls and Membranes, E. and F. N. Spon, Ltd., London, 1968.

5. H. A. Bladen and S. E. Mergenhagen, J. Bacteriol., 88, 1482 (1964).

6. J. W. Shands, J. Bacteriol., 90, 266 (1968).

7. J. W. Shands, Ann. N. Y. Acad. Sci., 133, 292 (1966).

8. W. Weidel, Annu. Rev. Microbiol., 12, 27 (1958).

9. S. de Petris, J. Ultrastruct. Res., 12, 247 (1965).

10. D. G. Bishop and E. Work, Biochem. J., 96, 567 (1965).

11. A. Taylor, K. Knox, and E. Work, Biochem. J., 99, 53 (1966).

12. K. W. Knox, M. Vesk, and E. Work, J. Bacteriol., 92, 1206 (1966).

13. K. W. Knox, J. Cullen, and E. Work, Biochem. J., 103, 192 (1967).

14. L. Rothfield and M. Pearlman-Kothencz, J. Mol. Biol., 44, 477 (1969).

15. T. Miura and S. Mizushima, Biochim. Biophys. Acta, 150, 159 (1968); ibid., 193, 263 (1969).

16. C. A. Schnaitman, J. Bacteriol., 104, 890 (1970).

17. M. J. Osborn, J. E. Gander, E. Parisi, and J. Carson, J. Biol. Chem., 247, 3962 (1972).

18. M. L. DePamphilis and J. Adler, J. Bacteriol., 105, 396 (1971).

19. C. A. Schnaitman, J. Bacteriol., 108, 553 (1971).

20. C. A. Schnaitman, J. Bacteriol., 108, 545 (1971).

21. D. A. White, W. J. Lennarz, and C. A. Schnaitman, J. Bacteriol., 109, 686 (1972).

22. O. Lüderitz, A. M. Staub, and O. Westphal, Bacteriol. Rev., 30, 192 (1966).

23. O. Lüderitz, K. Jann, and R. Wheat, in Comprehensive Biochemistry (M. Florkin and E. H. Stotz, eds.), Vol. 26A, pp. 105-228, Elsevier, Amsterdam, 1968.

24. O. Lüderitz, O. Westphal, A. M. Staub, and H. Nikaido, in Microbial Toxins (G. Weinbaum, S. Kadis, and S. J. Ajl, eds.), Vol. 4, pp. 145-233, Academic Press, New York, 1971.

25. H. Nikaido, Int. J. Syst. Bacteriol., 20, 383 (1970).

26. M. J. Osborn, Annu. Rev. Biochem., 38, 501-538 (1969).

27. A. Wright and S. Kanegasaki, Physiol. Rev., 51, 749 (1971).

28. J. W. Shands, Jr., in Microbial Toxins (G. Weinbaum, S. Kadis, and S. J. Ajl, eds.), Vol. 4, pp. 127-144, Academic Press, New York, 1971.

29. H. Nikaido, Advan. Enzymol., 31, 77 (1968).

30. L. Rothfield and D. Romeo, Bacteriol. Rev., 35, 14 (1971).

31. M. J. Osborn and L. I. Rothfield, in Microbial Toxins (G. Weinbaum, S. Kadis, and S. J. Ajl, eds.), Vol. 4, pp. 331-350, Academic Press, New York, 1971.

32. P. W. Robbins and A. Wright, in Microbial Toxins (G. Weinbaum, S. Kadis, and S. J. Ajl, eds.), Vol. 4, pp. 351-368, Academic Press, New York, 1971.

33. P. H. Mäkelä and B. A. D. Stocker, Annu. Rev. Genet., 3, 291 (1969).

34. B. A. D. Stocker and P. H. Mäkelä, in Microbial Toxins (G. Weinbaum, S. Kadis, and S. J. Ajl, eds.), Vol. 4, pp. 369-438, Academic Press, New York, 1971.

35. S. Kadis, G. Weinbaum, and S. J. Ajl (eds.), Microbial Toxins Vol. 5, Academic Press, New York, 1971.

36. O. Westphal and K. Jann, in Methods in Carbohydrate Chemistry (R. L. Whistler, ed.), Vol. 5, p. 83, Academic Press, New York, 1965.

37. C. Galanos, O. Lüderitz, and O. Westphal, Eur. J. Biochem., 8, 245 (1969).

37a. H. Kita and H. Nikaido, J. Bacteriol., 113, 672 (1973).

38. D. C. Ellwood, J. Gen. Microbiol., 60, 373 (1970).

39. R. J. Roantree, in Microbial Toxins (S. Kadis, G. Weinbaum, and S. J. Ajl, eds.), Vol. 5, pp. 1-37, Academic Press, New York, 1971.

40. H. Tauber and H. Russell, J. Biol. Chem., 235, 961 (1960).

41. A. P. Grollman and M. J. Osborn, Biochemistry, 3, 1571 (1964).

42. W. Dröge, V. Lehmann, O. Lüderitz, and O. Westphal, Eur. J. Biochem., 14, 175 (1970).

43. V. Lehmann, O. Lüderitz, and O. Westphal, Eur. J. Biochem., 21, 339 (1971).

44. M. J. Osborn, Proc. Nat. Acad. Sci. U. S., 50, 499 (1963).

45. P. Mühlradt, Eur. J. Biochem., 18, 20 (1971).

46. W. Dröge, E. Ruschmann, O. Lüderitz, and O. Westphal, Eur. J. Biochem., 4, 134 (1968).

47. R. Cherniak and M. J. Osborn, Fed. Proc. Fed. Amer. Soc. Exp. Biol., 25, 410 (1966).

47a. P. Mühlradt, Eur. J. Biochem., 11, 241 (1969).

48. D. Romeo, A. Girard, and L. Rothfield, J. Mol. Biol., 53, 475 (1970).

49. D. Malchow, O. Lüderitz, B. Kickhöfen, and O. Westphal, Eur. J. Biochem., 7, 239 (1969).

49a. O. Lüderitz, personal communication.

50. B. A. Key, G. W. Gray, and S. G. Wilkinson, Biochem. J., 120, 559 (1970).

51. D. T. Drewry, G. W. Gray, and S. G. Wilkinson, Eur. J. Biochem., 21, 400 (1971).

52. E. C. Heath, R. M. Mayer, R. D. Edstrom, and C. A. Beaudreau, Ann. N. Y. Acad. Sci., 133, 315 (1966).

53. A. J. Burton and H. E. Carter, Biochemistry, 3, 411 (1964).

54. J. Gmeiner, O. Lüderitz, and O. Westphal, Eur. J. Biochem., 7, 370 (1969).

55. J. Gmeiner, M. Simon, and O. Lüderitz, Eur. J. Biochem., 21, 355 (1971).

56. G. A. Adams and P. P. Singh, Biochim. Biophys. Acta, 187, 457 (1969).

57. G. A. Adams and P. P. Singh, Biochim. Biophys. Acta, 202, 553 (1970).

57a. W. Wober and P. Alaupovic, Eur. J. Biochem., 19, 340 (1971).

58. E. T. Rietschel, Ph.D. Thesis, Albert-Ludwig University, Freiburg, Germany (1971).

58a. M. Ikawa, J. B. Koepfli, S. G. Mudd, and C. Niemann, J. Amer. Chem. Soc., 75, 1035 (1953).

59. A. H. Fensom and G. W. Gray, Biochem. J., 114, 185 (1969).

60. G. A. Adams, T. G. Tornabene, and M. Yaguchi, Can. J. Microbiol., 15, 365 (1969).

61. M. J. Hewett, K. W. Knox, and D. G. Bishop, Eur. J. Biochem., 19, 169 (1971).

62. G. M. Gray, Biochim. Biophys. Acta, 65, 135 (1962).

63. C. H. Hsu and H. E. Conrad, Fed. Proc. Fed. Amer. Soc. Exp. Biol., 30, 1173 Abs. (1971).

64. D. E. Koeltzow, personal communication.

65. D. E. Koeltzow and H. E. Conrad, Biochemistry, 10, 214 (1971).

66. H. Nikaido, unpublished results.

66a. L. Leive, personal communication.

67. C. G. Hellerqvist, B. Lindberg, S. Svensson, T. Holme, and A. A. Lindberg, Carbohyd. Res., 9, 237 (1969).

68. O. Lüderitz, personal communication.

69. H. Nikaido, Eur. J. Biochem., 15, 57 (1970).

70. Y. Naide, H. Nikaido, P. H. Mäkelä, R. G. Wilkinson, and B. A. D. Stocker, Proc. Nat. Acad. Sci. U. S., 53, 147 (1965).

71. M. L. DePamphilis, J. Bacteriol., 105, 1184 (1971).

71a. V. Luzzati, F. Reiss-Husson, E. Rivas, and T. Gulik-Krzywicki, Ann. N. Y. Acad. Sci., 137, 409 (1966).

72. A. L. Olins and R. C. Warner, J. Biol. Chem., 242, 4994 (1967).

73. E. Ribi, R. L. Anacker, R. Brown, W. T. Haskins, B. Malmgren, K. C. Milner, and J. A. Rudbach, J. Bacteriol., 92, 1493 (1966).

74. H. van Vunakis, A. Ruffilli, and L. Levine, Biochem. Biophys. Res. Commun., 16, 293 (1964); J. Exp. Med., 121, 261 (1964).

75. P. T. Mora and B. G. Young, J. Gen. Microbiol., 26, 81 (1961).

76. J. A. Rudbach and A. G. Johnson, J. Bacteriol., 92, 892 (1966).

77. O. Westphal, J. Gmeiner, O. Lüderitz, A. Tanaka, and E. Eichenberger, in "Structure et effets biologiques des produits bactériens provenant de germes gram-négatifs. Coll. Paris, Oct. 1967," CNRS, Paris, 1969, p. 69.

78. N. Kasai and A. Nowotny, J. Bacteriol., 94, 1824 (1967).

79. S. A. Rooney and H. Goldfine, Fed. Proc. Fed. Amer. Soc. Exp. Biol., 30, 1173 Abs. (1971).

80. J. A. Nesbitt and W. J. Lennarz, J. Bacteriol., 89, 1020 (1965).

81. S. S. Taylor and E. C. Heath, J. Biol. Chem., 244, 6605 (1969).

82. H. C. Wu and T. C. Wu, J. Bacteriol., 105, 455 (1971); M. Sarvas, J. Bacteriol., 105, 467 (1971).

83. J. Gmeiner and H. Nikaido, unpublished results.

83a. G. O. Humphreys, I. C. Hancock, and P. M. Meadow, J. Gen. Microbiol., 71, 221 (1971).

84. L. Eidels and M. J. Osborn, Proc. Nat. Acad. Sci. U. S., 68, 1673 (1971).

85. P. Mühlradt, H. J. Risse, O. Lüderitz, and O. Westphal, Eur. J. Biochem., 4, 139 (1968).

86. T. Fukasawa and H. Nikaido, Virology, 11, 508 (1960).

87. T. Fukasawa and H. Nikaido, Biochim. Biophys. Acta, 48, 470 (1961).

88. H. Nikaido, Proc. Nat. Acad. Sci. U. S., 48, 1337 (1962).

89. D. Fraenkel, M. J. Osborn, B. L. Horecker, and S. M. Smith, Biochem. Biophys. Res. Commun., 11, 423 (1963).

90. T. Fukasawa, K. Jokura, and K. Kurahashi, Biochem. Biophys. Res. Commun., 7, 121 (1962); Biochim. Biophys. Acta, 74, 608 (1963).

91. T. A. Sundararajan, A. M. C. Rapin, and H. M. Kalckar, Proc. Nat. Acad. Sci. U. S., 48, 2187 (1962).

92. H. J. Risse, O. Lüderitz, and O. Westphal, Eur. J. Biochem., 1, 233 (1967).

93. H. Nikaido, Proc. Nat. Acad. Sci. U. S., 48, 1542 (1962).

94. M. J. Osborn, S. M. Rosen, L. Rothfield, and B. L. Horecker, Proc. Nat. Acad. Sci. U. S., 48, 1831 (1962).

95. L. Rothfield, M. J. Osborn, and B. L. Horecker, J. Biol. Chem., 239, 2788 (1964).

96. H. J. Risse, W. Dröge, E. Ruschmann, O. Lüderitz, and J. Schlosshardt, Eur. J. Biochem., 1, 216 (1967).

97. S. M. Rosen, M. J. Osborn, and B. L. Horecker, J. Biol. Chem., 239, 3198 (1964).

98. M. J. Osborn and L. D'Ari, Biochem. Biophys. Res. Commun., 16, 568 (1964).

99. M. J. Osborn, S. M. Rosen, L. Rothfield, L. D. Zeleznick, and B. L. Horecker, Science, 145, 783 (1964).

100. I. W. Sutherland, O. Lüderitz, and O. Westphal, Biochem. J., 96, 439 (1965).

101. R. D. Edstrom and E. C. Heath, Biochem. Biophys. Res. Commun., 16, 576 (1964).

102. R. D. Edstrom and E. C. Heath, J. Biol. Chem., 242, 3581 (1967).

102a. G. Bagdian and P. H. Mäkelä, Virology, 43, 403 (1971).

103. T. V. Subbaiah and B. A. D. Stocker, Nature (London), 201, 1298 (1964).

104. I. Beckmann, T. V. Subbaiah, and B. A. D. Stocker, Nature (London), 201, 1299 (1964).

105. A. M. C. Rapin and H. M. Kalckar, in Microbial Toxins (G. Weinbaum, S. Kadis, and S. J. Ajl, eds.), Vol. 4, p. 267, Academic Press, New York, 1971.

106. M. J. Osborn, Nature (London), 217, 957 (1968).

107. P. H. Mäkelä, M. Jahkola, and O. Lüderitz, J. Gen. Microbiol., 60, 91 (1970).

108. J. H. Johnston, R. J. Johnston, and D. A. R. Simmons, Biochem. J., 105, 79 (1967).

109. R. G. Wilkinson, N. A. Fuller, A. G. Lazen, and E. C. Heath, Bacteriol. Proc., p. 63 (1968).

110. S. Tamaki, T. Sato, and M. Matsuhashi, J. Bacteriol., 105, 968 (1971).

111. L. Rothfield and B. L. Horecker, Proc. Nat. Acad. Sci. U. S., 52, 939 (1964).

112. M. M. Weiser and L. Rothfield, J. Biol. Chem., 243, 1320 (1968).

113. L. Rothfield and R. W. Horne, J. Bacteriol., 93, 1705 (1967).

114. L. Rothfield, M. Takeshita, M. Pearlman, and R. V. Horne, Fed. Proc. Fed. Amer. Soc. Exp. Biol., 25, 1495 (1966).

115. L. Rothfield and M. Takeshita, Biochem. Biophys. Res. Commun., 20, 521 (1965); Ann. N. Y. Acad. Sci., 133, 390 (1966).

116. A. Endo and L. Rothfield, Biochemistry, 8, 3500 (1969).

117. E. Müller, A. Hinckley, and L. Rothfield, J. Biol. Chem., 247, 2614 (1972).

118. H. Sandermann, Jr. and J. L. Strominger, Proc. Nat. Acad. Sci. U. S., 68, 2441 (1971).

119. L. Rothfield and M. Pearlman, J. Biol. Chem., 241, 1386 (1966).

120. D. Romeo, A. Hinckley, and L. Rothfield, J. Mol. Biol., 53, 491 (1970).

121. A. Endo and L. Rothfield, Biochemistry, 8, 3508 (1969).

122. A. Hinckley, E. Müller, and L. Rothfield, J. Biol. Chem., 247, 2623 (1972).

123. L. Rothfield, D. Romeo, and A. Hinckley, Fed. Proc. Fed. Amer. Soc. Exp. Biol., 31, 12 (1972).

124. S. B. Levy and L. Leive, J. Biol. Chem., 245, 585 (1970).

125. R. D. Kornberg and H. M. McConnell, Proc. Nat. Acad. Sci. U. S., 68, 2564 (1971).

126. C. F. Kulpa, Jr. and L. Leive, personal communication.

127. R. D. Kornberg and H. M. McConnell, Biochemistry, 10, 1111 (1971).

128. H. Nikaido, K. Nikaido, T. V. Subbaiah, and B. A. D. Stocker, Nature (London), 201, 1301 (1964).

129. L. D. Zeleznick, S. M. Rosen, M. Saltmarsh-Andrew, M. J. Osborn, and B. L. Horecker, Proc. Nat. Acad. Sci. U. S., 53, 207 (1965).

130. H. Nikaido and K. Nikaido, Biochem. Biophys. Res. Commun., 19, 322 (1965).

131. H. Nikaido, Biochemistry, 4, 1550 (1965).

132. A. Wright, M. Dankert, and P. W. Robbins, Proc. Nat. Acad. Sci. U. S., 54, 235 (1965).

133. I. M. Weiner, T. Higuchi, M. J. Osborn, and B. L. Horecker, Proc. Nat. Acad. Sci. U. S., 54, 228 (1965).

134. J. S. Anderson, M. Matsuhashi, M. A. Haskin, and J. L. Strominger, Proc. Nat. Acad. Sci. U. S., 53, 881 (1965).

135. P. W. Robbins, A. Wright, and J. L. Bellows, Proc. Nat. Acad. Sci. U. S., 52, 1302 (1964).

136. A. A. Lindberg, T. Holme, C. G. Hellerqvist, and S. Svensson, J. Bacteriol., 102, 540 (1970).

137. M. Dankert, A. Wright, W. S. Kelley, and P. W. Robbins, Arch. Biochem. Biophys., 116, 425 (1966).

138. I. M. Weiner, T. Higuchi, M. J. Osborn, and B. L. Horecker, Ann. N. Y. Acad. Sci., 133, 391 (1966).

139. M. J. Osborn and I. M. Weiner, Fed. Proc. Fed. Amer. Soc. Exp. Biol., 26, 70 (1967).

140. M. J. Osborn and R. Y. Tze-Yuen, J. Biol. Chem., 243, 5145 (1968).

141. M. J. Osborn and I. M. Weiner, J. Biol. Chem., 243, 2631 (1968).

142. R. Yuasa, M. Levinthal, and H. Nikaido, J. Bacteriol., 100, 433 (1969).

143. S. Kanegasaki and A. Wright, Proc. Nat. Acad. Sci. U. S., 67, 951 (1970).

144. Y. Higashi, J. L. Strominger, and C. C. Sweeley, Proc. Nat. Acad. Sci. U. S., 57, 1878 (1967).

145. D. Bray and P. W. Robbins, Biochem. Biophys. Res. Commun., 28, 334 (1967).

146. P. W. Robbins, D. Bray, M. Dankert, and A. Wright, Science, 158, 1536 (1967).

147. J. L. Kent and M. J. Osborn, Biochemistry, 7, 4409 (1968).

148. J. L. Kent and M. J. Osborn, Biochemistry, 7, 4396 (1968).

149. A. Wright, M. Dankert, P. Fennessey, and P. W. Robbins, Proc. Nat. Acad. Sci. U. S., 57, 1798 (1967).

150. Y. Higashi, J. L. Strominger, and C. C. Sweeley, J. Biol. Chem., 245, 3697 (1970).

151. M. Lahav, T. H. Chin, and W. J. Lennarz, J. Biol. Chem., 244, 5890 (1969).

152. M. Scher, W. J. Lennarz, and C. C. Sweeley, Proc. Nat. Acad. Sci. U. S., 59, 1313 (1968).

153. J. G. Christenson, S. K. Gross, and P. W. Robbins, J. Biol. Chem., 244, 5436 (1969).

154. C. M. Allen, W. Alworth, A. MacRae, and K. Bloch, J. Biol. Chem., 242, 1895 (1967).

155. G. Siewert and J. L. Strominger, Proc. Nat. Acad. Sci. U. S., 57, 767 (1967).

156. P. H. Mäkelä, J. Bacteriol., 91, 1115 (1966).

157. N. A. Fuller and A. M. Staub, Eur. J. Biochem., 4, 286 (1968).

158. H. Nikaido, K. Nikaido, and P. H. Mäkelä, J. Bacteriol., 91, 1126 (1966).

159. H. Nikaido, M. Levinthal, K. Nikaido, and K. Nakane, Proc. Nat. Acad. Sci. U. S., 57, 1825 (1967).

160. T. Kuo, D. G. McPhee, V. Krishnapillai, and B. A. D. Stocker, personal communication.

161. T. Nakae and H. Nikaido, J. Biol. Chem., 246, 4386 and 4397 (1971).

162. T. Nakae, J. Biol. Chem., 246, 4404 (1971).

163. M. Levinthal and H. Nikaido, unpublished results.

164. M. Levinthal and H. Nikaido, J. Mol. Biol., 42, 511 (1969).

165. K. Bøvre and W. Szybalski, Virology, 38, 614 (1969).

166. H. A. Eisen, L. H. Pereira da Silva, and F. Jacob, C. R. Acad. Sci., Ser. D, 266, 1176 (1968); L. Reichardt and A. D. Kaiser, Proc. Nat. Acad. Sci. U. S., 68, 2185 (1971); H. Echols and L. Green, ibid., 68, 2190 (1971).

167. P. Gemski and B. A. D. Stocker, J. Bacteriol., 93, 1588 (1967).

168. E. M. Johnson, B. Krauskopf, and L. S. Baron, J. Bacteriol., 90, 302 (1965).

169. P. H. Mäkelä, J. Gen. Microbiol., 41, 57 (1965).

170. K. Kishi, Kitasato Med. J., 19, 129 (1969); P. H. Mäkelä, unpublished (quoted in ref. 34).

171. N. Kochibe, J. Biochem. (Tokyo), 68, 81 (1970).

172. F. Ørskov and I. Ørskov, Acta Pathol. Microbiol. Scand., 55, 99 (1962).

173. G. Schmidt, B. Jann, and K. Jann, Eur. J. Biochem., 10, 501 (1969).

174. G. Schmidt, I. Fromme, and H. Mayer, Eur. J. Biochem., 14, 357 (1970).

175. S. B. Formal, P. Gemski, Jr., and L. S. Baron, Bacteriol. Proc., p. 106 (1969).

176. V. G. Petrovskaya, personal communication.

177. D. A. R. Simmons, Bacteriol. Rev., 35, 117 (1971).

178. S. Matsui, Jap. J. Microbiol., 2, 153 (1958).

179. S. Iseki and S. Hamano, Proc. Japan Acad., 35, 407 (1969).

180. G. Giammanco, Ann. Inst. Pasteur, 114, 63 (1968).

181. H. Nikaido, Y. Naide, and P. H. Mäkelä, Ann. N. Y. Acad. Sci., 133, 299 (1966).

182. R. Yuasa, K. Nakane, and H. Nikaido, Eur. J. Biochem., 15, 63 (1970).

183. R. G. Wilkinson, P. Gemski, and B. A. D. Stocker, J. Gen. Microbiol., 70, 527 (1972).

184. G. Bagdian, O. Lüderitz, and A.-M. Staub, Ann. N. Y. Acad. Sci., 133, 405 (1966).

185. R. Losick, J. Mol. Biol., 42, 237 (1969).

186. D. Bray and P. W. Robbins, J. Mol. Biol., 30, 457 (1967).

187. G. Bagdian and P. H. Mäkelä, J. Gen. Microbiol., 57, xxix (1970).

188. R. Losick and P. W. Robbins, J. Mol. Biol., 30, 445 (1967).

189. H. Nikaido and P. H. Mäkelä, unpublished observations.

190. P. H. Mäkelä and O. Mäkelä, Ann. Med. Exp. Biol. Fenn., 44, 310 (1966).

191. F. Kauffman, J. Bacteriol., 41, 127 (1941); Die Bakteriologie der Salmonella Species, Munksgaard, Copenhagen, 1961.

192. P. W. Robbins and T. Uchida, Biochemistry, 1, 323 (1962).

193. T. Uchida, T. Makino, K. Kurahashi, and H. Uetake, Biochem. Biophys. Res. Commun., 21, 354 (1965).

194. A. Wright, Fed. Proc. Fed. Amer. Soc. Exp. Biol., 28, 658 (1969); J. Bacteriol., 105, 927 (1971).

195. T. Sasaki, T. Uchida, and K. Kurahashi, personal communication.

196. H. Nikaido, K. Nikaido, T. Nakae, and P. H. Mäkelä, J. Biol. Chem., 246, 3902 (1971).

197. K. Nikaido and H. Nikaido, J. Biol. Chem., 246, 3912 (1971).

198. M. Takeshita and P. H. Mäkelä, J. Biol. Chem., 246, 3920 (1971).

199. S. E. Luria and J. W. Burrous, J. Bacteriol., 74, 461 (1966).

200. V. D. Timakov, V. G. Petrovskaya, and V. M. Bondarenko, Ann. Inst. Pasteur, 118, 3 (1970).

201. R. Manson, D. A. R. Simmons, and V. G. Petrovskaya, Eur. J. Biochem., 17, 472 (1971).

201a. B. Lindberg, J. Lönngren, U. Rudén, and D. A. R. Simmons, Eur. J. Biochem., 32, 15 (1973).

202. P. W. Robbins, J. M. Keller, A. Wright, and R. L. Bernstein, J. Biol. Chem., 240, 384 (1965).

203. J. M. Keller, Ph.D. Thesis, Massachusetts Institute of Technology (1966).

204. E. M. Johnson, B. Krauskopf, and L. S. Baron, J. Bacteriol., 92, 1457 (1966).

205. F. Kauffman, Acta Pathol. Microbiol. Scand., 39, 299 (1956).

206. M. Berst, C. G. Hellerqvist, B. Lindberg, O. Lüderitz, S. Svensson, and O. Westphal, Eur. J. Biochem., 11, 353 (1969).

207. M. Sarvas, Ann. Med. Exp. Biol. Fenn., 45, 447 (1967).

208. M. Berst, O. Lüderitz, and O. Westphal, Eur. J. Biochem., 18, 361 (1971).

209. M. Sarvas and H. Nikaido, J. Bacteriol., 105, 1063 (1971).

210. A. G. Trejo, G. J. F. Chittenden, J. G. Buchanan, and J. Baddiley, Biochem. J., 117, 637 (1970).

211. H. Nikaido and M. Sarvas, J. Bacteriol., 105, 1073 (1971).

212. B. A. D. Stocker, A.-M. Staub, R. Tinelli, and B. Kopacka, Ann. Inst. Pasteur, 98, 505 (1960).

213. T. Uchida, P. W. Robbins, and S. E. Luria, Biochemistry, 2, 663 (1963).

214. M. Nurminen, C. G. Hellerqvist, V. V. Valtonen, and P. H. Mäkelä, Eur. J. Biochem., 22, 500 (1971).

215. F. A. Troy, F. E. Frerman, and E. C. Heath, J. Biol. Chem., 246, 118 (1971).

216. N. H. Behrens and L. F. Leloir, Proc. Nat. Acad. Sci. U. S., 66, 153 (1970).

217. N. H. Behrens, A. J. Parodi, and L. F. Leloir, Proc. Nat. Acad. Sci. U. S., 68, 2857 (1971).

218. W. Tanner, Biochem. Biophys. Res. Commun., 35, 144 (1969).

219. H. Kauss, FEBS Letters, 5, 81 (1969).

220. L. J. Douglas and J. Baddiley, FEBS Letters, 1, 114 (1968).

221. D. Brooks and J. Baddiley, Biochem. J., 115, 3017 (1969).

222. M. A. Cynkin and M. J. Osborn, Fed. Proc. Fed. Amer. Soc. Exp. Biol., 27, 293 (1968).

223. J. L. Kent and M. J. Osborn, Biochemistry, 7, 4419 (1968).

224. L. Leive, Proc. Nat. Acad. Sci. U. S., 53, 745 (1965).

225. L. Colobert, Ann. Inst. Pasteur, 95, 156 (1958).

226. G. W. Gray and S. G. Wilkinson, J. Gen. Microbiol., 39, 385 (1965).

227. L. Leive, V. K. Shovlin, and S. E. Mergenhagen, J. Biol. Chem., 243, 6384 (1968).

228. S. B. Levy and L. Leive, Proc. Nat. Acad. Sci. U. S., 61, 1435 (1968).

229. P. W. Robbins, Develop. Biol., Suppl. 2, 103 (1968).

230. L. Leive and D. Lawrence, Fed. Proc. Fed. Amer. Soc. Exp. Biol., 30, 1173 Abs. (1971).

231. M. Glaser, H. Simpkins, S. J. Singer, M. Sheetz, and S. J. Chan, Proc. Nat. Acad. Sci. U. S., 65, 721 (1970).

232. R. B. Taylor, W. P. H. Duffus, M. C. Raff, and S. de Petris, Nature New Biol., 233, 225 (1971).

233. M. J. Osborn, J. E. Gander, and E. Parisi, J. Biol. Chem., 247, 3973 (1972).

234. V. Braun and K. Rehn, Eur. J. Biochem., 10, 426 (1969).

235. V. Braun, K. Rehn, and H. Wolff, Biochemistry, 9, 5041 (1970).

236. V. Braun and U. Sieglin, Eur. J. Biochem., 13, 336 (1970).

237. V. Braun and H. Wolff, Eur. J. Biochem., 14, 387 (1970).

237a. E. H. Cota-Robles, J. Bacteriol., 85, 499 (1963).

238. M. E. Bayer, J. Gen. Microbiol., 53, 395 (1968).

238a. P. Mühlradt, personal communication.

239. R. J. Roantree, T. Kuo, D. G. MacPhee, and B. A. D. Stocker, Bacteriol. Proc., p. 79 (1969).

240. R. J. Roantree, T. Kuo, D. G. MacPhee, and B. A. D. Stocker, Clin. Res., 17, 157 (1969).

241. R. J. Roantree, personal communication.

242. J. L. Strominger, K. Izaki, M. Matsuhashi, and D. J. Tipper, Fed. Proc. Fed. Amer. Soc. Exp. Biol., 26, 9 (1967).

243. S. Schlecht and O. Westphal, Naturwissenschaften, 55, 494 (1968).

244. S. Schlecht and G. Schmidt, Zentr. Bakteriol. Parasitenk., Abt. I, Orig., 212, 505 (1969).

245. S. Schlecht and O. Westphal, Zentr. Bakteriol. Parasitenk., Abt. I, Orig., 213, 356 (1970).

246. G. Schmidt, S. Schlecht, and O. Westphal, Zentr. Bakteriol. Parasitenk, Abt. I, Orig., 212, 88 (1969).

246a. D. Papahadjopoulos and N. Miller, Biochim. Biophys. Acta, 135, 624 (1967).

246b. C. W. M. Haest, J. deGier, and L. L. M. van Deenen, Chem. Phys. Lipids, 3, 413 (1969).

247. J. W. Costeron, C. Forsberg, T. I. Matula, F. L. A. Buckmire, and R. A. MacLeod, J. Bacteriol., 94, 1764 (1967).

248. C. J. Scandella and A. Kornberg, Biochemistry, 10, 4447 (1971).

249. L. Leive, Biochem. Biophys. Res. Commun., 18, 13 (1965).

250. L. Leive, J. Biol. Chem., 243, 2373 (1968).

251. J. M. T. Hamilton-Miller, Biochem. Biophys. Res. Commun., 20, 688 (1965).

252. L. H. Muschel and L. Gustafson, J. Bacteriol., 95, 2010 (1968).

253. L. Leive and V. Kollin, Biochem. Biophys. Res. Commun., 28, 229 (1967).

254. M. J. Voll and L. Leive, J. Biol. Chem., 245, 1108 (1970).

255. L. A. Heppel, Science, 156, 1451 (1967).

256. J. Lopes, S. Gottfried, and L. Rothfield, J. Bacteriol., 109, 520 (1972).

257. W. R. Sistrom, Biochim. Biophys. Acta, 29, 579 (1958).

258. H. R. Kaback, Annu. Rev. Biochem., 39, 561 (1970).

259. D. C. Birdsell and E. H. Cota-Robles, J. Bacteriol., 93, 427 (1967).

260. J. P. Robbie and T. H. Wilson, Biochim. Biophys. Acta, 173, 234 (1969).

261. T. Kataoka, K. Inoue, O. Lüderitz, and S. C. Kinsky, Eur. J. Biochem., 21, 80 (1971).

262. H. Nikaido and T. Nakae, J. Inf. Dis., in press.

263. D. S. Feingold, J. N. Goldman, and H. M. Kurita, J. Bacteriol., 96, 2118 (1968).

264. R. G. Eagon and K. J. Carson, Can. J. Microbiol., 11, 193 (1965).

265. K. J. Carson and R. G. Eagon, Can. J. Microbiol., 12, 105 (1966).

266. S. Roseman, In Proceedings of the 4th International Conference on Cystic Fibrosis of the Pancreas (E. Rossi and E. Stoll, eds.), Part II, p. 244, S. Karger, New York, 1968.

267. H. Nikaido and W. Z. Hassid, Advan. Carbohyd. Chem. Biochem., 26, 352 (1971).

268. S. J. Singer and G. L. Nicolson, Science, 175, 720 (1972).

PART II

INTERACTION WITH THE ENVIRONMENT

Chapter 4

PRESENT STATUS OF BINDING PROTEINS THAT ARE RELEASED FROM GRAM-NEGATIVE BACTERIA BY OSMOTIC SHOCK

Barry P. Rosen and Leon A. Heppel

Department of Biological Chemistry
School of Medicine
University of Maryland
Baltimore, Maryland

and

Section of Biochemistry and Molecular Biology
Cornell University
Ithaca, New York

I. INTRODUCTION

A large and ever-growing number of binding proteins have been reported in the last half-dozen years. These are almost all proteins of relatively low molecular weight (below 50,000) and they possess the property of reversibly binding the substrates of specific transport systems. They are believed to be the carriers of the uptake systems that contain the initial recognition site for the material being transported. Extensive investigation has failed to show enzymatic activity for any of the binding proteins. Most of these proteins have been removed from gram-negative bacteria by the osmotic shock treatment [1,2], or by conversion of cells into spheroplasts [3]. These same procedures also remove a family of enzymes called periplasmic enzymes because they appear to be external to the protoplasmic membrane but internal to the outer membrane of the cell [4]. Modifications of the shock procedure have been used for other organisms, e.g., Neurospora crassa [5]. Binding proteins have also been obtained from animal tissues [6,7].

II. THE OSMOTIC SHOCK PROCEDURE

A. The Method as Applied to Gram-Negative Bacteria

Bacteria are first suspended in a hypertonic solution of sucrose containing EDTA. Then they are suddenly shifted to a cold solution of very low osmotic strength. The details of the procedure vary depending on the age of the culture.

Bacteria in stationary phase are harvested and washed several times with 0.01 M tris-HCl containing 0.03 M NaCl. One gram (wet weight) of cells is suspended in 80 ml of 20% sucrose containing 0.03 M tris-HCl (pH 7.3), at 23 °C. This corresponds to 10^{10} cells/ml. The suspension is treated with EDTA to give a concentration of 10^{-3} M, and gently shaken on a rotary shaker (180 rpm). After 5-10 min the mixture is centrifuged for 10 min at 13,000 g in the cold. The supernatant fluid is removed and the well-drained pellet is rapidly dispersed in 80 ml of cold water. After 5-10 min the mixture is centrifuged as before. The supernatant fluid (shock fluid) is removed.

For bacteria in exponential phase of growth the procedure is modified by using only 1×10^{-4} M EDTA and shocking with 0.5 mM $MgCl_2$ instead of water. For large-scale work, the ratio of EDTA to weight of cells is maintained constant, but the total volumes are reduced to 1 g (wet weight)/ 20-30 ml.

B. Comments on the Osmotic Shock Procedure

The osmotic shock procedure causes selective release of about 4% of the cell protein. Some 24 enzymes and 15 binding proteins have been tested and found to be released to the extent of 80% or greater [4]. Many other enzymes have been tested and found to remain entirely within the cell. Various aspects of the method have been evaluated. Among a number of amine buffers, only tris was found to be satisfactory [8]. The amount of EDTA must not be too great lest breakdown of RNA, loss of viability, and cell lysis ensue [2, 9, 10]. On the other hand, a certain level of EDTA is necessary, not only for good release of enzymes and binding proteins, but also to maintain viability in the face of a sudden osmotic transition. Sucrose may be replaced by another osmotic stabilizer (a substance obtainable in high osmolarity, which penetrates the bacterial cell very poorly). Sodium chloride [2,11] or glycylglycine will serve the purpose. The concentration of sucrose has been varied from 10 to 40% with similar results as far as selective release of enzymes is concerned.

Rosen and Hackette [12] have studied the release of certain enzymes by osmotic shock in *E. coli* 30E, an unsaturated fatty acid auxotroph, when the cultures were supplemented with either cis- or trans-unsaturated fatty acids. Cultures grown in oleate-supplemented medium release a large fraction of the cellular content of cyclic phosphodiesterase, acid hexose phosphatase, and 5'-nucleotidase, three typical periplasmic enzymes, following osmotic shock. Cultures grown in elaidate-supplemented medium release considerably less of these same enzymes after shock treatment. Cultures grown with either supplementation show total release of these enzymes upon conversion to spheroplasts, demonstrating that the enzymes are in the periplasmic space in both cases. When cells are changed from oleate to elaidate there is a rapid decrease in the ability of the cells to respond to osmotic shock, such that within a 25% increase in cell mass the culture responds to osmotic shock as would a culture grown overnight in elaidate-supplemented medium. The reverse experiment results in a gradual increase in the ability of the cells to respond to osmotic shock. This organism and similar mutants have been used to elucidate the mechanism of biogenesis of membrane transport systems [13, 14].

C. Localization of Binding Proteins

Evidence suggesting that the sulfate-binding protein is located internal to the cell wall, but external to the cell membrane has been reported by Pardee and Watanabe [15]. They made use of a reagent capable of permeating the cell wall but not the plasma membrane, i.e., diazo-7-amino-1,3-naphthalene disulfonate. Treatment of whole cells with this compound inactivates the sulfate binding protein. In addition, a specific antiserum inactivates the purified binding protein, but not the binding activity of intact cells. Nakane et al. [16] obtained evidence for surface localization of the leucine*-isoleucine-valine binding protein using a histochemical procedure of a special sort.

It is not clear whether the binding proteins exist free in the periplasmic space, or whether they are lightly attached to the outer surface of the protoplasmic membrane, so that they become detached by relatively gentle procedures such as osmotic shock. In any case, the binding proteins are not very firmly attached to membrane, as is the case for the M protein of Fox and Kennedy [17], the sugar-specific components of the Roseman phosphotransferase systems [18, 19], and the carrier protein for proline that was recently discovered in vesicles prepared by the Kaback procedure [20]. Membrane vesicles prepared according to Kaback were free of shock-releasable binding protein in the cases tested [21, 22].

An effort was made in this laboratory to demonstrate possible binding sites for the leucine-isoleucine-valine binding protein on membrane vesicles [23]. To this end *E. coli* were grown in the presence of radioactive leucine, and labeled binding protein was isolated and purified. Kaback vesicles were incubated with binding protein in the presence of Mg^{2+} at neutral pH. The mixture was centrifuged, and the membrane sediment was examined for radioactivity. No significant binding could be detected. We conclude that binding sites for the protein either do not exist on membrane vesicles, that conditions for binding were not realized, or that sites on the membrane were destroyed in the course of making vesicles.

---

*Throughout this chapter, the more common naturally occurring optical isomer is understood, unless specifically indicated otherwise.

D. Relationship between the Presence of a Shock-Releasable Binding Protein and the Effect of Shock on Active Transport in E. coli

Observations by a number of investigators have indicated that bacterial transport systems vary widely in their susceptibility to osmotic shock, and this usually can be correlated with the presence or absence of a shock-releasable binding protein. In the case of the following systems, the shock procedure causes a substantial decrease in active transport, and release of a binding protein of similar specificity properties has been demonstrated for the following: sulfate [24] ; leucine-isoleucine-valine transport system [25, 26] ; leucine-specific transport system [27] ; glutamine [28] ; cystine-diaminopimelic acid system [29] ; lysine-arginine-ornithine system [30-32] ; arginine-specific system [30-32] ; histidine [33, 34] ; phenylalanine [35, 36] ; galactose [26, 37-39] ; arabinose [40, 41] ; inorganic phosphate [42, 43] ; glucose-1-phosphate [44] ; thiamine [45, 46] ; riboflavin [45] ; glutamate [47] ; cyanocobalamine [48] . In these cases cells were usually grown on synthetic defined media.

Under special conditions of growth, as in the case of a rich bactotryptone-yeast extract medium, other transport systems also develop the property of being reduced by osmotic shock along with the release of specific binding activity, such as the cystine-specific system [49] and the lysine-specific system [50] . Anraku [51] pointed out that when E. coli W2243A were grown in 1% tryptone-salts-glycerol medium, binding activity specific for proline was released by osmotic shock. He also reported a reduction of alanine transport by shock, associated with release of an alanine-binding protein, in E. coli strain W3092, grown in synthetic medium. This is unusual behavior for the alanine transport system and may reflect strain differences.

In the case of the following E. coli transport systems no binding activity in shock fluid is usually demonstrated, and there is relatively little effect of shock on active transport. In addition, the system is present in vesicles prepared by Kaback's procedure, further evidence that the appropriate carrier protein is firmly attached to the cytoplasmic membranes: tryptophan in E. coli (T_3A) [52] ; proline [53] ; glycine [53] ; lysine-specific system [32] ; aspartic acid [53] ; asparagine [53] ; serine [53] ; alanine [53] ; tyrosine [53] ; β-galactosides [17, 54] .

As noted earlier, the list is generally correct for most tested strains of E. coli grown on synthetic medium, but unusual behavior may be encountered, depending on conditions of growth, strain of organism, etc.

In some instances transport is reduced by osmotic shock, but no binding protein can be detected. This was true when the uptake of glycylglycine was investigated [55] , perhaps because its poor affinity makes detection of binding activity difficult. As a result of this, discovery of a binding protein for glucose-6-phosphate was rendered difficult, and it was finally isolated by affinity chromatography [56].

It seems entirely reasonable to us that the binding proteins are the carrier proteins for shock-releasable transport systems and that they contain the initial recognition site for the substrate. The statement has been made that binding proteins may function instead as a kind of concentrating mechanism, designed to increase the concentration of substrate in the periplasmic space which is made available to the "true" carrier protein in the protoplasmic membrane. This is a possibility, but it is not very satisfying. First, this notion would require an efficient mechanism for release of substrate from binding protein to "true carrier," and some of the binding proteins have a very tight affinity indeed. Second, there should be evidence for the "true carrier" in Kaback vesicles. As indicated above, generally these vesicles are unable to transport substrates of shock-inhibitable systems where binding proteins have been discovered. Furthermore, in thoroughly poisoned vesicles lacking active transport it is possible to demonstrate binding by whole vesicles of compounds such as proline and lysine, which are taken up by normal vesicles. However, no binding can be demonstrated for compounds such as glutamine, which have shock-releasable binding proteins.

E. Methods for Measuring Binding Activity

1. Binding activity is most commonly measured by equilibrium dialysis using plexiglass chambers separated by a suitable membrane [57]. The apparatus is available commercially.

2. If the solute molecule is charged it can be bound to a suitable ion-exchange resin. If the resin had previously been equilibrated with a labeled solution of solute, addition of binding protein would shift the equilibrium and result in release of some of the labeled solute from the resin into supernatant fluid [24].

3. Sephadex G-25 columns are equilibrated with solute as described by Hummel and Dreyer. (The procedure, with references, is described in [37].) The binding protein is also equilibrated with the same concentration of solute and passed through the column. The large protein molecules are not retained and come off in an excluded peak that contains increased radioactivity of bound solute.

4. When available, antiserum binding to purified binding protein can be used as an assay in purifying later batches of the material and in experiments of various kinds.

5. For rapid scanning of chromatographic columns and other rough purposes, binding can be measured by interacting binding protein and

ligand in the presence of Mg^{2+} and buffer, filtering through Millipore or other nitrocellulose membrane, washing, and counting radioactivity retained on the filter. This is a convenient but treacherous and poorly understood assay. For some strange reason, the radioactive ligand appears to be irreversibly bound to protein immobilized on membrane, even though the binding is reversible in the equilibrium dialysis assay [29]. Labeled ligand bound "irreversibly" on the filter will nevertheless exchange with nonradioactive ligand.

III. GENERAL DESCRIPTION OF INDIVIDUAL BINDING PROTEINS

A. Proteins That Bind Inorganic Ions

1. Sulfate-Binding Protein from S. typhimurium

The first transport-related binding protein to be isolated was the sulfate-binding protein, reported by Pardee and Prestidge in 1966 [24]. It was first detected in cell-free extracts of Salmonella typhimurium LT_2, but was later found to be released by osmotic shock [58]. The purified protein has properties similar to most of the other shock-releasable binding proteins [58]. It has a molecular weight of 32,000 as determined by sedimentation equilibrium and velocity ultracentrifugation experiments. There is one binding site for sulfate per molecule of protein (0.84 moles SO_4^{2-}: 1.0 mole protein.) The protein is stable to a wide range of pH, temperature, and solvents. However, it is sensitive to variations in ionic strength. It is reversibly denatured by urea and irreversibly denatured by acetic anhydride. Substrate does not protect against this denaturation.

Interestingly, the protein does not contain any sulfur-containing amino acids [58]. It has been well characterized in terms of physical properties [59]. The protein has a Stokes radius of 24.7-24.9 Å, as determined by Sephadex gel filtration and the diffusion coefficient. The frictional ratio is 1.19, which leads to the picture of a prolate ellipsoid with dimensions of 112 Å long by 27 Å in diameter, when approximated by a cylinder. The protein has been crystallized [60], and X-ray diffraction studies have been begun [59].

The binding constant of the sulfate-binding protein appears to depend on the nature of the buffer system [58]. In 18 mM phosphate containing 20 mM KCl, the protein has a $K_d = 2 \times 10^{-8}$ M, but this value is increased 5-fold when minimal growth medium is used. The binding is inhibited by chromate (50% inhibition at a 50-fold excess of inhibitor).

The sulfate-binding protein has been localized within the periplasmic space [15]. Both binding and transport are inhibited by diazo-7-amino-1,3-naphthalene disulfonate, which cannot penetrate the cell membrane. Thus the binding protein is exposed to the external environment. Neither binding nor transport were inhibited by antiserum against the sulfate-binding protein. Since antibodies cannot penetrate the cell wall, the binding protein must not extend beyond the outer layer of the cell. These studies do not exclude the possibility that the binding protein exists within the protoplasmic membrane, but partially exposed to the outer surface.

The transport of sulfate in Salmonella typhimurium has been characterized by Dreyfuss [61]. Sulfate appears to be transported by a single system with a $K_m = 3.6 \times 10^{-5}$ M. At a sulfate concentration of 5×10^{-5} M, 2.5×10^{-5} M thiosulfate inhibits sulfate transport by 57%. Thus the system actually has more affinity for thiosulfate than for sulfate. Sulfite appears to be at least as good a substrate as sulfate itself [62]. Chromate is also a good transport inhibitor [62].

The sulfate transport system is regulated by at least two genes, the cysA and cysB genes, each of which consists of three complementation regions [63]. Additionally, mutations in other genes of the cysteine operon can lead to a loss of transport activity. There is a good correlation between the loss of sulfate transport activity and the sulfate-binding protein. Strain cysA20 has normal binding protein but lacks transport activity; the cysA gene is probably a regulatory gene for sulfate transport [63]. Mutants in the cysB gene, which is the regulatory gene of the cysteine operon, vary in their phenotypic expression. Strain N38a, a chromate-resistant mutant mapping in the cysB gene, lacks both transport and binding protein. However, strain R8, a revertant of N38a (in that it can now use sulfate as a source of sulfur), still has no binding protein, but shows better transport activity than the wild type, when both are derepressed. Pardee suggests that "... only a little binding activity may be necessary for full uptake activity, and the two activities are not proportional. If binding activity is involved in uptake, only a small fraction of the wild-type derepressed level is required [63]."

2. Phosphate-Binding Protein from E. coli

Medveczky and Rosenberg [42, 43] have described the isolation and purification of a phosphate-binding protein related to the phosphate transport system of E. coli. They have calculated that the binding protein is present at a level of about 2.5×10^4 molecules per cell. The protein is released by the normal osmotic shock procedure in about an 85% yield. If the tris-sucrose-EDTA step is left out, about 40% of the binding protein is still released. This cold shock also reduces phosphate transport. The transport of arginine and glutamine is not affected by cold shock alone, nor is there any release of the arginine specific or glutamine binding proteins [55]. Thus, this procedure does not appear to be as generally useful as the osmotic shock technique.

The purification of the phosphate-binding protein involves only three steps [43], and results in a nearly homogeneous preparation. The molecular weight of the protein has been estimated to be 42,000 by gel filtration and polyacrylamide gel electrophoresis. The protein yields the same molecular weight in the presence of 8 M urea, so that it would appear to be a single unit. This protein is the largest of the binding proteins yet isolated, nearly twice the molecular weight of the smallest (24,000 MW) -- it binds one molecule of phosphate per molecule of protein with a $K_d = 8 \times 10^{-7}$ M. Whereas there is a similarity between this protein and the R2 protein [42], the possibility of identity has been excluded. At least one strain of E. coli lacking the R2 protein has been shown to contain normal amounts of the phosphate binding protein. (The R2 protein is in the periplasmic space and is the product of one of the two regulator genes for alkaline phosphatase.)

The transport of phosphate in E. coli is achieved via two distinct systems: a high-affinity, nonspecific system and a low-affinity, specific system. The high-affinity system also transports arsenate [64, 65], which has been used to select for permease-less mutants [64, 65]. Two types of arsenate-resistant mutants have been isolated, one of which lacks the high-affinity transport activity, but has binding protein. The other type lacks both transport and binding protein. Since the K_m of this system is identical with the K_d of the binding protein, the isolation of a binding proteinless, transport-negative strain would strongly indicate that the phosphate binding protein is a component of the phosphate-arsenate transport system.

Medveczky and Rosenberg have reported the reconstitution of phosphate transport [43]. The addition of phosphate-binding protein to cold water-shocked cells causes a marked stimulation of both uptake and incorporation into acid-insoluble material. Phosphate transport in spheroplasts was also stimulated by the addition of binding protein. More important, reconstitution was effected in strain 10-1, which lacks both transport and binding protein, whereas strain B-14, which lacks transport but has binding protein, showed no such stimulation of its phosphate uptake. A number of other controls were performed, making this the best-documented example of removal and restoration of a component of an active transport system.

B. Carbohydrates

1. Galactose-Binding Protein from E. coli

Galactose can be transported into E. coli by at least five independent transport systems; however, only two of these, galactose permease and β - methylgalactoside permease, are induced by growth on galactose or its analog, D-fucose.

Galactose permease has a K_m for galactose of about 10^{-4} M and β-methylgalactoside permease has a K_m for galactose of 5×10^{-7}M. This last system is the one for which the galactose binding protein has been shown to be an essential component (for a review see Kalckar [66]).

This protein was first isolated by Anraku [67] who purified it to homogeneity and showed it to bind 1 mole of galactose or glucose per mole of protein with a binding constant of about 10^{-6}M [67]. He also provided evidence that this protein was involved in galactose transport [37, 67, 68].

Boos and co-workers have also isolated a galactose-binding protein from a strain of E. coli K_{12} closely related to the one used by Anraku [38]. This protein differed in two respects from the protein studied by Anraku, first in that it represented 1-0.4% of the cellular protein compared with 0.02% for Anraku's protein, and second, in its binding properties the protein isolated by Boos bound 2 mole of galactose per mole of protein with binding constants of 10^{-5} M and of 10^{-7} M. This preparation also showed a change in fluorescence in the presence of D-galactose or glucose with a K_d of 10^{-6} M. The galactose binding protein is coordinately regulated with β-methylgalactoside permease [69]. Both proteins are induced by galactose and D-fucose, repressed by growth on glucose and rendered constitutive either by a mutation that inactivates galactokinase, or by mutation in a gene called mglR which does not affect the synthesis of the enzymes of the galactose operon [69].

Two strains have been isolated, which contain structural gene mutations in the galactose-binding protein gene [70, 71], and a revertant of one of them has been found [71]. The first of these was reported by Hazelbauer and Adler [70] who have found that E. coli cells are chemotactic toward a large number of molecules, including galactose. The specificity of the galactose chemoreceptor is identical to that of the $P_{\beta g}$; moreover, the $P^-_{\beta g}$ strains, which lack binding protein are also taxis$^-$[70]. However, there are strains of the genotype taxis$^-$, permease$^+$, binding protein$^+$, and others which are taxis$^+$, permease$^-$, binding protein$^+$. It appears, then, that the galactose chemoreceptor and the $P_{\beta g}$ share one common component, the galactose binding protein. By selecting for taxis$^-$ strains, Hazelbauer and Adler were able to isolate a mutant that has a K_m of taxis 50-fold higher than the parent [70]. Boos found that the galactose binding protein cannot be measured in this strain, but that shock fluid contains antigenically cross-reacting material [71]. In addition, he isolated another strain that is $P^-_{\beta g}$, has no functional binding protein, but has CRM. Its revertant to $P_{\beta g}$ has normal binding protein [71].

Boos purified the galactose binding protein (or CRM) from each of the three strains [71]. Neither of the two cross-reacting proteins shows binding of galactose, up to a galactose concentration of 0.1 mM, as measured by equilibrium dialysis. The protein from strain EH3039 does show an increase of the tryptophan fluorescence upon addition of galactose. The native protein shows a half-maximal increase in fluorescence at 10^{-6} M galactose, whereas the mutant protein requires 3.4×10^{-3} M galactose for its half-maximal increase. Peptide maps of both proteins show differences from the normal galactose-binding protein, which demonstrates an actual structural change.

Although the mutation in the structural gene of the galactose-binding protein is accompanied by the loss of $P_{\beta g}$ activity, it was possible that this was the result of a double mutation. The isolation of a revertant of strain EH3039 has eliminated this possibility. At high galactose concentrations, the low-affinity galactose permease transports enough galactose even in $P_{\beta g}^{-}$ strains to allow for good growth with galactose as the sole carbon source. At a galactose concentration of 10^{-6} M, only $P_{\beta g}^{+}$ strains can transport enough galactose to grow. Using this as a method for selection, Boos isolated a revertant of strain EH3039, which shows 65% of the $P_{\beta g}$ activity present in the wild type [71]. The purified galactose-binding protein from this strain has regained its binding capacity, although there are still slight differences between the kinetic parameters of the revertant protein and those of the wild-type protein. In addition, the peptide map of this protein shows the same pattern as that of strain EH3039. The isolation of a mutant with a structural alteration in the gene for a binding protein demonstrates that such binding proteins are indeed components of transport systems; this represents an important contribution to the field of active transport.

2. Glucose-1-Phosphate

Fukui and Miyairi [44] noted the presence of an energy-dependent active transport system for glucose-1-phosphate in *Agrobacterium tumefaciens*. The transport system was formed inducibly by growing the organisms on a glucose-1-phosphate or sucrose medium. It had a high specificity for glucose-1-phosphate with a value for K_m of 4.5×10^{-6} M at pH 8.2. A glucose-1-phosphate binding factor was found in shock fluid prepared from these cells.

3. L-Arabinose

A binding protein for L-arabinose has been reported by Hogg and Englesberg [40] and by Schleif [41]. Hogg and Englesberg reported that the protein consists of a single polypeptide chain of 32,000 MW, with 0.81 binding sites per molecule in their most highly purified preparation.

Although they find a dissociation constant of 5.7×10^{-6} M, this value may not represent the ratio of k_{off}/k_{on}, because the binding does not appear to be freely reversible. In fact, by adding radioactive arabinose to the ammonium sulfate fraction, this unusual binding provided a means of following the protein through the individual steps of the purification procedure. The bound arabinose is exchangeable with added free arabinose.

Schleif [41] reported that the binding is temperature-dependent, with $K_{d(4°C)} = 2 \times 10^{-7}$ M, $K_{d(24°C)} = 7 \times 10^{-7}$ M, and $K_{d(37°C)} = 2 \times 10^{-6}$ M. Moreover, the protein has no binding at 70° C, although cooling to 24° C restores binding. (Temperature effects on binding may not be as marked for other binding proteins, since the K_d of arginine binding to the arginine-specific binding protein is the same at 4°, 23°, and 37°C [57].)

It is difficult to establish with which L-arabinose transport system the arabinose-binding protein is involved. Four transport systems have been described for arabinose [41, 72, 73], although two are minor systems seen only in *araC*⁻ strains [73], which probably represent the transport of arabinose via other systems such as the β-methylgalactoside system. The other two systems are the arabinose-specific system [72] and a higher-affinity system [41].

The arabinose-specific permease is a high-capacity, low-affinity system with a K_m of 5×10^{-5} M [41] to 1.3×10^{-4} M [72]. This system is induced by L-arabinose and depends on a functional *araC* (regulatory) gene product. Mutants in the *araE* gene, which is unlinked to the genes of the arabinose operon, lack this transport activity. Strains of *araC*⁻ also lack this system, and, additionally, lack the arabinose-binding protein. By adding arabinose-binding protein antiserum to agar plates, Hogg [74] has been able to screen arabinose-resistant colonies for the presence of binding protein and has, in this way, isolated a number of *araC*⁻ mutants.

Although the arabinose binding protein and the arabinose-specific permease have similar specificities (weak affinity for D-fucose and L-xylose [40, 72]), there are several reasons for suspecting that the binding protein is not an obligatory component of that system. First, the K_m for arabinose uptake via the arabinose-specific system is at least 10-fold, and possibly 50-fold, higher than the K_d of arabinose binding. Second, a mutant of *E. coli* B/r, strain UP1009, lacks the arabinose-binding proteins but has greater than 60% of normal induced transport [40]. The lesion is in neither the *araC* locus nor the *araE* locus. Strains with mutations in the *araE* gene have only about 20% of normal induced transport, but have normal levels of binding protein.

A low-capacity, high affinity ($K_m = 3 \times 10^{-6}$ M) arabinose transport system has been observed in *araC*^{c} strains [41]. It is not seen in inducible strains, nor in constitutive strains grown in the presence of arabinose. It may be masked under these conditions due to the high internal pools of arabinose, which could exit and dilute the radioactive arabinose in the uptake assay. Strain ara90 [41], which has lost 90% of its concentrative ability, shows only one transport system and has normal binding protein. This mutation does not map in the araE locus, and the K_m of the remaining system is not given, so that it is not possible to say which system is absent. One suspects that the arabinose-specific system is missing in this strain, since it is the high-capacity system and does not require the arabinose-binding protein. If this is so, then the ara90 and araE loci would both be involved in the arabinose-specific permease.

Schleif [41] has suggested that the binding protein might be part of the high affinity system. The fact that strain UP1009 is deficient in binding protein but has only slightly reduced transport suggests that the system with which the binding is associated is of low capacity. Opposing this is the fact that Schleif [41] has isolated a mutant, ara56, which lacks 90% of the binding protein and 92% of its concentrative capacity, but is still $araC^c$, suggesting that the binding protein and the arabinose-specific permease are related. However, the binding protein remaining in this strain has normal binding properties, which allows for the possibility of a gene that regulates only arabinose transport, and is separate from the araC gene. Such a relationship exists between the cysA and cysB genes [52]. Mutants in either gene reduce the activity of the sulfate-transport system and in some, but not all, cases, reduce the amount of sulfate-binding protein. Thus, the high-affinity arabinose-transport system has not been investigated in enough detail to determine its relationship to the arabinose-binding protein.

4. Ribose and Maltose

Hazelbauer and Adler [70] reported the isolation of binding proteins for ribose and maltose. The ribose-binding protein has a $K_d = 2 \times 10^{-6}$ M. Both the ribose-binding protein and the ribose chemoreceptor are induced by growth on ribose; neither is inhibited by galactose, glucose, or maltose. Likewise, the maltose-binding protein ($K_d = 5 \times 10^{-6}$ M) and the maltose chemoreceptor are induced by growth on maltose, and neither is inhibited by glucose or galactose. Their role in chemotaxis has been suggested. It is not known if there is any relationship between these binding proteins and the corresponding transport systems, although the interrelationship of the galactose-binding protein, the β-methylgalactoside permease, and the galactose chemoreceptor indicates the need for such a study.

Aksamit and Koshland [75] also studied the ribose-binding protein because of their interest in chemotaxis. The protein is obtained from the shock fluid of *S. typhimurium* LT-2. Their final, purified preparation is homogeneous by disc gel and sodium dodecyl sulfate gel electrophoresis and by electrofocusing. Its molecular weight is 36,000 and it has a binding constant of 3.3×10^{-7} M, determined by equilibrium dialysis. One mole of D-ribose is bound per mole of protein and the binding is highly specific. The binding protein restored chemotaxis to shocked cells, a result similar to that obtained by Adler with the galactose-binding protein.

C. Vitamins

1. Thiamine and Riboflavin

A binding protein for riboflavin has been described [45]. The protein has been partially purified, exhibits a binding constant of 3×10^{-5} M, and has a molecular weight of 48,000. Two laboratories have reported the isolation of a thiamine-binding protein [45, 46]. The protein has a relatively high affinity for thiamine ($K_d = 3 \times 10^{-7}$ M) and has a molecular weight of 42,000 [45]. Thiamine pyrophosphate competes for both transport and binding [45]. The level of binding protein and the activity of the transport system are reduced when the cells are grown on thiamine-containing medium, and derepressed during growth on adenine [46].

2. Cyanocobalamine

A binding protein for tritiated cyanocobalamine was isolated by Taylor et al. [48] from _E. coli_ B. The factor was released both by treatment with tris-EDTA and by osmotic shock. Its molecular weight by gel filtration was 22,000 but a very large binding molecule (molecular weight in excess of 200,000) was also found. Both proteins formed reversible complexes with dissociation constants of 0.6×10^{-8} M (measured at 4°C) and 0.5×10^{-8} M (measured at 31°C). The exact relationship between the two proteins has not been determined; neither is readily converted to the other. Both proteins are distinct from the B_{12} apoenzyme.

D. Amino Acids

1. Leucine-Binding Proteins from E. coli

Piperno and Oxender [25] were the first to discover an amino acid binding protein. They observed that when _E. coli_ are subjected to osmotic shock the transport of leucine and isoleucine is reduced, but the uptake of alanine and proline are unaffected. A binding protein for leucine, isoleucine, and valine is found in the shock fluid, whereas binding activity for alanine and proline is not found. It is important to include study of a transport system that is unaffected by shock; it serves as a useful control to rule out nonspecific injury to the cells. This control is missing in a number of investigations on transport in shocked cells.

Piperno and Oxender found that the dissociation constants for the binding protein are similar to the values for K_m of transport for leucine, isoleucine, and valine. The protein is termed the leucine-isoleucine-valine protein (LIV-binding protein.) It was purified by DEAE-cellulose chromatography and crystallized from 2-methyl-2,4-pentanediol [76]. Its molecular weight is between 34,000 and 36,000. Treatment with 6 M urea causes a substantial and reversible conformational change [76, 77]. Anraku [26, 37, 68] independently purified the LIV-binding protein, and his results generally agree with those of Oxender and his associates.

Furlong and Weiner [27] observed that _E. coli_ strain 7 has a dual transport system for leucine; approximately one-half of the uptake is inhibitable by excess of nonradioactive isoleucine or valine, while the other half of leucine uptake appears to be via a leucine-specific system. Trifluoroleucine is a specific inhibitor of the leucine-specific transport system. Thus it was found that unlabeled isoleucine did not completely inhibit the initial rate of uptake of labeled leucine; the remainder could be inhibited by trifluoroleucine. Similarly, excess of trifluoroleucine also yielded only partial inhibition of leucine uptake; the residual transport was then inhibited by isoleucine. These workers then discovered a leucine-specific binding protein, which was purified and crystallized. It had a molecular weight of 36,000 and a K_d of 7×10^{-7} M. No binding of any other amino acid could be demonstrated.

2. Glutamine-Binding Protein from _E. coli_

The glutamine transport system is highly specific for L-glutamine. In _E. coli_ strain 7, which was the most intensively studied, it has a K_m of 8×10^{-8} M and a V_m of 10 nmole/min/mg protein [78]. No other natural amino acid competes for active transport. Of a variety of analogs listed, only γ-glutamylhydrazide and γ-glutamylhydroxamate are found to compete for entry; they are competitive inhibitors. In the presence of azaserine, an inhibitor of γ-glutamyl transfer reactions, 85% of the glutamine taken up remains as free glutamine at an internal concentration of 4×10^{-3} M. This represents a 400-fold concentration of glutamine over the external medium.

Osmotic shock causes a reduction of 90% or more in the initial uptake of glutamine, while the transport of certain other amino acids such as proline and glycine is relatively unaffected. The shock fluid contains a protein that specifically binds L-glutamine. It was tested for a number of enzymatic activities and none was found. The binding of glutamine is reversible and the bound substrate is chemically unaltered.

The glutamine-binding protein was purified by means of DEAE-cellulose chromatography and isoelectric focusing. It appears to be homogeneous as judged by its behavior in the ultracentrifuge, and by the fact that it migrates as a single band in polyacrylamide disc gel electrophoresis run at several pH values and over a wide range of cross-linking of gels. Antibody to the binding protein gives only one band in Ouchterlony double diffusion plates.

Approximately 1 mole of L-glutamine is bound per mole of protein, and the K_d for the binding reaction is 3×10^{-7} M as determined by equilibrium dialysis. A similar value is obtained from fluorescence measurements. The protein exhibits native tryptophan fluorescence when excited at 280 nm, with an emission maximum at 336 nm. Glutamine specifically causes a spectral shift toward the blue as well as a drop in amplitude. By measuring the extent of quenching of fluorescence at different concentrations of glutamine, a dissociation constant of 3×10^{-7} M is obtained. Still another method to measure K_d can be used. With a stopped-flow apparatus the fluorescence change is measured as a function of time and rate constants for the binding reaction are determined. The value of k_{on} is 9.8×10^7 M^{-1} sec^{-1} with a standard deviation of 0.82×10^7, and k_{off} is 16 sec^{-1} with a standard deviation of 1.7. This gives a K_d directly of 0.16 μM, which again is in good agreement with the K_d determined by equilibrium dialysis.

Indirect evidence strongly supports the idea that the glutamine binding protein functions in active transport. Thus, osmotic shock caused a 90% decrease in initial rate of transport accompanied by the release of binding protein. The glutamine-binding protein is relatively abundant, a fact that can be correlated with the observation that this is the most active of the amino acid-transport systems. In a mutant, GLNP1, which can use glutamine as a sole carbon source, the initial rate of transport was increased three-fold, together with a comparable threefold increase in level of binding protein. A second mutant, (strain GH20) was derived from GLNP1, by selecting for resistance to γ-glutamylhydrazide. Strain GH20 had only 3% of the initial rate of uptake and about 5% of the binding protein of strain GLNP1. In other experiments, growth of strain 7 on a rich medium (3% dehydrated tryptone and 4% yeast extract) caused both the initial rate of glutamine uptake and the level of binding protein to fall to one-third of the values obtained with minimal salts medium. Finally, γ-glutamylhydrazide and γ-glutamylhydroxamate competitively inhibit both binding and transport with similar K_i values.

3. Glutamic Acid-Binding Protein from E. coli

Halpern and co-workers [47] have described a highly specific, energy-dependent active transport system for glutamate in E. coli K12. This system is several-fold elevated in mutants able to use glutamate as sole carbon source. The supernatant fluid obtained after centrifuging spheroplasts contains a protein-binding for L-glutamate with a K_d of 6.7×10^{-6} M. This binding is competitively inhibited by L-glutamate-γ-methyl ester, which is also a competitive inhibitor of glutamate transport. Noncompetitive inhibitors of glutamate uptake were also examined; they are L-alanine and α-ketoglutarate. These compounds also act as noncompetitive inhibitors of the binding reaction.

Spheroplasts show reduced uptake of glutamate (10-30%): addition of a crude glutamate-binding fraction completely restores the uptake capacity of spheroplasts. This activity is precipitated between 90% and 100% saturation with ammonium sulfate. The binding protein is sensitive to pronase.

In later experiments (unpublished) it was shown that the glutamate-binding protein is released by osmotic shock, a treatment that also reduces the capacity of E. coli for glutamate uptake. Preliminary experiments indicated that shocked cells also showed restoration of uptake upon addition of a crude glutamate-binding preparation. It is presumed that this shock-releasable system is different from the glutamate transport system present in vesicles prepared by Kaback's procedure.

The glutamate-binding protein has been purified 22-fold by means of ammonium sulfate fractionation and DEAE-Sephadex chromatography. This fraction also fully restores the uptake capacity of spheroplasts.

4. Phenylalanine

The organism, Comamonas sp. (11299a), formerly Pseudomonas sp., has an inducible phenylalanine uptake system, with a K_m of 2×10^{-5} M, and which is competitively inhibited by tyrosine and phenylalanine [35]. From this organism, Kuzuya et al. [35] isolated a binding protein by osmotic shock. It was purified to homogeneity by ammonium sulfate fractionation, chromatography on carboxymethyl-Sephadex in the presence of phenylalanine, and passage through a column of Sephadex G-100. The K_d is 1×10^{-7} M and other aromatic amino acids can also bind. Reagents that are known to react with the ϵ-amino groups of lysine interfere with the binding reaction. The protein has an isoelectric point of 9, and extent of binding is independent of pH between 5 and 9. There is stimulation by NaCl and strong inhibition by urea. A substantial fraction of the activity remains after 3 min of boiling. The molecular weight as determined by

Sephadex G-100 filtration is spurious unless 1×10^{-5} M phenylalanine is present. Then a value of 24,000 is obtained, which agrees with the results derived from use of sodium dodecyl sulfate gels. Induced cells contain about two times as much binding material as uninduced cells, but the time course of induction of binding material is quite different from that for induction of transport.

Klein et al. obtained a phenylalanine-binding protein from E. coli by osmotic shock [36]. Unfortunately purification has proven to be very difficult because of the extreme instability of this protein.

5. Histidine-Binding Protein from S. typhimurium

A histidine-binding protein was reported by Rosen and Vasington [33], and also by Ames and Lever [34].* In a later paper [79] the purification to homogeneity of this protein was described. Conventional techniques were applied to concentrated osmotic shock fluid. Purity of the final preparation was indicated by disc gel electrophoresis, sedimentation velocity, and equilibrium ultracentrifugation. It has a molecular weight of $25,250 \pm 300$. It is stable to variations in temperature, ionic strength, and pH. The protein binds histidine with a K_d of 1.5 μM and binds arginine weakly. The simultaneous release of the histidine-binding protein and the reduction of transport caused by osmotic shock, as well as the similarity of kinetic constants and specificity of the binding protein and the histidine permease system suggested to Rosen and Vasington that the histidine-binding protein is a functional component of the histidine permease system. The purification and some of the properties of the wild-type J protein have also been described by Lever [80].

Ames and her associates [34] have extensively studied the transport of histidine in Salmonella typhimurium. There are two major systems for the entry of this amino acid: the histidine-specific system (K_m of 8×10^{-8} M) and a general aromatic amino acid-transport system for the uptake of histidine (K_m of 1.1×10^{-4} M), as well as of phenylalanine, tyrosine, and tryptophan. Mutants defective in each of these systems were described. Ames and Lever [34] showed that uptake via the histidine-specific transport system involves at least two components, J and K, acting in parallel. Each of these two components requires a second protein, the same for both, so that transport through either J or K depends on the integrity of another protein specified by the hisP gene. The protein corresponding to the hisJ gene is a histidine-binding protein.

---

*Dr. G. Ames (personal communication) believes that her J protein is different from the binding protein purified by Rosen and Vasington.

A consideration of the elegant genetic analyses of Ames and co-workers may be summarized as follows.

1. The P protein is essential for histidine transport by the J protein; it is coded for by the hisP gene [34], but no biochemical identification has as yet been established. The P protein is also necessary for the function of the K-P permease (K_m about 10^{-7} M).

2. Osmotic shock releases two proteins that bind L-histidine, of which the J protein comprises 95% and a K protein, 5% of the total. Mutants selected for increased utilization of D-histidine show a fivefold increase in the J-binding protein, associated with a similar increase in the transport of both D- and L-histidine. These mutants show increased sensitivity to the analog, 2-hydrazino-3-(4-imidazolyl)propionic acid. These mutants (dhuA) are probably regulatory mutants with increased levels of normal J protein.

3. Strains with a mutation in the hisJ gene show altered binding protein and this gene would appear to be the structural gene for the J protein. The J protein from a particular mutant strain shows a different elution profile on DEAE-cellulose from wild-type J protein, has a higher dissociation constant (2 μM compared with 0.1 μM) and is inactivated to the extent of 87% after 10 min at 100°C compared with 30% for wild-type protein. The mutant protein also decays more rapidly at 4°C than wild-type protein. The mutant with this altered J protein shows an increased temperature sensitivity of histidine transport. It is unable to grow on histidine at 40°C, and histidine uptake is greatly reduced at this temperature. This provides powerful evidence that the J binding protein is concerned in the active transport of histidine.

More indirect evidence is also cited by Ames and Lever. Thus only those compounds that affect histidine transport are able to inhibit histidine binding. The affinity of the J protein for a variety of compounds correlates well with the affinity of these same compounds for the J-P permease in intact cells.

6. Cystine-Binding Proteins from E. coli

Leive and Davis [81] found two transport systems for cystine in E. coli W, one shared with the structural analog diaminopimelic acid (DAP) and the other specific for cystine. Recently the uptake of cystine has been studied by Berger and Heppel [49] using the mutants supplied by Dr. Leive. The "general," or DAP-inhibitable system shows a K_m of 3×10^{-7} M, and is inhibited by a variety of cystine and DAP analogs. The "specific system," which is DAP-resistant, has a K_m of 2×10^{-8} M and is relatively more selective. A mutant, D_2W, has only the general system and in increased amount.

The two transport systems exhibit drastically different sensitivities towards osmotic shock [29, 49]. Thus, in strain W grown on minimal medium, shock causes a reduction of only the general system. There is no significant change in that fraction of active transport of cystine which is not competed for by DAP, so that cystine uptake falls by only 40%. In strain D_2W, which has only the "general" system to begin with, osmotic shock reduces cystine uptake by 90% or more. These data suggested to us that the shock fluid might contain binding protein active toward both cystine and DAP, but would be free of cystine-specific binding activity. This proved to be the case. Osmotic shock of both W and D_2W yielded a cystine-binding protein whose binding was completely inhibited by excess of non-radioactive DAP.

The cystine-binding protein has been purified to homogeneity by DEAE-cellulose chromatography at pH 7.2, followed by preparative gel electrophoresis and electrofocusing with a pH gradient of 4-6. The purified protein has a molecular weight of 28,000 and an isoelectric point of pH 4.8. Its stability properties closely resemble those reported for other shock-releasable binding proteins. It binds cystine with a K_d of 1×10^{-8} M. When cells were grown in a very rich medium, the general transport system was very greatly reduced and no DAP-inhibitable binding of cystine in the shock fluid could be detected.

A wide variety of cystine and DAP analogs were examined for the ability to inhibit cystine transport and binding. It was found that compounds containing both α and α' amino and carboxyl groups are effective as inhibitors, provided that the "bridge" part of the molecule is of proper length. Compounds containing bridges of 3 or 4 units were potent inhibitors, whereas those with shorter or longer groups did not compete. The list of compounds active with the "specific" system was smaller than the number active with the "general" system.

The existence of analogs which competitively inhibit transport proved useful in the selection of mutants with altered transport properties. Selenocystine, which can enter the cell by either system, is also a potent growth inhibitor. Strain D_2W was used in the selection of mutants able to grow in its presence. Several of the resistant mutants were found to be almost totally devoid of cystine uptake. Upon osmotic shock, these mutants released very low amounts of cystine binding protein.

7. Basic Amino Acid-Binding Proteins from E. coli

Chromatography on DEAE-cellulose of crude shock fluid proteins has been found to yield three peaks of arginine-binding activity [31, 32]. Wilson and Holden [31] reported that none of the peaks bound lysine or leucine, nor was the binding of arginine inhibited by canavanine. Rosen, on the other hand, has purified each of the three proteins, and has found that one of them (the LAO-binding protein) binds a number of other basic amino acids in addition and is probably not related to arginine transport [32]. This protein is described below. In addition, Rosen and Vasington [33] have reported the existence of a histidine-binding protein from Salmonella typhimurium, which elutes from a DEAE-cellulose column at the same position as the first of the arginine binding proteins. The histidine-binding protein has a low affinity for arginine as well, and both are basic proteins [33], so that it is possible that this arginine-binding protein is actually associated with histidine transport and not arginine transport.

The third of the arginine-binding proteins is specific for arginine ($K_d = 3 \times 10^{-8}$ M), and exhibits stability to extremes of pH and ionic strength, as do most other binding proteins [82]. The purified arginine-specific binding protein has a molecular weight of 27,700 by sedimentation equilibrium ultracentrifugation, and antiserum prepared against the protein does not react with the LAO-binding protein. An amino acid analysis of both the arginine specific and LAO-binding proteins shows that each of them contains two half-residues of cysteine, which are probably in the disulfide form [82]. This is of interest, since none of the other binding proteins released by osmotic shock has been shown to contain cysteine or cystine. The affinity of the arginine-specific binding protein is not altered at 4°, 23°, or 37° C [5, 7], unlike the arabinose-binding protein, which changes affinity with changes in temperature [41].

The data indicate that the arginine-specific binding protein is a component of the arginine-specific transport system, although the correlation is mostly indirect. The release of the binding protein is associated with a nearly complete loss of arginine specific transport (95%) in E. coli K12 [32]. However, in E. coli W, arginine transport appears to be only slightly reduced [30]. The K_d of the arginine-specific binding protein is similar to that of the transport system ($K_m = 2.6 \times 10^{-8}$ M) and neither is inhibited by lysine, arginine analogs, or arginine precursors [32, 55]. The other two binding proteins with affinity for arginine have neither specificities nor affinities in common with the arginine-transport system. Growth on enriched medium causes a 70-80% reduction in the activity of the transport system and in the amount of shock releasable binding protein [50]. Wilson and Holden have reported the reconstitution of arginine transport in shocked cells by the addition of binding protein [31]; however, no correction for adsorption of protein to the filter was made. Since the arginine-binding proteins, and, indeed, many of the binding proteins adsorb to nitrocellulose filters [32, 41], the small increase in radioactive arginine on the filter may be due to such effects.

A protein that binds lysine, arginine, and ornithine (the LAO-binding protein) has been purified by Rosen [32]. The protein has been found to bind, in order of affinity, arginine, lysine, ornithine, citrulline, and canavanine [32, 82]. The dissociation constants range from 2×10^{-7} M for arginine to greater than 10^{-4} M for canavanine. In addition, N-α-acetylornithine, trans-4,5-dehydrolysine, and 4-oxalysine are inhibitors of binding (but have not been tested directly for binding.) However, N-α-acetylglutamate, another arginine precursor, does not inhibit the binding of arginine.

The properties of the LAO-binding protein are similar to those of other binding proteins, in that the protein is stable to extremes of pH, ionic strength, and temperature, and is reversibly denatured by urea. The protein has a molecular weight of 26,200 as determined by sedimentation equilibrium ultracentrifugation, and a pI of 5.1 from isoelectric focusing.

The LAO-binding protein has been related to a transport system for lysine and ornithine [32]). The affinity of the transport system for lysine and ornithine is similar to the affinity of the LAO-binding protein for those two substrates. Arginine uptake was found to be completely insensitive to ornithine and lysine, as mentioned above. Moreover, the K_i of arginine inhibition or ornithine or lysine uptake is an order of magnitude greater than the K_m of arginine uptake, whereas the K_i's and K_m's for the other substrates were approximately the same. For these reasons Rosen [32, 82] has postulated that the arginine-specific and LAO-transport systems are distinct, and that arginine is an inhibitor of the LAO system, but not a substrate.

Little information has been obtained on the genetics of the transport systems for these amino acids. Schwartz et al. have isolated a canavanine-resistant strain of *E. coli*, which has reduced transport for arginine, lysine, and ornithine [83]. From this, they postulated a single system for the uptake of all three amino acids. In view of the work of Rosen [32] this does not appear likely, but more work is necessary to resolve this question.

E. Binding Proteins from Microorganisms Other Than Bacteria

1. Neurospora crassa

By varying the conditions for osmotic shock from those used for *E. coli*, Wiley [5] was able to reduce tryptophan transport in *Neurospora* by 90% without an appreciable loss of cell viability and this was accompanied by the release of a tryptophan-binding protein and alkaline phosphatase. Unfortunately he did not report whether or not some other transport system, *not* associated with a releasable binding protein was unimpaired; this would have strengthened the significance of his observation. The specific activity of binding (0.034 nmole/mg protein) was very low compared with E. coli systems. The activity was purified 20-fold by treatment with Sephadex. The following evidence was presented to indicate a relation between this binding protein and the transport of tryptophan. (1) It appears to be localized near the cell surface; (2) a transport-negative mutant shows reduced binding; (3) specificity of the binding reaction is similar to the specificity of transport, and different substrates show affinities which rank in the same order; (4) the dissociation constant for the binding reaction (8×10^{-5} M) was approximately the same as the K_m for active transport.

Stuart and DeBusk [84] report that simple washing of the conidia with dilute KCl releases a number of glycoproteins, several of which bind arginine. At the same time, the washed cells show a 30% reduction in arginine transport activity. When the crude protein solution is passed through a Sephadex G-25 column to which arginine has been chemically linked, three peaks of protein were eluted with a pH gradient. All three peaks bound both arginine and phenylalanine. The elution pattern from transport negative mutants was examined. In strains lacking basic amino acid transport, peaks B and C were missing. In strains lacking neutral amino acid transport, only peak C was missing. Peak A was present in all cases.

2. Saccharomyces

Schwencke et al. [85] have modified the osmotic shock procedure for Saccharomyces chevalieri. A number of catabolic enzymes are released into the shock fluid, including invertase (68%), acid phosphatase (18%), 5'-nucleotidase (71%) and alkaline pyrophosphatase (11%). These same enzymes are released upon conversion to spheroplasts. Proline uptake is reduced 70-90% in osmotically shocked cells, and the possibility of the release of a proline-binding protein is being investigated.

Bussey and Umbarger [86] have reported the isolation of a leucine-binding protein from sonic extracts of Saccharomyces cerevisiae and another undefined strain of Saccharomyces. The protein in both cases exhibits a K_d 100-fold lower than the K_m of leucine transport. The K_d's and K_m's are considerably different between the two strains. In strain 60615 both the transport system and the binding protein are induced by growth in the presence of leucine. Strain 60615tfl$_2$, a trifluoroleucine-resistant mutant, shows derepressed transport activity and binding protein levels, demonstrating that the two are regulated by a common gene.

F. Carrier Proteins That Are Not Released by Osmotic Shock

This review is limited to shock-releasable binding proteins, but brief mention should nevertheless be made of the important studies on firmly bound bacterial carrier proteins. In 1965, Fox and Kennedy [17] found a membrane-localized protein which they called the M protein. It is essential for the operation of the β-galactoside transport system in E. coli and can be extracted from membrane preparations by using detergents. The M protein has two binding sites for the sugar being transported.

Another very significant development of recent years has been the discovery of the phosphotransferase system by Roseman and his associates. This system is responsible for the active uptake of a number of sugars and alcohols in E. coli, M. pyogenes, and certain other bacteria. The reader is referred to the important recent papers from Roseman's laboratory in order to get started on the fairly extensive literature [11, 18, 19, 88-91].

Mention should also be made of the very interesting studies of calcium-binding proteins from mammalian cells. Wasserman and his colleagues [92-95] discovered a calcium-binding protein in the intestinal mucosa of chicks. This resulted from their studies on the mechanism of action of vitamin D on the intestinal uptake of calcium. The calcium-binding activity was missing from intestines of rachitic animals and appeared after vitamin D administration, along with restored capacity for calcium absorption. The protein was purified to homogeneity. Its molecular weight is 25,000-28,000. It binds 1 mole of calcium per mole of protein.

Lehninger [96] found that when rat liver mitochondria were exposed to distilled water they lost their high-affinity binding sites for calcium. Soluble, heat-labile binding activity could be recovered from the distilled water extract. This factor probably functions in a calcium-transport system located in the inner membrane of rat liver mitochondria.

Recently, binding activities for amino acids have been removed from bacterial membrane preparations with detergents. The solubilization of a protein fraction from membrane vesicles with binding properties for proline was reported by Gordon et al. [21]. Membrane vesicles were partially solubilized with the nonionic detergent Brij 36-T and the solubilized material was fractionated by Sephadex G-100 chromatography carried out in the presence of detergent. Three fractions with proline-binding activity were obtained; one of them has a relatively low molecular weight and high specific activity. This fraction has no D-lactate dehydrogenase activity. The binding is inhibited by p-chloromercuribenzoate and the inhibition is reversed by dithiothreitol. These compounds have similar effects on proline transport by intact membrane vesicles. The fraction also binds serine, glycine, lysine, and tyrosine. The authors feel that their fraction may be the "carrier protein" of the D-lactate dehydrogenase-coupled-amino acid transport system.

Friedberg [23] has also obtained an amino acid-binding protein, of the tightly associated kind, from membrane vesicles made by Kaback's procedure. He was able to remove lysine-binding activity by repeated washing with low ionic strength buffer, but no other binding activities were removed in this fashion. An attempt was made to measure binding by whole membranes using equilibrium dialysis in the presence of 1×10^{-5} M carbonyl p-trifluoromethoxyphenylhydrazone. By this test a 60-70% reduction in binding for whole membrane vesicles was observed, and the removed lysine binding was recovered in the washings.

The lysine-binding capacity in the washings showed no inhibition in the presence of excess nonradioactive arginine or ornithine. No binding of ^{14}C-arginine or ^{14}C-ornithine was detected. The specificity of binding therefore agreed with the specificity of the lysine uptake system present in Kaback vesicles, being lysine-specific and uninhibited by other basic amino acids. The washings were centrifuged at 140,000 x g for 6 h at 4° C and the supernatant fraction was placed on a Sephadex G-100 column. Activity was largely recovered in a peak of low molecular weight (under 50,000). The lysine binding capacity was reduced by N-ethylmaleimide and p-hydroxymercuric benzoate. Inhibition by the latter was reversed by dithiothreitol. The apparent K_D of binding for L-lysine was 8×10^{-6} M. This is similar to the K_m of the lysine-specific uptake system. It is unfortunate that similar data are not presented for the fraction of Gordon et al. Binding activity for labeled L-lysine is present in the presence of a very large excess of nonradioactive D-lysine, which rules out nonspecific binding.

IV. GENERAL CONCLUSIONS

Most of the evidence concerning a relationship between binding proteins and active transport is indirect, but in the aggregate it is impressive. Direct evidence has been very slow in coming. However, the recent experiments of Ames and Lever on a mutation in the structural gene for a histidine-binding protein are quite convincing. The experiments of Boos are also quite important, although there are some reservations because his strains are not strictly isogenic. Efforts at reconstitution of transport using binding proteins have been generally discouraging, but positive results are claimed by at least three laboratories. We are especially impressed by the fact that osmotic shock selectively destroys those transport systems for which a binding protein can be isolated, and the procedure has little effect on systems in which the carrier protein is tightly bound to the membrane. In general, membrane vesicles prepared by Kaback's procedure show transport corresponding to those systems for which no binding protein is released. It is our opinion that the case for binding proteins is strong, and experiments should be designed to obtain a better understanding of how they function rather than to prove or disprove their role in the transport system.

The indirect lines of evidence have been summarized a number of times [52, 97-101]. They may be enumerated as follows.

1. Osmotic shock causes a simultaneous loss of transport activity and binding activity in the shock fluid. This line of evidence has been attacked, and the loss in transport activity upon shock has been attributed to nonspecific cell injury. However, if the shock procedure is properly carried out, there is no loss in cell viability and shocked cells resume growth in nutrient medium after a lag period not significantly longer than for cells that were simply washed and chilled. Most important, many transport systems are unaffected by the shock treatment and in the investigated cases, such systems are _not_ associated with shock-releasable binding proteins.

2. Binding proteins have been localized in the cell envelope, where transport factors would be expected to reside.

3. Similarities in kinetic constants: In general, K_D of binding is similar to K_m of transport and substrates, and inhibitors of the same system fall in the same order when ranked in both processes. The values of K_D and K_m are not usually identical and this is not surprising, for the binding reaction is probably not measured under true _in vivo_ conditions.

4. Co-regulation: In many cases there is a parallel repression of transport activity and synthesis of binding protein or a parallel induction of both.

5. There are numerous instances where competitive inhibitors of the binding reaction were found also to be competitive inhibitors of active transport.

6. Reversion of transport-negative-binding protein-negative mutants to transport-positive results in all cases thus far noted to a binding-protein-positive state.

REFERENCES

1. H. C. Neu and L. A. Heppel, _J. Biol. Chem._, _240_, 3685 (1966).

2. N. G. Nossal and L. A. Heppel, _J. Biol. Chem._, _241_, 3055 (1966).

3. R. Repaske, _Biochim. Biophys. Acta_, _30_, 225 (1958).

4. L. Heppel, in _Structure and Function of Biological Membranes_ (L. I. Rothfield, ed.), p. 223, Academic Press, New York, 1971.

5. W. R. Wiley, J. Bacteriol., 103, 656 (1970).

6. A. N. Taylor and R. H. Wasserman, Arch. Biochem. Biophys., 119, 536 (1967).

7. A. L. Lehninger, Biochem. Biophys. Res. Commun., 42, 312 (1971).

8. H. C. Neu, J. Gen. Microbiol., 57, 215 (1969).

9. H. C. Neu, D. F. Ashman, and T. D. Price, J. Bacteriol., 93, 1360 (1967).

10. L. Leive and V. Kollin, Biochem. Biophys. Res. Commun., 28, 229 (1967).

11. W. Kundig, F. D. Kundig, B. Anderson, and S. Roseman, J. Biol. Chem., 241, 3243 (1966).

12. B. P. Rosen and L. Hackette, J. Bacteriol., 110, 1181 (1972).

13. H. U. Schairer and P. Overath, J. Mol. Biol., 44, 209 (1969).

14. G. Wilson and C. F. Fox, J. Mol. Biol., 55, 49 (1971).

15. A. B. Pardee and K. Watanabe, J. Bacteriol., 96, 1049 (1968).

16. P. K. Nakane, G. E. Nichoalds, and D. L. Oxender, Science, 161, 182 (1968).

17. C. F. Fox and E. P. Kennedy, Proc. Nat. Acad. Sci. U. S., 54, 891 (1965).

18. W. Kundig, S. Ghosh, and S. Roseman, Proc. Nat. Acad. Sci. U. S., 52, 1067 (1964).

19. W. Kundig and S. Roseman, J. Biol. Chem., 246, 1407 (1971).

20. H. R. Kaback in Current Topics in Membranes and Transport (A. Kleinzeller and F. Bronner, eds.), Academic Press, New York, 1971.

21. A. S. Gordon, F. J. Lombardi, and H. R. Kaback, Proc. Nat. Acad. Sci. U. S., 69, 358 (1972).

22. L. A. Heppel, private communication.

23. I. F. Friedberg, private communication.

24. A. B. Pardee and L. S. Prestige, Proc. Nat. Acad. Sci. U. S., 55, 189 (1966).

25. J. R. Piperno and D. L. Oxender, J. Biol. Chem., 241, 5732 (1966).

26. Y. Anraku, J. Biol. Chem., 243, 3116 (1968).

27. C. E. Furlong and J. H. Weiner, Biochem. Biophys. Res. Commun., 38, 1076 (1970).

28. J. H. Weiner, C. E. Furlong, and L. A. Heppel, Arch. Biochem. Biophys., 142, 715 (1971).

29. E. A. Berger and L. A. Heppel, private communication.

30. O. H. Wilson and J. T. Holden, J. Biol. Chem., 244, 2737 (1969).

31. O. H. Wilson and J. T. Holden, J. Biol. Chem., 244, 2743 (1969).

32. B. P. Rosen, J. Biol. Chem., 246, 3653 (1971).

33. B. P. Rosen and F. D. Vasington, Fed. Proc. Fed. Amer. Soc. Exp. Biol., 29, 342 (1970).

34. G. F. Ames and J. Lever, Proc. Nat. Acad. Sci. U.S., 66, 1096 (1970).

35. H. Kuzuwa, K. Bronwell and G. Gurloff, J. Biol. Chem., 246, 6371 (1971).

36. W. L. Klein, A. S. Dahms, and P. D. Boyer, Fed. Proc. Fed. Amer. Soc. Exp. Biol., 28, 341 (1970). No. 540.

37. Y. L. Anraku, J. Biol. Chem., 243, 3123 (1968).

38. W. Boos, Eur. J. Biochem., 10, 66 (1969).

39. W. Boos and M. Sarvas, Eur. J. Biochem., 13, 526 (1970).

40. R. W. Hogg and E. Englesberg, J. Bacteriol., 100, 423 (1969).

41. R. Schleif, J. Mol. Biol., 46, 185 (1969).

42. N. Medveczky and H. Rosenberg, Biochim. Biophys. Acta, 192, 369 (1969).

43. N. Medveczky and H. Rosenberg, Biochim. Biophys. Acta, 211, 158 (1970).

44. S. Fukui and S. Miyairi, J. Bacteriol., 101, 685 (1970).

45. T. W. Griffith, C. Carraway, and F. R. Leach, Fed. Proc. Fed. Amer. Soc. Exp. Biol., 30, 363 abs. (1971).

46. A. Iwashima, A. Matsumura, and Y. Nose, J. Bacteriol., 108, 1419 (1971).

47. H. Barash and Y. S. Halpern, Biochem. Biophys. Res. Commun., 45, 681 (1971).

48. R. T. Taylor, S. A. Norrell, and M. L. Hanna, Arch. Biochem. Biophys., 148, 366 (1972).

49. L. A. Heppel, B. P. Rosen, I. Friedberg, E. Berger, and J. H. Weiner in The Molecular Basis of Biological Transport (J. F. Woessner and F. Huijing, eds.), Academic Press, New York, 1972.

50. B. P. Rosen, Fed. Proc. Fed. Amer. Soc. Exp. Biol., 30, 1061 abs. (1971).

51. Y. Anraku, J. Biochem (Japan), 70, 855 (1971).

52. D. L. Oxender, Ann. Rev. Biochem., 41, 777 (1972).

53. H. R. Kaback and L. S. Milner, Proc. Nat. Acad. Sci. U. S., 66, 1008 (1970).

54. E. M. Barnes, Jr., and H. R. Kaback, Proc. Nat. Acad. Sci. U. S., 66, 1190 (1970).

55. B. P. Rosen, J. H. Weiner, and L. A. Heppel, personal communication.

56. G. Dietz, personal communication.

57. C. E. Furlong, R. G. Morris, M. Kandrach, and B. P. Rosen, Anal. Biochem., 47, 514 (1972).

58. A. B. Pardee, J. Biol. Chem., 241, 5886 (1966).

59. A. B. Pardee, Science, 156, 1629 (1967).

60. R. Langridge, H. Shinagawa, and A. B. Pardee, Science, 169, 59 (1970).

61. J. Dreyfuss, J. Biol. Chem., 239, 2292 (1964).

62. A. B. Pardee, L. S. Prestidge, M. B. Whipple, and J. Dreyfuss, J. Biol. Chem., 241, 3962 (1966).

63. N. Ohta, P. R. Galsworthy, and A. B. Pardee, J. Bacteriol., 105, 1053 (1971).

64. R. L. Bennett and M. H. Malamy, Biochem. Biophys. Res. Commun., 40, 496 (1970).

65. N. Medveczky and H. Rosenberg, Biochim. Biophys. Acta, 241, 494 (1971).

66. H. M. Kalckar, Science, 174, 557 (1971).

67. Y. Anraku, J. Biol. Chem., 242, 793 (1967).

68. Y. Anraku, J. Biol. Chem., 243, 3128 (1968).

69. J. Lengeler, K. O. Hermann, H. J. Unsold, and W. Boos, Eur. J. Biochem., 19, 457 (1971).

70. G. L. Hazelbauer and J. Adler, Nature New Biol., 230, 101 (1971).

71. W. Boos, J. Biol. Chem., 247, 5414 (1972).

72. C. P. Novotny and E. Englesberg, Biochim. Biophys. Acta, 117, 217 (1966).

73. J. Singer and E. Englesberg, Biochim. Biophys. Acta, 249, 498 (1971).

74. R. W. Hogg, J. Bacteriol., 105, 604 (1971).

75. R. R. Aksamit and D. E. Koshland, Jr., Fed. Proc., 31, 1363 abs. (1972).

76. W. R. Penrose, G. E. Nichoalds, J. R. Piperno, and D. L. Oxender, J. Biol. Chem., 243, 5921 (1968).

77. W. R. Penrose, R. Zand, and D. L. Oxender, J. Biol. Chem., 245, 1432 (1970).

78. J. H. Weiner and L. A. Heppel, J. Biol. Chem., 246, 6933 (1971).

79. B. P. Rosen and F. D. Vasington, J. Biol. Chem., 246, 5351 (1971).

80. J. Lever, J. Biol. Chem., 247, 4317 (1972).

81. J. Leive and B. D. Davis, J. Biol. Chem., 240, 4362 (1965).

82. B. P. Rosen, J. Biol. Chem., 248, 1211 (1973).

83. J. Schwartz, W. Maas, and E. Simon, Biochim. Biophys. Acta, 32, 582 (1959).

84. W. D. Stuart and A. G. DeBusk, Arch. Biochem. Biophys., 144, 512 (1971).

85. J. S. Schwencke, G. Farias, and M. Rojas, Eur. J. Biochem., 21, 137 (1971).

86. H. Bussey and H. E. Umbarger, J. Bacteriol., 103, 277 (1970).

87. W. Kundig and S. Roseman, J. Biol. Chem., 246, 1393 (1971).

88. M. H. Saier, Jr., W. Scott Young III, and S. Roseman, J. Biol. Chem., 246, 5838 (1971).

89. B. Anderson, N. Weigel, W. Kundig, and S. Roseman, J. Biol. Chem., 246, 7023 (1971).

90. M. H. Saier, Jr., B. U. Feucht, and S. Roseman, J. Biol. Chem., 246, 7819 (1971).

91. M. H. Saier, Jr. and S. Roseman, J. Biol. Chem., 247, 972 (1972).

92. R. H. Wasserman and A. N. Taylor, Science, 152, 791 (1966).

93. R. H. Wasserman in Metabolic Transport (L. E. Hokin, ed.), Vol. VI, Academic Press, New York, 1972.

94. R. H. Wasserman, R. A. Corradino, and A. N. Taylor, J. Biol. Chem., 243, 3978 (1968).

95. R. H. Wasserman and A. N. Taylor, J. Biol. Chem., 243, 3987 (1968).

96. A. L. Lehninger, Biochem. Biophys. Res. Comm., 42, 312 (1971).

97. A. B. Pardee, Science, 162, 632 (1968).

98. L. A. Heppel, J. Gen. Physiol., 54, 95s (1969).

99. H. R. Kaback, Ann. Rev. Biochem., 39, 561 (1970).

100. Lin, E. C. C., in Membrane Structure and Function (L. I. Rothfield, ed.), Academic Press, New York, 1971.

101. H. M. Kalckar, Science, 174, 557 (1971).

Chapter 5

BACTERIAL TRANSPORT MECHANISMS

H. R. Kaback

The Roche Institute of Molecular Biology
Nutley, New Jersey

I. INTRODUCTION

The mechanisms involved in active transport and their relationship to the cell membrane represent two major obstacles to understanding membrane structure and function. Until recently, progress in this area was hampered by the lack of adequately defined experimental systems. Investigations were limited to kinetic studies of uptake or application of physical techniques to systems in which there was little possibility of correlating structural observations with defined membrane phenomena. Within a relatively short period of time, however, rapid advances have taken place on a more biochemical level as a result of the development of experimental systems that allow the investigator to probe on a deeper level. Some of these systems and the advances gained in utilizing them are discussed elsewhere in this volume. The present discussion emphasizes another aspect of these developments, particularly those derived from studies with isolated bacterial membrane vesicles, an experimental system developed primarily in the author's laboratory.

It was first reported in 1960 that subcellular preparations from E. coli catalyze the uptake of the amino acid glycine [1]. Subsequent studies confirmed the initial observation and further demonstrated that cell-free preparations of membrane vesicles, essentially devoid of cytoplasmic constituents, catalyze the uptake of glycine [2,3] and the active transport of proline [4, 5]. These studies established a model system that allows examination of the problems outlined above.

II. ISOLATION OF BACTERIAL MEMBRANE VESICLES

A. Preparation and Homogeneity

An electron micrograph of a longitudinal section through an intact E. coli cell is shown in Fig. 1. The cell is rod-shaped with two trilaminar "unit membrane" structures bordering its exterior. The outer membrane, present in gram-negative bacteria only, is the lipopolysaccharide layer of the cell wall; the inner membrane is the plasma membrane. Located between these membranes, in the periplasmic space, is the peptidoglycan layer of the cell wall; this structure, however, cannot be seen in Fig. 1. The rigid peptidoglycan layer is responsible for the shape of bacterial cells, allows the concentration of solutes against large concentration gradients, and prevents the bacterium from bursting in hypotonic environments [6]. Within the inner cell membrane are the ribosomes, nucleoplasm, and most of the "soluble" components of the cell. When such a cell is grown in the presence of penicillin or is treated with enzymes that attack the rigid layer of the cell wall (e.g., lysozyme with most organisms or lysostaphin with Staphylococci), the rigid peptidoglycan layer is outgrown or degraded. As a result, the cell becomes sensitive to changes in osmolarity and will burst in sufficiently dilute media. This manipulation is the basis for the preparation of bacterial membrane vesicles [7].

The structures shown in Figs. 2 and 3 are obtained by osmotic lysis of penicillin- or lysozyme-EDTA-treated E. coli against large osmotic gradients in the presence of EDTA, DNase, and RNase, followed by extensive washing and differential centrifugation. Details of the procedure have been described previously [7]. When the ML strain of E. coli is subjected to this procedure (Fig. 2), the structures consist of intact "unit membrane"-bound sacs varying from 0.5 to 1.5 μ in diameter. Most of the sacs are surrounded by a single 65-70 Å membrane. The sacs are empty and without apparent internal structure. Vesicles prepared from cells made osmotically sensitive by either lysozyme-EDTA (Fig. 2A), or penicillin (Fig. 2B), cells are morphologically identical.

When E. coli W is subjected to the same procedures (Fig. 3A, B), intact, membrane-bound sacs are also obtained, but the vesicles are more heterogeneous. The diameters of the sacs vary from 0.1 to 1.5 μ, and they are surrounded by one to five or six membrane layers. As in E. coli ML, these vesicles are also empty and devoid of internal structure, each membrane layer is 65-70 Å thick, and there is no significant difference between vesicles prepared from either type of osmotically sensitized cell (compare Figs. 3A and B). Vesicles prepared from other strains of E. coli (K12, W2244, and K_2lt) are indistinguishable from E. coli W. The reason for this morphological difference between ML membranes and

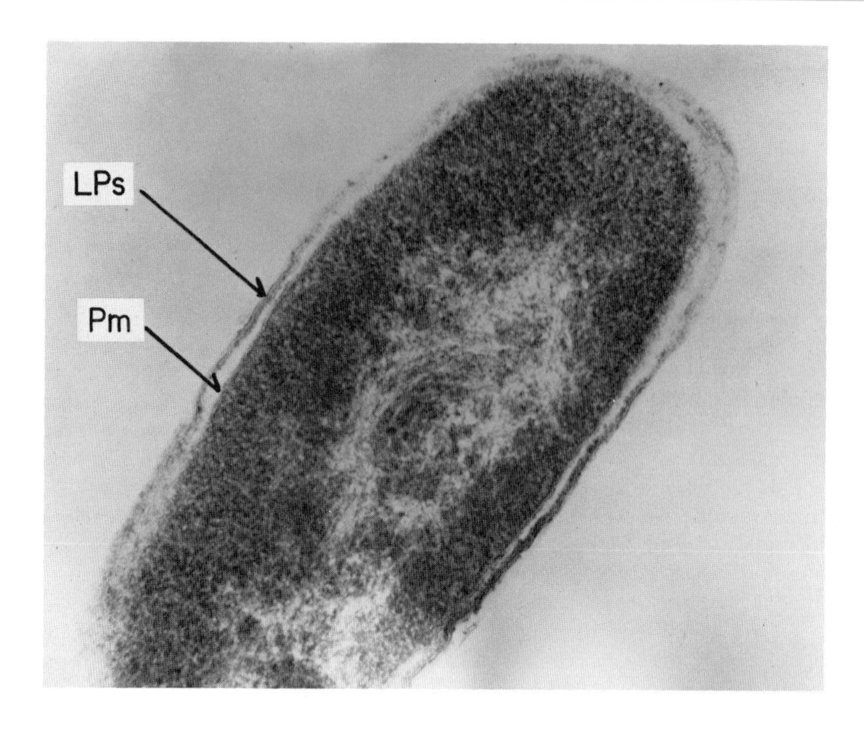

FIG. 1. Electron micrograph of a longitudinal section through an intact E. coli cell. LPS, lipopolysaccharide; PM, plasma membrane. This micrograph was taken by Dr. Samuel Silverstein of the Rockefeller University. From H. R. Kaback [8].

membranes prepared from other strains of E. coli is probably related to the observation that ML vesicles contain negligible quantities of lipopolysaccharide, whereas the other vesicle preparations contain significant quantities of this cell-wall component, as discussed below. In any case, there are no demonstrable physiological differences between membranes prepared from the ML strain and those prepared from other strains of E. coli as judged by each of the transport systems thus far studied. Moreover, vesicles from gram-positive organisms that have no lipopolysaccharide also catalyze active transport.

The purity and homogeneity of the vesicle preparations have been established by a variety of other criteria [2, 4, 7, 8] :

1. Membrane vesicles can be separated from whole cells physically by density centrifugation, and the preparations contain negligible quantities of whole cells or partially lysed forms.

2. The vesicles contain less than 5% of the nucleic acids, 10-15% of the protein, and at least 70% of the phospholipids of the whole cells from which they are prepared. Less than 1 or 2% of the activities of glutamine synthetase, β-galactosidase, fatty acid synthetase, leucine-activating enzyme, or each of the "periplasmic enzymes" found in the parent cells are retained by the membrane vesicles.

3. Although 10-15% of the total cellular protein remain in the vesicles, almost all of the "soluble" proteins are lost, as demonstrated by acrylamide disc gel electrophoresis.

4. Vesicles retain 10% or less of the diaminopimelic acid of the osmotically sensitized form, indicating that very little peptidoglycan is left in the preparations. Regarding lipopolysaccharide, ML vesicles have less than 3% (by dry weight), whereas vesicles from a number of other strains of E. coli and S. typhimurium have from 7 to 17%.

Expressed as a function of dry weight, the vesicles are approximately 60-70% protein, 30-40% phospholipid, and 1% carbohydrate.

B. Physical Properties

One essential property of any system to be used to study transport is that it must have a continuous surface (i.e., it must be able to retain transported substrate). Although the sectioned material presented in Figs. 2 and 3 suggests that the vesicles are closed structures, other techniques must be utilized to substantiate this impression. The electron micrographs shown in Figs. 4 and 5 were obtained using negative staining (Figs. 4A, B) or freeze-etching techniques (Fig. 5) so that the surface of the vesicles can be observed. The micrographs show typical vesicles prepared from E. coli ML 308-225 (Figs. 4A and 5) and E. coli W (Fig. 4B). In both cases, there are no gross defects in the surface of the vesicles. It is also significant that the stain (i.e., phosphotungstic acid) does not penetrate the interior of the vesicles (Fig. 4).

More convincing evidence for membrane continuity demonstrates that the vesicles are osmotically intact. They shrink and swell appropriately when the osmolarity of the medium is increased or decreased, as shown by light scattering [5] or by measurement of the dextran-impermeable intramembranal space [8]. Moreover, there is a diffusion barrier to phosphoenolpyruvate, which can be overcome transiently by osmotically shocking the vesicles in the presence of this highly charged phosphorylated compound [8].

A significant number of vesicles do not become inverted during lysis (i.e., the vesicles do not turn inside out). This has been ascertained by

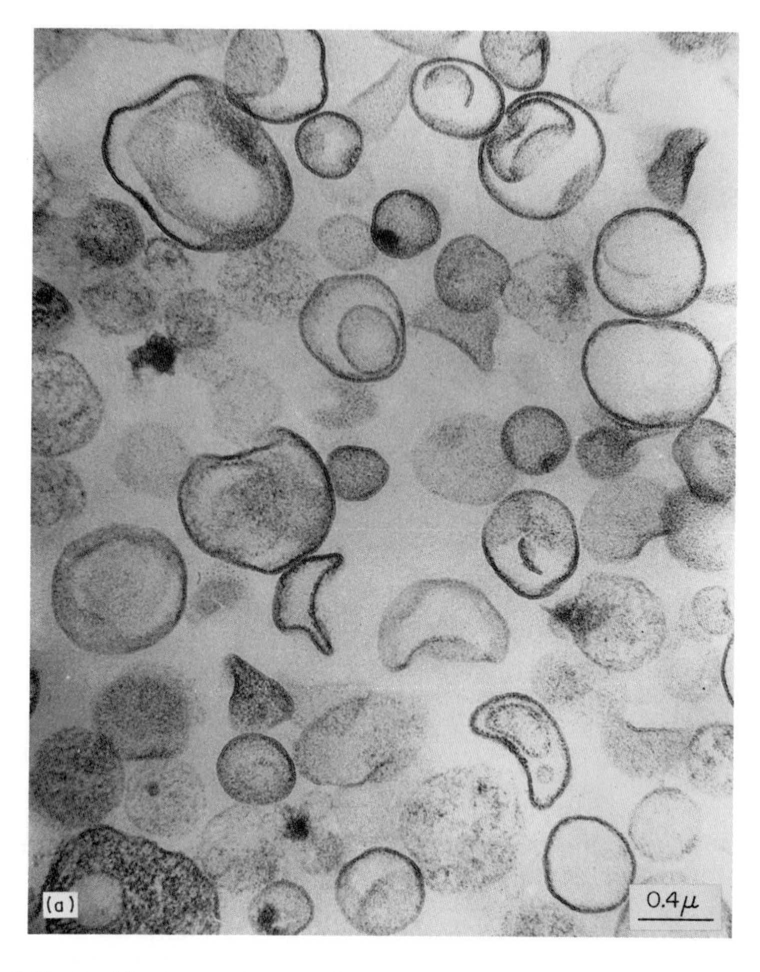

FIG. 2. Electron micrograph of membrane vesicle preparations from E. coli ML 308-225. The membranes were prepared from (A) lysozyme-EDTA-induced spheroplasts and (B) penicillin-induced spheroplasts. The micrographs were taken by Dr. Vincent Marchesi of the National Institute of Arthritis and Metabolic Disease. From H. R. Kaback [7, 8].

observing lysis under phase-contrast microscopy [8a]. In addition, vesicles prepared from B. megaterium grown into the late logarithmic phase of growth retain at least some α-hydroxybutyrate granules [8a].

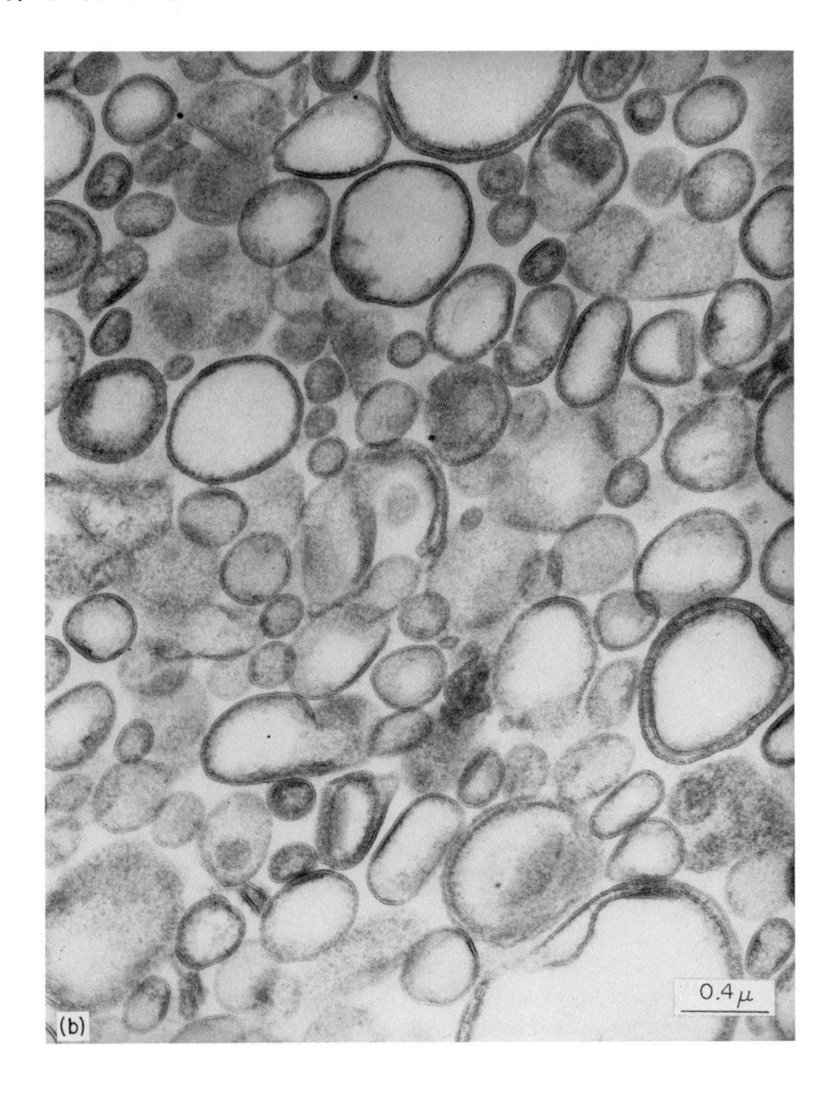

Optical rotatory dispersion (ORD) and circular dichroism (CD) studies [8b] on membrane vesicles from E. coli ML 308-225 and K12 are essentially identical to those found with B. subtilis membranes and human red blood cell membranes [9] and Erlich's ascites cell membranes [10].

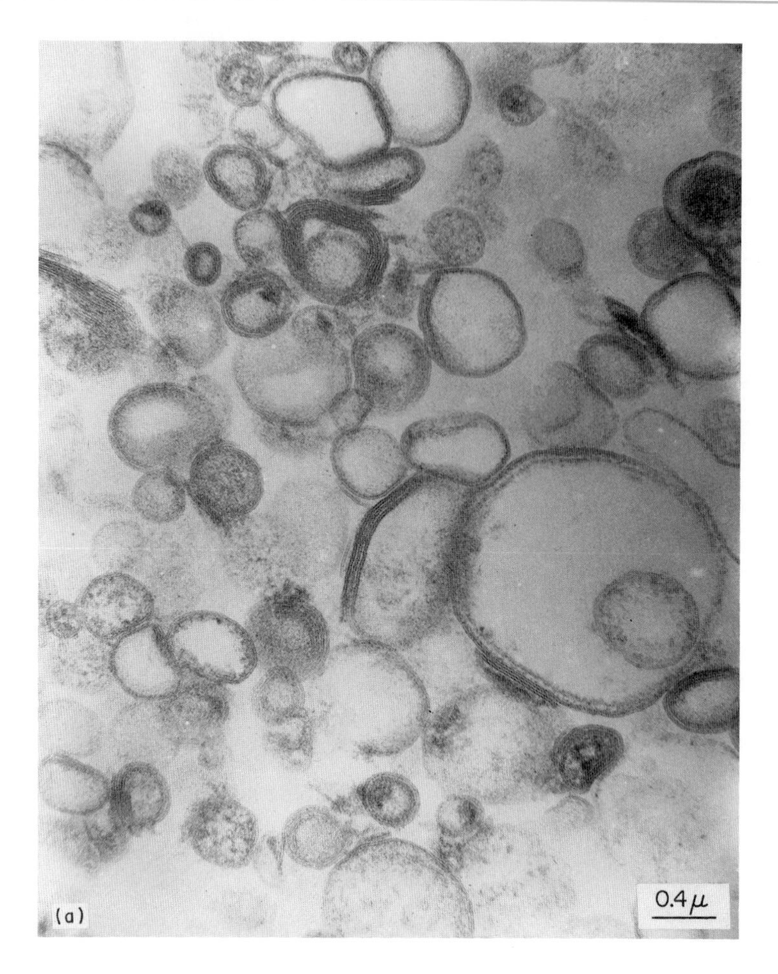

FIG. 3. Electron micrograph of membrane vesicle preparations from E. coli W. The membranes were prepared from (A) lysozyme-EDTA-induced spheroplasts and (B) penicillin-induced spheroplasts. The micrographs were taken by Dr. Vincent Marchesi of the National Institute of Arthritis and Metabolic Disease. Micrograph B is from H. R. Kaback [7, 8].

III. TRANSPORT PHENOMENA — DEFINITIONS

Before discussing experimental observations related to transport per se, several mechanisms by which substances cross cell membranes should be defined, since their clear distinction is important to the discussion.

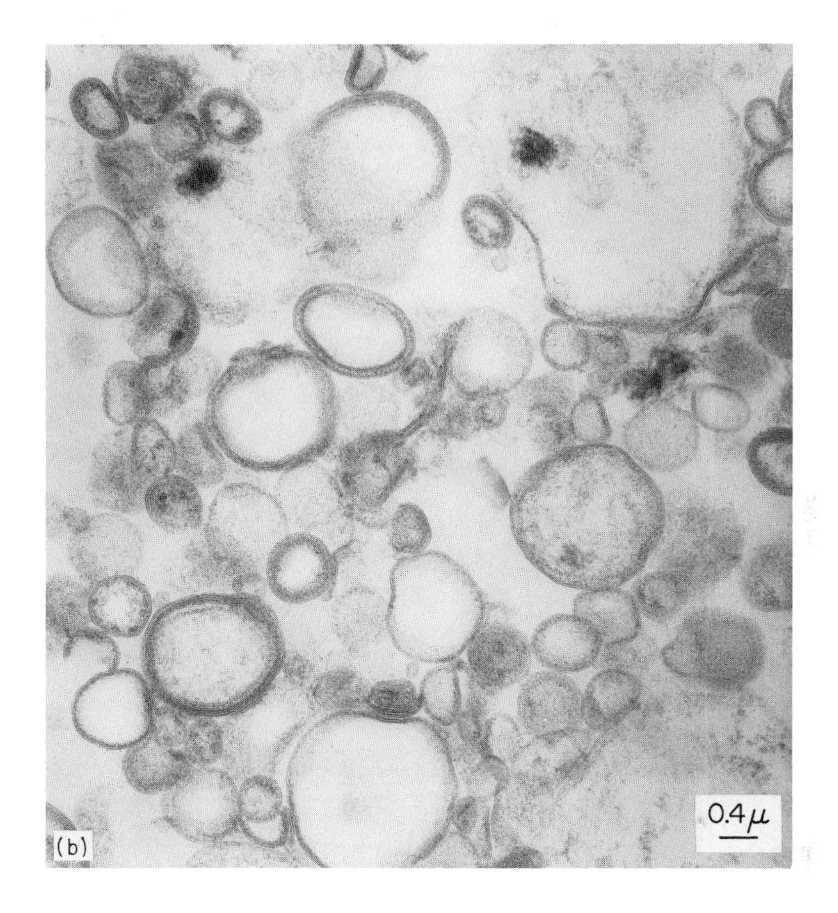

Some of these mechanisms, which are based on kinetic and thermodynamic considerations, are discussed in detail in the literature [11].

A. Passive Diffusion

By passive diffusion a substance crosses a membrane as a result of random molecular motion. The transported solute does not interact specifically with any molecular species in the membrane. Passive diffusion mechanisms may be modified by solvent drag (in which the penetrating substance is swept through aqueous pores in the membrane by bulk water flow), membrane charge, and by the degree of hydrophobicity of the diffusion barrier.

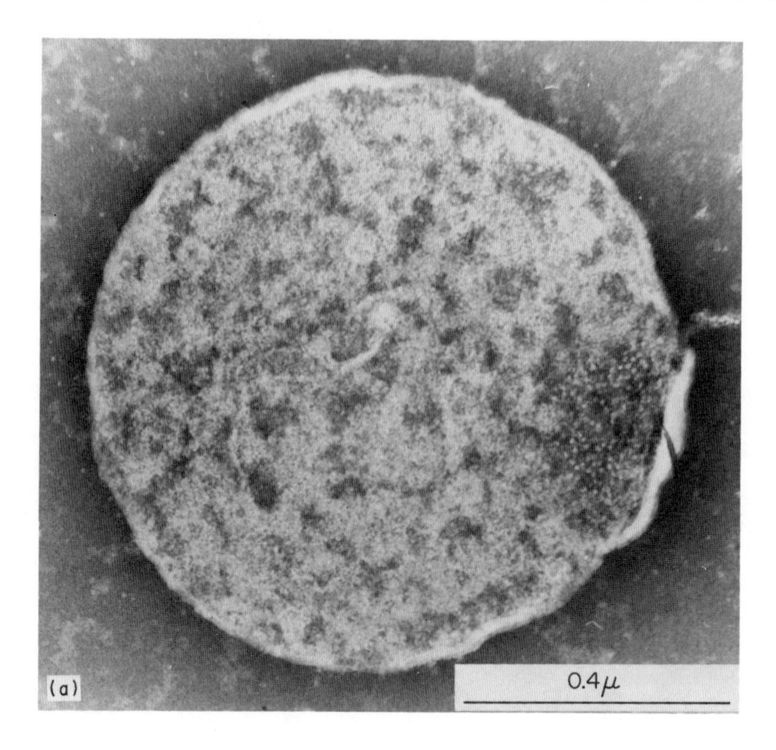

FIG. 4. Electron micrograph of E. coli ML 308-225 (A), and E. coli W (B) membrane vesicles negatively stained with phosphotungstic acid. Micrographs were taken by Dr. Vincent Marchesi of the National Institute of Arthritis and Metabolic Disease. From H. R. Kaback [7, 8].

B. Facilitated Diffusion

In facilitated diffusion the transported solute is presumed to combine reversibly with a specific "carrier" in the membrane. The carrier or carrier-substrate complex oscillates between the inner and outer surfaces of the membrane, releasing and binding molecules on either side. Because of the short distances covered, it is postulated that thermal energy or molecular deformation or both resulting from binding and release of substrate can account for the small amount of motion needed.

Neither of these mechanisms requires metabolic energy nor do they lead to concentration against a gradient.

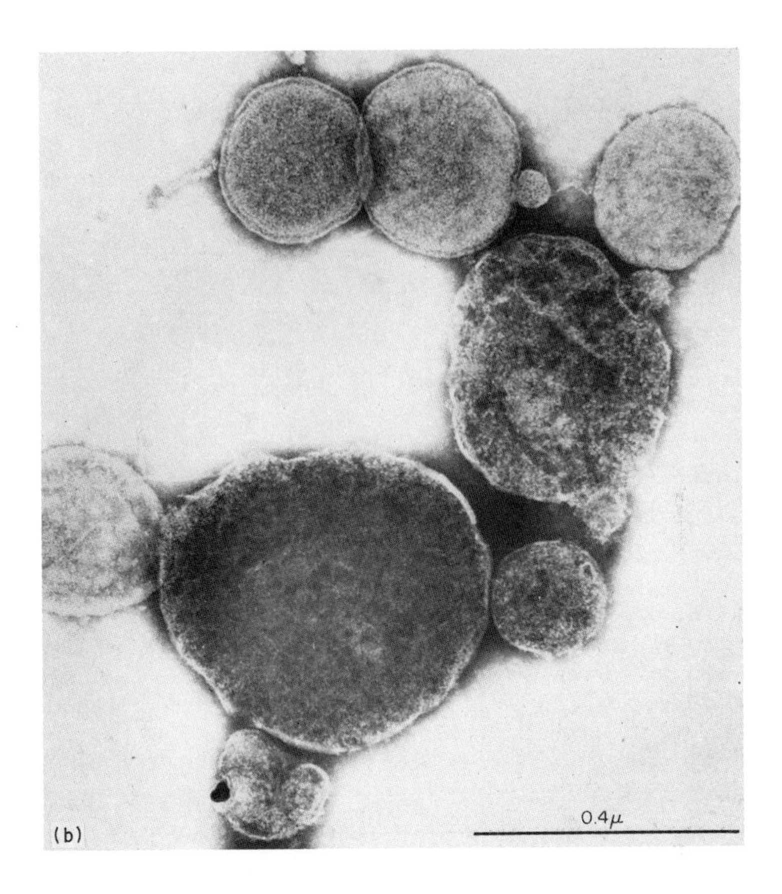

C. Active Transport

By means of active transport, solute is accumulated against an electrochemical or osmotic gradient. This mechanism requires metabolic energy on the part of the cell, as well as a specific membrane-carrier molecule. Most models for this mechanism postulate that the penetrating species combines with a carrier and that the carrier or the carrier-substrate complex is then subjected to modification in the membrane. The carrier-substrate complex formed on the outside surface of the membrane is modified in such a way that the carrier has a lower affinity for substrate. Substrate is released into the interior of the cell, the high-affinity form of the carrier is regenerated, and the cycle is repeated.

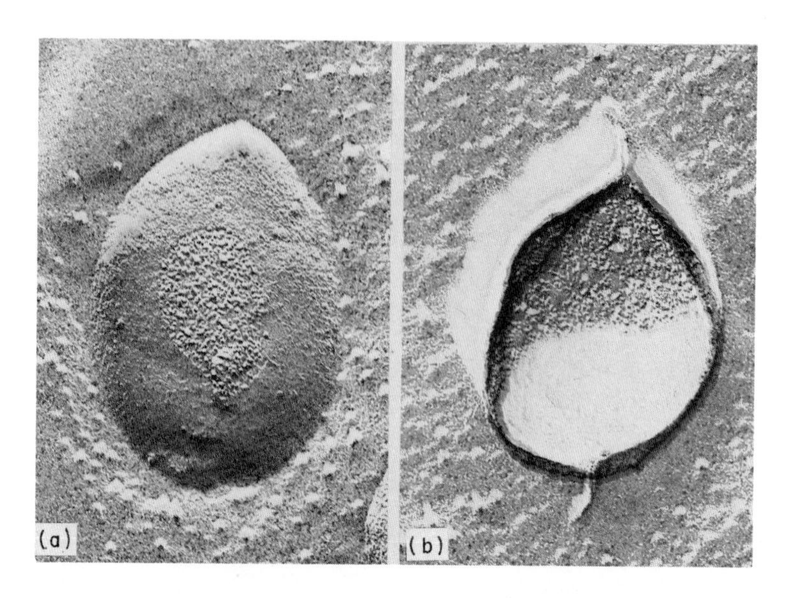

FIG. 5. Electron microscopy of freeze-etched membrane vesicles from E. coli ML 308-225. The micrographs were taken by Dr. Thomas Tillach and Dr. Vincent Marchesi of the National Institute of Arthritis and Metabolic Disease. (A) Outer surface; (B) Inner surface. Magnification about X 140,000. From H. R. Kaback [7].

D. Group Translocation

Another transport mechanism that is metabolically dependent is group translocation. In this process a covalent change is exerted upon the transported molecule such that the reaction itself results in the passage of the molecule through the diffusion barrier. Group translocation is not a "classic" active transport mechanism because the transported solute is modified chemically. One such group translocation mechanism, called vectorial phosphorylation [8, 12], leads to the translocation and accumulation of a molecule within the cell by phosphorylation during passage through the membrane.

IV. ACTIVE TRANSPORT

A. Coupling of a Membrane-Bound D-Lactic Dehydrogenase (D-LDH) to Amino Acid and Sugar Transport in E. coli Membrane Vesicles

1. Amino Acid Transport

Recent studies [13-15] have defined the energetics of amino acid transport in detail. Addition of D-lactate to membrane vesicles dramatically stimulates proline uptake with a 20- to 30-fold increase over baseline levels. Only succinate, L-lactate, D, L-α-hydroxybutyrate, and NADH replace D-lactate to any extent whatsoever, and each is much less effective than D-lactate. Furthermore, NAD in the presence of D-lactate causes no additional stimulation of proline transport.

^{14}C-D-Lactate or ^{14}C-succinate are converted stoichiometrically to pyruvate or fumarate, respectively [13-16], and neither pyruvate nor fumarate has any effect on proline transport. These results indicate that the concentrative uptake of proline involves electron transfer, and more specifically, that a membrane-bound lactic dehydrogenase with a high degree of specificity toward D-lactate is tightly coupled to proline transport.

Since the rate and extent of conversion of D-lactate to pyruvate is much greater than can be accounted for by proline transport alone, the effect of D-lactate on the transport of other amino acids was investigated. The results of these experiments [13-15] demonstrate that the conversion of D-lactate to pyruvate also markedly stimulates the initial rates of uptake and the steady-state levels of accumulation of glutamic acid, aspartic acid, tryptophan, serine, glycine, alanine, lysine, phenylalanine, tyrosine, cysteine, leucine, isoleucine, valine, and histidine. The transport of glutamine, arginine, cystine, methionine, and ornithine is stimulated only marginally. Virtually all of the radioactivity taken up by the vesicles is recovered as the unchanged amino acid [13]. Furthermore, the steady-state concentration of each of the amino acids taken up by the vesicles is many times higher than that of the medium. Primarily D-lactate and, to a much lesser extent, succinate, L-lactate, D, L-α-hydroxybutyrate, and NADH are the only energy sources that initiate the uptake of any of the amino acids. However, the relative effects of these compounds on a particular amino acid transport system vary [14]. If effective, succinate is one-third to one-half as active as D-lactate, and L-lactate only one-tenth to one-fifth as effective.

When vesicles are prepared from cells grown on either glycerol or enriched media, a membrane-bound α-glycerol-P dehydrogenase is induced which is coupled to amino acid transport about as effectively as succinic dehydrogenase [17, 18]. Similar observations have also been made for formate dehydrogenase [18].

Competition experiments indicate that there are at least nine discrete amino acid transport systems, each one of which is specific for a structurally related group of amino acids [19]. Genetic evidence in some cases corroborates the assignment of a particular group of amino acids to one transport system. The data presented in Table 1 are a summary of kinetic constants obtained for the transport of various amino acids by *E. coli* ML 308-225 membrane vesicles. The amino acids listed have been grouped according to their affinity for a common transport system as determined by competitive and/or genetic studies [19]. The observation that the transport systems for histidine and valine have two sets of kinetic constants is consistent with studies carried out with whole cells [20, 21].

2. β-Galactoside Transport

Although the β-galactoside transport system in E. coli has been examined in detail, the mechanism of coupling of metabolic energy to active galactoside transport has been a mystery. A role for ATP in lactose transport in E. coli was suggested [22]; however, studies on anaerobic thiomethylgalactoside (TMG) uptake [23] indicate that uncouplers of oxidative phosphorylation block TMG accumulation but do not alter ATP levels. Fox and Kennedy [24] demonstrated the existence of "M protein," the product of the y gene [25], but data suggesting that the P-enolpyruvate-P-transferase system catalyzed TMG uptake in E. coli [26] raised the possibility that the M protein might be an inactivated form of enzyme II (cf. Section V, A). This topic has been discussed in detail in previous reviews [8, 12].

Since much of the interest in this laboratory over the past few years was directed toward the role of the P-enolpyruvate-P-transferase system in sugar transport, and since all attempts to implicate this system in the transport of galactosides were uniformly negative, the effect of D-lactate on the uptake of β-galactosides by E. coli vesicles was investigated [16]. Conversion of D-lactate to pyruvate in membrane vesicles prepared from cells containing a functional y gene markedly stimulates the initial rate of transport of β-galactosides, and in a short time, the vesicles accumulate these sugars to concentrations at least 100-times higher than the medium. All of the galactosides accumulated are recovered as the unchanged substrates. As demonstrated for amino acid transport, only D-lactate, and to a lesser extent, D, L-α-hydroxybutyrate, succinate, L-lactate, and NADH increase lactose transport above endogenous levels. α-Glycerol-P and formate also stimulate lactose transport in membranes prepared from cells grown on appropriate media [18, 26].

Vesicles prepared from E. coli GN-2 [27], a mutant lacking enzyme I of the P-transferase system (cf. Section V, A), but constitutive for lac, transport β-galactosides in the presence of D-lactate, but are completely unable to vectorially phosphorylate α-methylglucoside (α-MG) [7, 12, 28]. D-Lactate does not stimulate α-MG uptake. Moreover, P-enolpyruvate does not stimulate lactose, isopropylthiogalactoside (IPTG), or TMG uptake, nor is lactose-P, IPTG-P, or TMG-P detected. Finally, membranes that transport β-galactosides fail to exhibit phosphatase activity toward TMG-P, and the addition of lactose to vesicles incubated with ^{32}P-enolpyruvate does not accelerate the appearance of $^{32}P_i$, as might be expected if a lactose-P P-hydrolase were involved [12]. Clearly, β-galactoside transport in E. coli does not involve the P-enolpyruvate-P-transferase system.

TABLE 1

K_m and V_{max} Values for Transport of Various Amino Acids by E. coli ML 308-225 Membrane Vesicles in the Presence of 20 mM D-lactate

L-Amino acid	K_m (μM)	V_{max} (nmole/mg protein/min)
Proline	1.0	1.3
Glutamic acid	11.0	4.0
Aspartic acid	2.9	1.1
Serine	2.6	4.0
Threonine	5.4	1.4
Glycine	1.4	0.75
Alanine	8.4	1.0
Phenylalanine	0.42	2.6
Tyrosine	0.68	3.2
Tryptophan	0.33	0.7
Histidine	0.15	0.20
	4.0	0.7
Leucine	1.1	0.25
	18.0	1.1
Isoleucine	1.7	0.25
	21.0	0.60
Valine	2.0	0.2
	29.0	1.1
Cysteine	38.0	18.5

3. Coupling of Other Sugar Transport Systems to D-LDH

The transport systems for galactose [29], arabinose, glucuronic acid, gluconic acid, and glucose-6-P are coupled to D-LDH in a manner identical to that described for β-galactosides [15, 19, 30]. Transport of these sugars by vesicles requires induction of the parent cells, does not involve the P-enolpyruvate-P-transferase system, and is inhibited by the same conditions that affect amino acid and β-galactoside transport (see below). Kinetic constants for the transport of most of these sugars by the vesicles listed are given in Table 2.

D-Galactose transport is exhibited by _lac_ y^- vesicles prepared from galactose-induced _E. coli_ ML 3 and ML 35, and _E. coli_ ML 32400 [31] and W3092cy$^-$ [32], which transport galactose constitutively. Of particular interest is the observation that "galactose binding protein" is totally absent from the vesicles, as is a high-affinity, low V_{max} galactose transport system present in the whole cells [29]. These findings, in addition to the observation that the vesicles do not transport β-methylgalactoside, indicate that the galactose-transport system retained by the vesicles is the so-called "_gal_ permease" system [33]. Moreover, it is obvious that this system does not require the galactose-binding protein for activity.

Recent experiments carried out collaboratively with Dr. Maurizio Iaccarino of the International Institute of Genetics and Biophysics in Naples indicate that a similar situation may exist for the leucine-isoleucine-valine transport system in _E. coli._ Mutants defective in this transport system [34] contain normal amounts of periplasmic binding protein for these amino acids [34a]. On the other hand, membrane vesicles prepared from one of the mutants manifest the transport defect present in the whole cells (i.e., the vesicles do not transport leucine, isoleucine, or valine in the presence of D-lactate or ascorbate-PMS) [34b]. Thus, although current evidence indicates that the periplasmic binding proteins play a role in transport (cf. Chapter IV), these experiments with the membrane vesicle system suggest that the binding proteins may not be involved in the _translocation_ of solute across the membrane.

4. Activity of Membrane Vesicles Compared to Whole Cells

Although previous data suggest that the rates of sugar and amino acid transport by membrane vesicles is 10-20% of that found in whole cells, recent experiments show that vesicles lose significant activity due to excessive manipulation during preparation. Vesicles prepared using gentle conditions (i.e., avoiding vigorous homogenization in tight-fitting homogenizers) and assayed for lactose or proline transport prior to freezing have specific activities 3 to 5 times higher than whole cells (comparing

TABLE 2

Kinetic Constants for Sugar-Transport Systems in Membrane Vesicles

Sugar	Vesicles (*E. coli*)	Electron donor	K_m (mM)	V_{max} (nmole/mg protein/min)
Lactose	ML 308-225[a]	D-Lactate	0.2	78
		ASC-PMS	0.2	240
Glucose-6-P	GN-2[a]	D-Lactate	0.25	50
Glucuronic acid	ML 30[a]	D-Lactate	0.03	83
Galactose	ML 35	D-Lactate	0.05	12
		ASC-PMS	0.05	18
Arabinose	ML 30	D-Lactate	0.14	12

[a] Membranes prepared under gentle conditions (i.e., avoiding the use of tight-fitting homogenizer) and assayed prior to freezing in liquid nitrogen. See Section IV, A.

initial rates of uptake per milligram of protein). However, quantitative comparisons between vesicles and whole cells are extremely difficult to interpret, especially when the transport activity manifested by whole cells may be the composite of more than one system (e.g., galactose and leucine, isoleucine, and valine transport).

5. Source of D-lactate in *E. coli*

Escherichia coli has two distinct D-LDH's — a soluble, pyridine nucleotide-dependent enzyme, which catalyzes the conversion of pyruvate to D-lactate [34, 35], and a membrane-bound, flavin-linked enzyme, which catalyzes the reverse reaction [37]. Apparently the soluble enzyme produces D-lactate, which may then be utilized by the membrane-bound enzyme to drive many transport systems and perhaps other cellular processes.

B. Substrate Oxidation by Membrane Vesicles

There is no relationship between the oxidase activity of vesicles toward various substrates and their ability to stimulate transport [17]. With vesicles prepared from succinate-grown cells, succinate is oxidized much faster than D-lactate and NADH is oxidized approximately as fast, yet D-lactate is markedly more effective as an electron donor for transport.

D-Lactate-dependent transport by vesicles is inhibited by anoxia. Moreover, the electron-transfer inhibitors cyanide, 2-heptyl-4-hydroxyquinoline-N-oxide (HOQNO), amytal, and the specific D-lactic dehydrogenase inhibitor oxamic acid effectively block transport [17]. Inhibition of D-lactate oxidation by these compounds is similar to the inhibition of transport [17, 19], and these inhibitors also effectively block succinate and NADH oxidation. Investigations of the respiratory chain of *E. coli* [38] have identified the amytal-sensitive site as a flavoprotein between D-LDH and cytochrome b_1. HOQNO acts between cytochrome b_1 and cytochrome a_2, perhaps at a quinone-containing component, and cyanide blocks cytochrome a_2. Thus, each of the dehydrogenases studied is coupled to oxygen via a membrane-bound respiratory chain.

The D-lactate-dependent transport systems are not significantly inhibited by high concentrations of arsenate or oligomycin [23-25]. Moreover, vesicles have been prepared with arsenate buffer rather than phosphate buffer throughout the procedure, and subsequently transport the assayed in arsenate buffer in the absence of inorganic phosphate. Little or no inhibition of transport was observed. It is apparent from these studies that the effect of D-lactate is not mediated by the production of stable high-energy phosphate compounds. This conclusion is supported by many observations, among which are the absence of transport in the presence of ATP under conditions in which ATP is demonstrably accessible to reactive sites within the membrane [39], the failure of ADP or other nucleoside-diphosphates to stimulate transport in the presence of D-lactate or other electron donors, and the observation that the membrane preparations do not carry out oxidative phosphorylation [40, 62]. In addition, the ATP content of vesicles is below the limits of the luciferin-luciferase assay and does not increase when vesicles are incubated with electron donors [41]. Vesicles prepared from mutants that are "uncoupled" for oxidative phosphorylation [42] transport a variety of substrates normally in the presence of D-lactate or ascorbate-PMS [42a]. Finally, proline transport in *M. phlei* respiratory particles occurs in the absence of oxidative phosphorylation [43].

Transport and D-lactate oxidation by the vesicles are inhibited by the sulfhydryl reagents N-ethylmaleimide (NEM) and *p*-chloromercuribenzoate (PCMB) [16, 17, 30]. These reagents are discussed in greater detail subsequently.

Dinitrophenol (DNP), carbonyl cyanide m-chlorophenylhydrazone (CCCP), and azide do not significantly affect D-lactate oxidation, despite profound inhibition of transport [15-17]. This finding is not surprising since most bacterial electron transfer systems are not subject to respiratory control.

Taken as a whole, the observations indicate that the specificity of the transport systems for D-LDH cannot be accounted for solely on the basis of its presence in the vesicles (to the exclusion of other dehydrogenases), and furthermore, that the coupling of D-LDH to transport involves the flow of electrons through a respiratory chain to oxygen as the terminal electron acceptor.

C. Site of Energy Coupling between D-LDH and Transport

Difference spectra between D-lactate-, succinate-, NADH-, L-lactate-, or dithionite-reduced samples and oxidized samples are indistinguishable [15, 17]. Furthermore, difference spectra between D-lactate-reduced and NADH-, succinate-, L-lactate- and dithionite-reduced samples show no absorption bands. Since the rate of reduction of cytochrome b_1 by these substrates is directly proportional to their rates of oxidation, each dehydrogenase must be coupled to the same cytochrome chain. Thus the site of energy-coupling between D-LDH and transport must lie between the primary dehydrogenase and cytochrome b_1, the first cytochrome in the *E. coli* respiratory chain.

In vesicles that oxidize succinate more rapidly than D-lactate, the addition of increasing concentrations of succinate in the presence of saturating concentrations of D-lactate results in progressive inhibition of proline uptake. Under the same experimental conditions, succinate inhibits the conversion of ^{14}C-D-lactate to pyruvate. On the other hand, succinate does not inhibit the isolated, partially purified D-LDH [17] nor the membrane-bound enzyme when dichloroindophenol (DCI, an artificial electron acceptor that accepts electrons directly from flavins) is used rather than oxygen as an electron acceptor. A reasonable mechanism for this inhibition is that succinic dehydrogenase, when it is more active than D-LDH, is able to saturate cytochrome b_1 kinetically, and thus inhibit D-LDH. In order for transport to be inhibited by this means, the site of energy coupling must be proximal to cytochrome b_1.

Direct evidence for the location of the energy-coupling site is provided by experiments in which the effects of NEM and PCMB on D-lactate and NADH oxidation were investigated [15, 17]. Both of these sulfhydryl reagents markedly inhibit D-lactate oxidation at concentrations that block transport. Moreover, PCMB inhibition is reversed by dithiothreitol, providing further evidence for sulfhydryl involvement in D-lactate oxidation. The effect of these thiol reagents on D-lactate oxidation is not expressed at the level of the primary dehydrogenase. Neither the D-lactate:DCI reductase activity of the intact vesicles nor that of the solubilized, partially purified preparation of this enzyme is sensitive to NEM or PCMB. NADH oxidation is also insensitive to NEM or PCMB. Thus neither the primary dehydrogenase nor the cytochrome chain contains a reactive sulfhydryl group, and the site(s) of inhibition of D-lactate oxidation by NEM and PCMB must lie between D-LDH and the cytochromes.

An abbreviated schematic representation of the sequence of events thought to occur is presented in Fig. 6. Electrons from D-lactate flow through the "carriers" or something closely aligned with the "carriers" before entering the cytochrome chain, after which they ultimately reduce oxygen to water. Each of the electron-transfer inhibitors shown inhibits solute transport by interrupting the flow of electrons before (oxamate), at (PCMB, NEM), or after (amytal, HOQNO, cyanide, or anoxia) the site of energy coupling.

D. Mechanism of Energy Coupling of D-LDH to Transport

The findings presented above, especially when considered in conjunction with those to be discussed, indicate that the "carrier" components of these transport systems reflect the redox state of the respiratory chain between D-LDH and cytochrome b_1.

D-LDH activity and the initial rates of sugar and amino acid transport respond identically to temperature, and both phenomena have the same activation energy of 8400 cal/mole [30]. A similar activation energy for transport is also obtained with an artificial electron donor system [43a] — ascorbate-phenazine methosulfate — which reduces the respiratory chain above D-LDH (cf. Section IV, E). On the other hand, the D-lactate:DCI reductase activity of the vesicles exhibits a markedly different temperature profile and has an activation energy of approximately 30,000 cal/mole [43a]. Thus some component of the membrane that is common to both the D-lactate oxidation and transport apparently determines the activation energy for D-lactate oxidation.

Evidence has been presented [30] that demonstrates that the steady-state levels of solute accumulation at temperatures ranging from 0° to 53° C represent equilibrium states in which there is a balance between influx and efflux. Moreover, efflux induced at 45° C is a saturable, "carrier"-mediated phenomenon with a much lower affinity (i.e., higher K_m) than the influx system, but a very similar maximum velocity [30]. Anoxia, cyanide, HOQNO, CCCP, and DNP also induce the efflux of solute and the kinetics of cyanide-induced efflux manifest the same apparent K_m

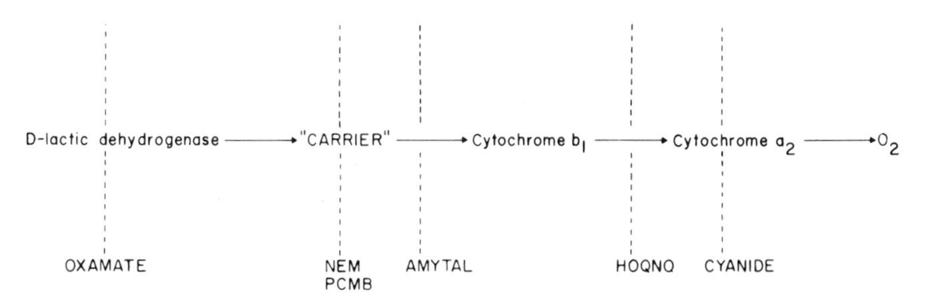

FIG. 6. Electron-transfer pathway from D-lactic dehydrogenase to oxygen showing sites of inhibition of electron-transfer inhibitors.

as temperature-induced efflux and the same maximum velocity as the influx process. Thus efflux occurs at rates that are comparable to those of influx, but much higher concentrations of substrate are required to saturate the "carrier" on the inside of the membrane. Furthermore the rate of efflux responds to temperature in a manner that is essentially the inverse of the response of the steady-state level of lactose accumulation to temperature [30].

The β-galactoside transport system in E. coli is inhibited by sulfhydryl reagents [44-46], and the ability of two substrates of this system to protect against sulfhydryl inactivation led to the identification and purification of M protein, the product of the y gene [24, 25, 47, 48]. However, very little evidence has been presented that has any bearing on the mechanistic role of sulfhydryl groups in galactoside transport. Evidence discussed above demonstrates that D-lactate-induced respiration is inhibited by PCMB and NEM, and that the site(s) of action of these compounds is between D-LDH and cytochrome b_1, i.e., at the site of energy coupling. Virtually every D-LDH-coupled transport system is inhibited by PCMB and NEM, and PCMB inhibition is reversed by dithiothreitol [30]. Moreover, all "carrier"-mediated aspects of transport, even those that are independent of D-lactate oxidation, are inhibited by PCMB and NEM; and in each case, the inhibition by PCMB is reversed by dithiothreitol [30]. Temperature-induced efflux, exchange of external lactose with ^{14}C-lactose present in the intramembranal pool, and efflux induced by the addition of 2,4-dinitrophenol or cyanide are all blocked by NEM and PCMB, and the inhibition by PCMB is completely reversed with dithiothreitol. Thus D-lactate oxidation and "carrier" activity are both dependent upon functional sulfhydryl groups, which are located in a specific region of the respiratory chain.

Electron-transfer inhibitors whose sites of action are well documented (cf. Fig. 6) were studied with respect to their effect on the ability of the vesicles to retain accumulated solute [30]. Each inhibitor used blocks D-lactate oxidation and the initial rate of influx by at least 70-80% [15, 17, 18].

Only anoxia and those inhibitors which block electron transfer after the site of energy-coupling (i.e., anoxia, cyanide, HOQNO, and amytal) cause efflux. Thus reduction of the electron-transfer chain between D-LDH and cytochrome b_1 is responsible for efflux. Strikingly, oxamate, which inhibits electron transfer before the site of energy coupling, does not induce efflux, nor does PCMB or NEM. Since oxamate (and PCMB or NEM) prevent reduction of the respiratory chain, and anoxia, cyanide, HOQNO, and amytal cause reduction of a segment of the respiratory chain between D-LDH and cytochrome b_1 (cf. Fig. 6), the rate of efflux must be determined by the redox state of the respiratory chain at the site of energy coupling.

These experimental findings are consistent with the conceptual model presented in Fig. 7. Simplistically the "carriers" are depicted as obligatory electron-transfer intermediates between D-LDH and cytochrome b_1. In the oxidized state, the "carrier" has a high-affinity site for ligand which it binds on the exterior surface of the membrane. Electrons coming ultimately from D-lactate through one or possibly more flavoproteins reduce a critical disulfide in the "carrier" resulting in a conformational change. Concomitant with this conformational change, the affinity of the "carrier" for ligand is markedly reduced, and ligand is released on the interior surface of the membrane. The reduced "sulfhydryl" form of the "carrier" is then oxidized by cytochrome b_1 and the electrons flow through the cytochrome chain to reduce molecular oxygen to water. The reduced form of the "carrier" can also "vibrate" and catalyze a low-affinity, "carrier"-mediated, nonenergy-dependent transport of ligand across the membrane.

The effect of anoxia and electron-transfer inhibitors on the time course of uptake is consistent with this conceptual model [30]. Since the removal of oxygen or the presence of electron-transfer inhibitors, which inhibit after the site of energy-coupling, cause reduction of the energy-coupling site (cf. Fig. 6), membranes incubated under these conditions exhibit profound inhibition of uptake throughout the time course of the experiment. On the other hand, inhibitors that work before or at the site of energy coupling prevent reduction of the energy-coupling site (cf. Fig. 6), and vesicles incubated under these conditions exhibit markedly diminished initial rates of uptake but eventually accumulate significant quantities of solute.

The proposed mechanism implies that there are functionally heterogenous intermediates between different D-LDH molecules and cytochrome b_1. For each transport system, there should be an intermediate with a binding site that is specific for a particular transport substrate. Supportive evidence for this prediction is provided by studies of lactose transport in the presence of structurally unrelated substrates that are also transported

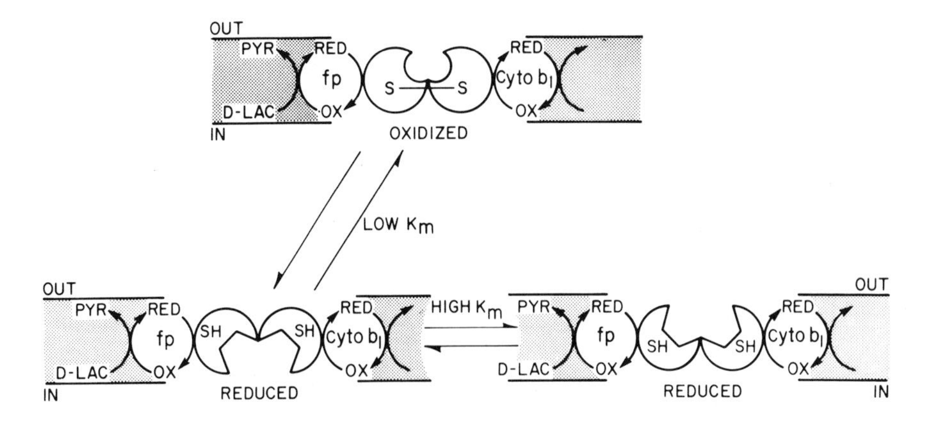

FIG. 7. Conceptual working model for D-lactic dehydrogenase-coupled transport systems. D-LAC, D-lactate; PYR, pyruvate; fp, flavoprotein; cyto b_1 , cytochrome b_1; ox, oxidized; red, reduced. OUT signifies the outside surface of the membrane; IN signifies the inside surface. The spheres located between fp and cyto b_1 represent the "carrier"; ◡◡ , a high-affinity binding site and ╲⁄╲╱, a low-affinity binding site. The remainder of the cytochrome chain from cytochrome b_1 to oxygen has been omitted. From H. R. Kaback and E. M. Barnes, Jr. [47].

by D-LDH-coupled systems [30]. Little or no inhibition of either the initial rate of lactose uptake or the steady-state level of lactose accumulation is observed. Moreover the sum of the V_{max} values of the known D-LDH-coupled transport systems in a particular membrane preparation is equal to or less than the V_{max} of D-LDH in the same membrane preparation.

Although no direct evidence has been presented which demonstrates that the "carriers" undergo oxidation-reduction and it is almost certain that the "carriers" are not direct electron-transfer intermediates in the respiratory chain, this formulation is consistent with much of the experimental data currently available, and is certainly the simplest conception possible. D-lactate oxidation and the D-lactate-dependent and -independent aspects of transport are inhibited by PCMB and NEM. Furthermore PCMB inhibition is reversed by dithiothreitol. Since the site of energy-coupling lies between the primary dehydrogenase and cytochrome b_1, and the site(s) of inhibition by sulfhydryl reagents is also between D-LDH and cytochrome b_1, the proposed mechanism is supported by more than simplicity of conception. However, the model does not account for the action of "uncoupling agents" (i.e., DNP, CCCP) or azide on the system. These reagents abolish transport without inhibition of D-lactate oxidation [14-17], indicating that the "carriers" are not direct electron-transfer intermediates in the respiratory chain.

The observations and the conceptual model to which they have led do not conflict with studies in whole cells [49-51]. Furthermore, the proposed concept would explain some apparent inconsistencies. Only TDG and melibiose protect galactoside transport against inhibition by NEM [47]. Since many other galactoside analogs are transported via this system, it was proposed that M protein has two binding sites and that TDG and melibiose bind to one site only. No evidence has been presented to substantiate this suggestion. According to the concept proposed here, M protein has two sites, but one site is involved in oxidation-reduction and the other in binding substrate. Since presumably all galactosides bind to M protein by virtue of their galactose moieties, it does not seem unlikely that TDG and melibiose, due to their size (TDG) or shape (melibiose is an α-galactoside), sterically protect a sulfhydryl group that is not in the binding site.

The proposed mechanism could also explain so-called "energy-uncoupled" galactoside-transport mutants [52, 53]. These mutants have increased amounts of M protein, but do not catalyze the concentrative uptake of galactosides as well as the parent. Uptake of TMG by membrane vesicles prepared from one of these mutants [52] is stimulated by D-lactate, but only about one third as well as the parent preparation. D-lactate-dependent transport of proline and P-transferase-mediated uptake of α-methylglucoside, on the other hand, are identical [53a]. Possibly the defect in these mutants is related to redox properties of M protein.

E. Coupling of Ascorbate-Phenazine Methosulfate (PMS) to Transport

Konings, Barnes, and Kaback [54] demonstrated that an artificial electron donor system --ascorbate-PMS-- can be coupled to transport in vesicle preparations. The initial rate of lactose transport in *E. coli* membrane vesicles is stimulated about three times more effectively by ascorbate-PMS than D-lactate, the best physiological electron donor. However, comparative stimulation of various transport systems by ascorbate-PMS or D-lactate differs. The significance of this observation is discussed below.

The effect of ascorbate-PMS on transport is inhibited by anoxia, and by cyanide, HOQNO, PCMB, NEM, and amytal in a manner identical to that described for D-lactate. Significantly, oxamic acid has no effect, since this inhibitor acts at the level of D-LDH. Thus ascorbate-PMS reduces the respiratory chain of the vesicles at a redox level below that of cytochrome b_1. This conclusion is substantiated by spectrophotometric data [54] that demonstrate that ascorbate-PMS reduces only 21% of the membrane-bound flavoprotein, whereas D-lactate reduces 63%. These results place the site of energy-coupling within a portion of the respiratory chain above 80% of the membrane-bound flavoprotein but below cytochrome b_1. Moreover, since PCMB and NEM inhibit transport in the presence of ascorbate-PMS, the postulated sulfhydryl component(s) of the respiratory chain, which is(are) common to both transport and D-lactate oxidation, must lie within the segment of the respiratory chain reduced by ascorbate-PMS.

Amytal inhibits the effect of ascorbate-PMS on transport, which suggests that there are flavoproteins between the "carriers" and cytochrome b_1 or possibly that the "carriers" are flavoproteins themselves. No evidence has been presented to resolve these possibilities. Since the methods used in these experiments do not distinguish between flavoprotein and non-heme iron [38], it is also possible that non-heme iron protein is present in this segment of the respiratory chain. The iron chelator enterochelin [55] inhibits lactose transport with D-lactate <u>or</u> ascorbate-PMS as electron donors. Moreover, enterochelin does not cause efflux of solute accumulated in the intramembranal pool. Since Bragg [56] has shown that ascorbate-PMS reduces nonheme iron in respiratory particles from <u>E. coli</u>, this component must be near the site of ascorbate-PMS reduction either before or at the site of energy coupling.

Regarding the proposed mechanism (cf. Fig. 7), it is interesting that the relative effects of ascorbate-PMS or D-lactate differ depending upon the transport system studied. Preliminary observations indicate that although most transport systems are stimulated more by ascorbate-PMS than by D-lactate, some are stimulated more by D-lactate, and others are stimulated about equally by both. Possibly there are small differences in the redox potentials of various energy-coupling sites such that a few are not completely reduced by ascorbate-PMS.

F. A Kinetic Model for the Redox Transport Mechanism

The conceptual model presented in Fig. 7 may be expressed kinetically [19] and Dr. F. J. Lombardi has derived the following differential equation to describe the proposed mechanism.

$$v_{net} = v_{in} - v_{out}$$

$$v_{net} = \left[\frac{V^{\uparrow}_{max} S^{o}}{K^{\uparrow}_{m} + S^{o}}\right](1-\alpha) - \left[\frac{\dfrac{V^{eff}_{max}}{K^{i}_{m}}(S^{i} - S^{o})}{1 + \dfrac{S^{o}}{K^{o}_{m}} + \dfrac{S^{i}}{K^{i}_{m}}}\right]\alpha$$

where v = velocity; $V^{\uparrow}_{max}$ = maximum velocity of energy-dependent active uptake; $K^{\uparrow}_{m}$ = apparent Michaelis constant for energy-dependent active uptake; S^{o} = external substrate concentration; α = fraction of carriers in the reduced form under steady-state conditions; V^{eff}_{max} = maximum velocity of efflux; K^{i}_{m} = apparent Michaelis constant for efflux; S^{i} = internal substrate concentration; and K^{o}_{max} = apparent Michaelis constant for energy-independent efflux.

Each term in this equation can be measured experimentally. With the exception of the term α, the determination of the other terms in the equation should be obvious from the previous discussion. The term α is measured in the following manner. The rate of efflux at the steady state is a function of the amount of "reduced carrier." Thus, vesicles are incubated with a radioactive transport substrate in the presence of D-lactate. After substrate has been accumulated to steady-state levels, an excess of nonradioactive substrate is added, and the initial rate of efflux is measured. In this manner, the specific activity of the external transport substrate is markedly decreased, and unidirectional efflux is observed. Expressing the above equation in the form v=[f] $(1-\alpha)$-[g] α, the rate of efflux under these conditions will be given by [g]. The same experiment is then carried out but cyanide is added in addition to nonradioactive substrate in order to reduce the respiratory chain, and convert all of the "carrier" to the "reduced form." Under these conditions, α will be equal to 1, and according to the abbreviated equation written above, the rate of efflux is given by [g]. The ratio of these rates is equal to α.

Using data from experiments carried out at various temperatures with the equation shown, a computer was programmed to generate uptake curves as a function of time. The computer-generated curves are indistinguishable from experimental data [30]. Although this type of treatment does not discriminate between mechanisms in which the "carrier" is or is not an electron transfer intermediate, it provides strong evidence for the general type of redox mechanism presented.

G. Solubilization and Partial Purification of "Carriers"

Recently Dr. A. S. Gordon solubilized and partially purified a fraction from vesicles which apparently contains many of the amino acid "carriers" [57]. Membrane vesicles are extracted with non-ionic detergents, and the extract is subjected to gel filtration in the presence of detergent. Three 280-nm absorbance peaks are eluted from Sephadex G-100 columns: (1) excluded protein (peak I); (2) included protein of relatively high molecular weight (peak II); and (3) included protein of low molecular weight (peak III).

Material in each of the three protein-containing fractions binds proline. However, peak III exhibits a much higher specific activity than the two other fractions. Proline is not altered chemically by each fraction nor by the unfractionated extract, no detectable D-LDH activity is associated with peak III, and succinic dehydrogenase and NADH dehydrogenase activities are not detected in the column effluent.

The uv absorption spectrum of peak III indicates that this fraction is composed predominantly of protein. Moreover, when extracts from ^{32}P-labeled membranes are chromatographed, no significant radioactivity is recovered in peak III. Thus very little, if any, phospholipid is associated with this fraction.

The proline-binding activity of peak III is highly specific. Only proline itself inhibits the binding of the radioactive amino acid. In contrast, binding is not affected by a mixture of structurally unrelated amino acids.

Proline binding by peak III is inhibited by NEM and PCMB, and PCMB inhibition is reversed by dithiothreitol. These results are analogous to the behavior of the transport system [30, 57]. In contrast to inhibition of binding by sulfhydryl reagents, electron transfer, and general metabolic inhibitors such as amytal, HOQNO, CCCP, and DNP do not inhibit.

Peak III also exhibits binding activity for lysine, serine, tyrosine, and glycine. These amino acids, like proline, are transported by systems with low apparent K_m's suggesting that the "carriers" have high affinities. Binding activity for each of these amino acids is inhibited reversibly by PCMB; moreover, binding of these amino acids is not altered by the presence of structurally unrelated amino acids. Binding of amino acids other than those mentioned has not been tested.

These solubilized membrane components are distinctly different from the "binding proteins" isolated by cold osmotic shock [58-60]. The latter are water-soluble and localized in the periplasmic space, whereas these components are associated with cytoplasmic membrane and are soluble in detergent only. In addition, as discussed previously, the vesicles contain no galactose-binding protein. Finally, the periplasmic binding proteins are insensitive to sulfhydryl reagents and contain no cysteine.

Although the "carrier" components of other D-LDH-coupled transport systems may also be present in peak III, many of them would be difficult to detect by binding assays. It is unlikely, for instance, that M protein could be detected, since the β-galactoside transport system has a relatively high K_m [17, 19, 30].

H. General Importance of Dehydrogenase-Coupled Transport

The use of ascorbate-PMS has extended the study of respiration-coupled transport to membrane vesicles prepared from a wide variety of bacteria [54]. This artificial system markedly stimulates the transport of amino acids by membrane vesicles prepared from E. coli, S. typhimurium, Ps. putida, P. mirabilis, B. megaterium, B. subtilis, M. denitrificans, and S. aureus.

Although the physiological electron donor(s) for the transport systems in many organisms is(are) not known at present, the following observations are significant.

1. Short et al. [41] have demonstrated that the transport of 16 amino acids by membrane vesicles prepared from S. aureus is coupled to a membrane-bound α-glycerol-P dehydrogenase.

2. Konings and Freese have shown that the concentrative uptake of a variety of amino acids, in addition to L-serine [61], by B. subtilis membranes is coupled primarily to α-glycerol-P and NADH-dehydrogenase, and also, to some extent, to L-lactic dehydrogenase [62].

3. Membrane vesicles from M. denitrificans accumulate glycine, alanine, glutamine, and asparagine in the presence of D-lactate [62a].

4. Barnes [62b] has recently observed that glucose transport by membrane vesicles prepared from the obligate aerobe Azotobacter vinlandii is markedly stimulated by ascorbate-PMS and L-malate (in the presence of flavin adenine dinucleotide).

These and other observations demonstrate that active transport systems that are basically similar to the D-LDH-coupled systems in E. coli are present in a variety of other organisms. The coupling of particular dehydrogenases to transport may be important with regard to the ecology of various bacterial species.

I. Valinomycin-Induced Rubidium Transport

The cyclic depsispeptide antibiotic valinomycin [63, 64] greatly increases membrane permeability, specifically to potassium, rubidium, and cesium [65, 67]. Membranes altered in permeability by this ionophore include bacterial [65, 68, 69], erythrocyte [66, 70, 71], mitochondrial [72], and artificial black lipid films [67]. Studies with black lipid films indicate that valinomycin forms one-to-one complexes with potassium ions [71] which then diffuse across the film. By this means; potassium is sheltered from the hydrophobic interior of the film within the cyclic valinomycin molecule [64], the exterior of which is soluble in the matrix of the film. Valinomycin facilitates the net movement of potassium in mitochondria [72], which has led to the hypothesis that this ionophore may be a prototype for natural potassium carriers.

Recently, Bhattacharyya, Epstein, and Silver [73] reported that addition of valinomycin to E. coli membrane vesicles results in the accumulation of K^+ or Rb^+ by a temperature- and energy-dependent process. Moreover, these workers demonstrated that vesicles prepared from mutants altered in K^+ transport show defects manifested by the intact cells. Since the vesicles establish what appears to be a proton gradient [74], it seemed likely that the extrusion of protons might be the driving force for this transport system and possibly others.

Dr. F. J. Lombardi of this laboratory and Dr. John P. Reeves of Rutgers University have shown that valinomycin-induced Rb^+ uptake is analogous in nearly all respects to the respiration-linked transport of sugars and amino acids. D-Lactate and ascorbate-PMS are the most effective electron donors in E. coli and M. denitrificans vesicles, whereas α-glycerol-P and ascorbate-PMS are most effective in membranes from S. aureus. In E. coli vesicles, 2 mole of Rb^+/mole of D-lactate oxidized, are transported and both D-lactate-dependent Rb^+ uptake and D-lactate oxidation are blocked by anoxia, oxamate, amytal, HOQNO, and cyanide. Studies of oxygen utilization and spectrophotometric evidence indicate that the energy-coupling site for Rb^+ transport in E. coli membranes is located between D-LDH and cytochrome b_1. Agents that block electron transport beyond this site (anoxia, amytal, HOQNO, and cyanide) cause rapid efflux of accumulated Rb^+, whereas oxamate produces no efflux. These results provide strong evidence that valinomycin does not simply enhance the passive flux of Rb^+ across the membrane, since the vesicles maintain a large Rb^+ concentration gradient in the presence of oxamate despite inhibition of D-lactate oxidation and Rb^+ influx.

J. Proton and Potential Gradients

Chemiosmotic coupling has been suggested by Mitchell [75-77] as a mechanism for oxidative phosphorylation in mitochondria, and Harold [78] and West [79] have applied this theory to active transport in bacteria. By this means it is postulated that the active efflux of protons or sodium ions results in a transmembrane potential which is positive outside, negative inside. Under these conditions valinomycin would facilitate the passive movement of K^+ through the membrane to neutralize the electrical charge across the membrane (Fig. 8, electrogenic efflux). Alternatively, as shown on the right in Fig. 8 (i.e., electrogenic influx), valinomycin could induce active K^+ uptake resulting in a transmembrane potential negative outside, positive inside, and by this means, cause the passive efflux of protons or sodium with the electrical gradient established and possibly against their own chemical gradient. The following evidence indicates that the latter alternative is probably the case for valinomycin-induced K^+ transport in this system, and that proton or potential gradients are not the primary driving force for the D-LDH-coupled transport systems.

1. The transport of each ionic species in the reaction mixtures was measured in the presence of D-lactate or ascorbate-PMS, and the vesicles do not accumulate magnesium, sulfate, phosphate, potassium (in the absence of valinomycin), sodium, chloride, or pyruvate.

2. Although the pH of lightly buffered membrane suspensions decreases on addition of D-lactate and other electron donors under certain conditions [74], a proton gradient is probably not established under most conditions. Thus, membrane vesicles treated with phospholipase A-B [81] such that they retain the catalytic activities associated with transport (i.e., P-enolpyruvate-dependent phosphorylation of α-methylglucoside and D-lactate oxidation), but are unable to retain transported solute [8, 12, 14, 15, 19] exhibit similar pH effects. Moreover, so-called "proton conductors" such as DNP and CCCP have the same effect on normal and phospholipase-treated vesicles. Since these preparations are devoid of a diffusion barrier, it is highly unlikely that the observed pH changes can be the result of proton gradients.

3. Vesicles containing standard concentrations of Na^+ or K^+ (i.e., 50 mM) incubated in the presence of D-lactate exhibit no change in the rate or extent of acidification on addition of valinomycin and K^+ or valinomycin, respectively.

4. The addition of transport substrates (i.e., lactose plus a mixture of amino acids) has no effect on the rate or absolute amount of acidification.

5. Under conditions in which the vesicles catalyze active transport (i.e., in 50 mM potassium, sodium, or choline phosphate buffers), they do not take up the lipid-soluble weak acid 5,5-dimethyloxazolidine-2,4-dione (DMO) in the presence of D-lactate or other electron donors. DMO is a weak acid, metabolically inert, which diffuses passively across many

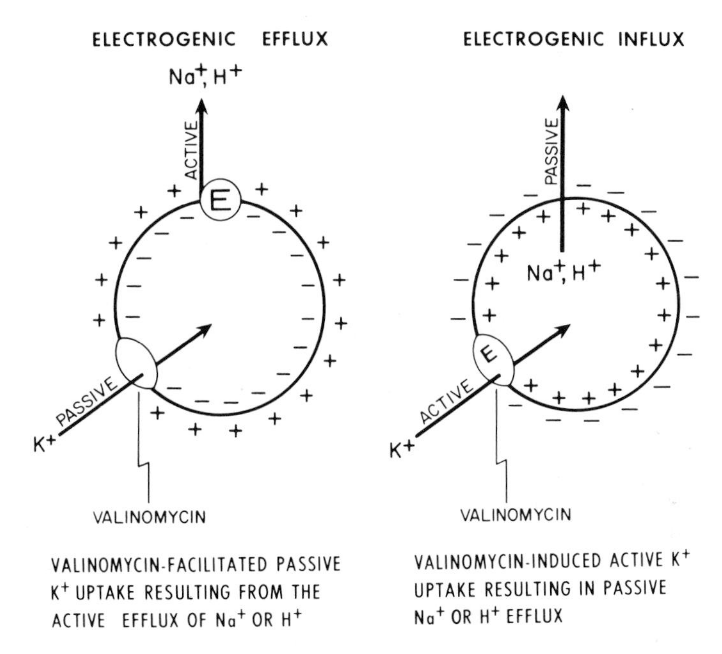

FIG. 8. Possible mechanisms for valinomycin-induced active potassium transport.

biological membranes. The permeability coefficient of the uncharged acid is generally much greater than that of the anion. The distribution of DMO is therefore a function of pH [78, 80], and these experiments indicate that the intravesicular pH is not alkaline with respect to the medium.

6. When vesicles are loaded with ^{22}Na, addition of valinomycin, D-lactate, and RbCl or KCl results in the movement of ^{22}Na out of the intravesicular space against its own concentration gradient. ^{22}Na efflux is obliterated by the omission of any one of these components. Moreover, D-lactate alone, or D-lactate plus a mixture of lactose and amino acids does not induce ^{22}Na efflux, nor does acidification of the external medium. Finally, addition of D-lactate, in addition to tetraphenylarsonium chloride, a lipophilic cation, has no effect on ^{22}Na movements.

These results taken as a whole indicate that the extrusion of Na^+ or protons from the vesicles is the result of (rather than the cause of) the active accumulation of Rb^+ or K^+.

K. ANS Fluorescence Studies

1-Anilino-8-naphthalene sulfonic acid (ANS) is an anion that becomes highly fluorescent when it interacts with certain proteins and membranes [82-88]. The behavior of this fluorescent probe has been postulated to reflect membrane potential changes in mitochondria [89]. When D-lactate is added to membrane vesicles in the presence of ANS, there is a rapid decrease in fluorescence that is maintained until the reaction mixture becomes anaerobic at which time there is a large increase in fluorescence. The rate at which the fluorescence decrease progresses is most marked with D-lactate, and slower with succinate, L-lactate, or NADH. Addition of electron-transfer inhibitors prior to D-lactate inhibits the initial decrease in fluorescence. If added subsequent to D-lactate, however, the inhibitors that act after the site of energy coupling (i.e., cyanide, HOQNO, and amytal) reverse the initial decrease in fluorescence, whereas oxamate, NEM, and PCMB, which inhibit before or at the site of energy-coupling cause little or no change in fluorescence. These effects are obviously analogous to certain aspects of the transport systems. As demonstrated with the pH measurements, the behavior of phospholipase-treated membranes is similar to that of untreated membranes with regard to each of the fluorescent studies mentioned. Thus, it is unlikely that membrane potentials are the primary driving force for transport.

More definitive evidence that the fluorescence changes described are due to structural changes in the membrane has recently been obtained from preliminary studies of ANS fluorescence in the energy-transfer mode. If a fluorescent probe like ANS can absorb energy from another fluorescing species in its environment, the ANS molecule becomes excited and emits photons. This phenomenon is called energy transfer [90], and by this means, information can be obtained with regard to the primary fluorescent source. In this case, the tryptophan residues in the membrane proteins are excited by light at 292 nm, and the fluorescence emission of ANS is recorded at 480 nm. Since tryptophan emission and ANS absorption spectra overlap, it is possible to study ANS in an energy-transfer mode. It has been demonstrated directly in this system that the ANS fluorescence is due to absorption from tryptophan by the decrease in tryptophan emission. Moreover, the ratio of the 290-370 nm <u>excitation</u> maxima of ANS is increased when energy transfer is the mode of excitation. In any case, the effects of various electron donors and electron-transfer inhibitors on ANS fluorescence in the energy-transfer mode are qualitatively similar to the findings described above, and phospholipase-treated membrane vesicles exhibit similar phenomena. These studies provide strong support for the contention that conformational changes in components of the membrane are intimately involved in the mechanism of active transport.

V. GROUP TRANSLOCATION

A. The P-Enolpyruvate-P-Transferase System

A bacterial phosphotransferase system was described in 1964 [91], which catalyzes the transfer of phosphate from P-enolpyruvate to various carbohydrates according to the following overall reaction.

$$\text{Sugar} + \text{P-enolpyruvate} \longrightarrow \text{sugar-P} + \text{pyruvate} \qquad (1)$$

The individual reactions involved in the overall P-transferase system are shown below.

$$\text{P-enolpyruvate} + \text{HPr} \underset{\text{Enzyme II, } Mg^{2+}}{\overset{\text{Enzyme I, } Mg^{2+}}{\rightleftharpoons}} \text{pyruvate} + \text{P-HPr} \qquad (2)$$

$$\text{P-HPr} + \text{sugar} \xrightarrow[\text{(Factor III)}]{} \text{sugar-P} + \text{HPr} \qquad (3)$$

$$\text{P-enolpyruvate} + \text{sugar} \xrightarrow[\text{HPr, } Mg^{2+}\text{, Enzyme II, (Factor III)}]{\text{Enzyme I}} \text{sugar-P} + \text{pyruvate} \qquad (4)$$

HPr, a heat-stable, low molecular weight protein that has been purified to homogeneity [92] and Enzyme I are predominantly soluble proteins, whereas Enzyme II is membrane bound. Enzyme II is responsible for specificity with respect to the various sugars studied. Recently, Enzyme II activity has been solubilized and partially purified [93], and it has been shown that at least two protein fractions (one of which is sugar-specific) and phosphatidylglycerol are required for Enzyme II activity. There is also evidence in some systems for the involvement of a fourth protein (Factor III) [94, 95] the function of which is unknown. Subsequent to the initial description of the P-transferase system, biochemical and genetic evidence was presented by a number of laboratories indicating that this system might be involved in bacterial carbohydrate metabolism or transport or both. For an extensive review of this work, the reader is directed to previous reviews [12, 96, 97].

Studies utilizing isolated bacterial membrane vesicles have demonstrated that the P-transferase system catalyzes the vectorial phosphorylation of glucose and related monosaccharides in E. coli, S. typhimurium, and B. subtilis [8, 12, 28, 98]. By this means, free sugar in the medium is transported into the vesicles as sugar-P without passing through an intramembranal free sugar pool. Uptake and phosphorylation of αMG, glucose, fructose, or mannose by membrane vesicles from E. coli, S. typhimurium, and B. subtilis are almost totally dependent on the presence of P-enolpyruvate and only P-enolpyruvate. There is little or no dephosphorylation or subsequent metabolism of the sugar-P's formed, and membrane vesicles prepared from various HPr or Enzyme I mutants are not able to take up or phosphorylate α-MG with or without P-enolpyruvate. Thus, the effect of P-enolpyruvate must be mediated by the P-enolpyruvate-P-transferase system. In addition, there is a stoichiometric relationship between ^{32}P-enolpyruvate disappearance and the appearance of ^{32}P in α-MG-P, suggesting that P-enolpyruvate provides energy for the simultaneous uptake and phosphorylation of these sugars.

Direct evidence for this hypothesis is derived from double isotope experiments in which the intravesicular pool is preloaded with ^{14}C-glucose under conditions in which there is no phosphorylation. After removal of the external isotope, the preloaded vesicles are exposed to ^{3}H-glucose in the presence of P-enolpyruvate. The vesicles show an almost absolute preference for ^{3}H-glucose added to the outside simultaneously with P-enolpyruvate.

The possibility that the P-enolpyruvate-P-transferase system phosphorylates sugars that enter an intramembranal pool by facilitated or passive diffusion is also inconsistent with at least three more observations. (1) The initial rates of α-MG uptake and phosphorylation exhibit saturation kinetics with an apparent K_m of about 4×10^{-6} M, whereas the rate of uptake of free α-MG into the intramembranal pool exhibits linear kinetics over a millionfold α-MG concentration range and is independent of the presence of P-enolpyruvate. (2) The intramembranal pool contains only one-third of the free α-MG concentration of the external pool. (3) The rate of appearance of α-MG-P in the intramembranal pool precedes that of free α-MG. These experiments provide strong, if not unequivocal, evidence that the P-enolpyruvate-P-transferase system is responsible for the translocation of sugars and their concentration within the membrane vesicles as phosphorylated derivatives.

Membrane vesicles take up little glucose-P when incubated with or without P-enolpyruvate. Moreover, the diffusion-limited rates of appearance of glucose-6-P or free glucose into the intramembranal pool are similar. In light of the previous observations, this finding implies that external sugar reaches a catalytic site within the membrane and is translocated as a result of phosphorylation. In addition, these experiments suggest that the P-transferase system does not function as a trap since there would be little advantage in trapping sugar as sugar-P in order to decrease its outward diffusions through the membrane.

1. Regulation of the P-transferase System

Membrane preparations from E. coli, S. typhimurium, or B. subtilis concentrate sugar-P to different levels [8, 28, 98], and the ability of the membranes to catalyze P-enolpyruvate-dependent phosphorylation is inversely related to the ability of the membrane preparation to retain phosphorylated sugar (i.e., transport) [8]. Moreover, the intramembranal concentration of α-MG-P in membrane preparations incubated under leaky conditions (i.e., 46°C) and in the presence of high concentrations of α-MG oscillates with respect to time. Both the period and the amplitude of the oscillations vary with the rate of leakage of α-MG-P from the intramembranal pool and the intramembranal concentration of α-MG-P [98a]. These observations suggested that the P-transferase activity of the vesicles is regulated by the intravesicular concentration of sugar-P.

Further experiments [8, 99] demonstrate that the glucose-transport mechanism is subject to very rigorous control. Glucose or α-MG transport and phosphorylation are noncompetitively inhibited by glucose-6-P, glucose-1-P, and by a variety of related hexose-P's. The inhibitory sites for the 6-P and 1-P esters are separate, distinct, and accessible from both sides of the membrane. The inhibitory effects of glucose-6-P and glucose-1-P on glucose uptake by the membranes varies independently in different membrane preparations; moreover, inhibition of glucose transport by 1-P esters is antagonized by glucose-6-P, and vice versa. Finally, glucose-1-P specifically inhibits the uptake of fructose and galactose. These experiments, especially when considered in conjunction with the independent experiments of Lowry et al. [99a], provide a strong indication that glucose-6-P and glucose-1-P may be important regulatory elements in the metabolism of carbohydrates transported via the P-transferase system.

2. Effect of HPr and Enzyme I

Since HPr and Enzyme I are predominantly soluble proteins, and because the vesicles retain minute quantities of these components, the effects of exogenous HPr and Enzyme I on transport and phosphorylation were investigated [8, 12, 98]. The results are as follows. (1) The initial rate of α-MG uptake and phosphorylation by E. coli ML 308-225 membrane vesicles increases linearly over a large concentration range of P-enolpyruvate in the absence of HPr and Enzyme I. (2) Under the standard assay conditions (i.e., in the presence of NaF), HPr and Enzyme I cause slight inhibition of uptake over the same range of P-enolpyruvate concentrations. (3) In the absence of NaF, however, HPr and Enzyme I cause sigmoid kinetics with marked stimulation of the initial rate of α-MG uptake at very low concentrations of P-enolpyruvate. Moreover, α-MGP recovered from the intravesicular pool closely mirrors total uptake demonstrating that HPr and Enzyme I stimulate uptake and phosphorylation simultaneously. (4) When filtrates from reaction mixtures containing HPr, Enzyme I, and NaF are chromatographed, considerable amounts of α-MGP are found in the external medium. Thus, NaF prevents the stimulatory effect of HPr and Enzyme I on transport with relatively minor effects on total phosphorylation. It is possible that Enzyme II can release sugar-P either into the vesicles or into the external medium depending upon the nature of the complex formed with exogenously added HPr and Enzyme I. In the presence of these components, Enzyme II may undergo a conformational change such that the complex formed is oriented for vectorial phosphorylation (i.e., transport). NaF, for unknown reasons, may prevent an orientation of the reconstituted complex which is essential for vectorial phosphorylation.

Although these data show that NaF abolishes the stimulatory effect of exogenously added HPr and Enzyme I on α-MG transport, subsequent experiments demonstrate that other variables, as yet undefined, are also important. Except for the case of E. coli ML 308-225 described above, although phosphorylation of α-MG by membrane preparations from a variety of organisms and HPr or Enzyme I mutants of E. coli and S. typhimurium is stimulated by appropriate additions of HPr, Enzyme I, or both, transport is not stimulated despite the absence of NaF. Moreover, vesicle preparations from E. coli ML 308-225 (the organism used for the original studies) do not behave consistently as described above. Addition of exogenous HPr and Enzyme I virtually always enhances phosphorylation but does not stimulate transport. Conceivably, the complex formed between HPr, Enzyme I, and Enzyme II must assume a highly specific conformation to achieve vectorial phosphorylation. If this is the case, the conformation of such a complex will almost certainly depend upon the immediate environment (i.e., the milieu within the membrane), and changes in the membrane resulting from differences in growth conditions or genetics of the parent cells might be responsible for the variable ability of HPr and Enzyme I to stimulate transport. Such changes in the membrane have been shown to occur [8, 12]. In any case, these data, taken as a whole, indicate that the reconstituted systems currently used to assay P-transferase activity [96, 100] represent only a crude reflection of transport since the vectorial nature of the intact system is missing. Perhaps these observations account for the failure of sugar-P's to regulate the activity of the reconstituted P-transferase system, and for the kinetic results recently presented by Rose and Fox [101].

3. General Importance of the P-transferase-Mediated Transport Systems

In S. aureus, the P-transferase system catalyzes the transport of many carbohydrates [94, 102-106]. In most other organisms, however, this does not appear to be the case. In E. coli, for instance, this system is involved in the transport of relatively few sugars, i.e., glucose, mannose, fructose, and mannitol [107]. It is not involved in amino acid transport or in most inducible sugar transport systems in E. coli, as discussed above, nor is it apparently involved in glucose transport in obligate aerobes [108].

A conceptual model that has been proposed as a possible mechanism for vectorial phosphorylation [8, 12], as well as experiments related to the role of phosphatidylglycerol in the P-transferase system [109], are not discussed here.

B. Adenine Phosphoribosyltransferase

Evidence for a group translocation mechanism for purine transport has recently been presented [110]. The uptake of adenine by membrane vesicles from E. coli is accompanied by its conversion to AMP and is stimulated by P-ribose-PP. The enzyme mediating this activity—adenine phosphoribosyltransferase — is required for uptake, and variations in enzyme activity are reflected by changes in adenine transport.

VI. APPLICATIONS TO THE STUDY OF MEMBRANE STRUCTURE

A. Functional Separation of Transport from Barrier Function

Because of the unique features of the P-enolpyruve-P-transferase system, this vectorial phosphorylation mechanism may be used to study certain functional aspects of membrane structure. As opposed to the "carrier"-mediated active transport systems discussed above, sugars transported via the P-transferase system appear in the intravesicular pool as phosphorylated derivatives. Since the vesicles do not dephosphorylate sugar-P's, nor can they transport sugar-P's [8, 12, 28, 99] either by active transport or by facilitated diffusion unless a specific sugar-P transport system is induced in the parent cells (cf. Section IV, A, 3), the retention of substrates transported via the P-transferase system is a reflection of the passive permeability of the membrane. By means of simple experimental manipulations [8, 12, 14, 15, 19, 98], the permeability barrier of the vesicles can be studied as an isolated functional entity.

Mouse duodenal phospholipase (A and B activities) [81],* snake venom phospholipase A, a variety of detergents (i.e., Triton X-100, Tween 40, and tergetol-4), toluene, acetone extraction, and Surfactin [111], a bacteriolytic substance produced by B. subtilis, all produce markedly increased rates of α-MGP leakage with little or no change in the ability of the vesicles to phosphorylate α-MG in the presence of P-enolpyruvate [8, 12, 14, 15, 19, 98]. Moreover, with mouse duodenal phospholipase and snake venom phospholipase A, the loss of barrier function is the direct result of hydrolysis of phosphatidylethanolamine (the predominant phospholipid in E. coli), indicating that this phospholipid plays an important role in membrane barrier function [19]. Phospholipase-D (cabbage) specifically inhibits the ability of the membranes to carry out phosphorylation by means of its ability to hydrolyze phosphatidylglycerol [109]. Treatment of the vesicles with this enzyme has no effect on passive efflux of α-MGP. Proteases (e.g., chymotrypsin, trypsin, or pronase) and protein fixatives (e.g., glutaraldehyde) have little or no effect on α-MGP efflux, despite the fact that these agents completely inhibit sugar and amino acid transport and cause marked changes in the membrane proteins as demonstrated by SDS acrylamide gel electrophoresis [8, 12, 14, 15].

B. Temperature Transitions

When vectorial phosphorylation of α-MG is studied as a function of temperature, two optima are observed — one for the initial rate of uptake which is at about 46° C with every preparation, and another for the maximum amount of α-MGP accumulated (uptake in 30 min), which varies from 30° to 46° C depending upon the carbon source used for the growth of the parent cells [8, 12, 14, 15, 19, 98]. Studies of α-MGP efflux as a function of temperature demonstrate that there is no leakage of sugar-P between 0° C and the 30-min temperature optimum for the particular vesicle preparation studied. At precisely the 30-min temperature optimum, the membranes begin to lose sugar-P. Above this critical temperature, the rate of efflux increases and becomes maximal at about 55° C. The effects of incubation at temperatures up to and including 46° C are completely and instantaneously reversed by lowering the temperature

* This enzyme preparation was graciously contributed by Dr. A. Ottolenghi, Department of Pharmacology, Duke University, Durham, North Carolina.

of the reaction mixtures. Preliminary observations indicate that the critical leakage temperature of B. subtilis vesicles is correlated with the average chain length of the fatty acids in the membrane phospholipids [8, 12, 98]. E.coli vesicles, on the other hand, show an apparent increase in the diphosphatidylglycerol (cardiolipin) content of the membrane, which is correlated with an increase in the critical leakage temperature [111a].

C. Structural and Functional Correlations

Over the past few years, various spectroscopic techniques have been applied to the study of biological membrane structure [9, 10, 84, 112-118]. In most cases, however, little or no attempt was made to correlate these studies with biochemical processes in a defined biological membrane system. In view of the ability of highly purified membrane vesicles to catalyze the active transport of a large number of solutes, and the demonstration of at least two types of temperature-induced transitions in transport-related functions, this system seems ideal for such structure-function investigations. The studies to be described have been initiated recently by Dr. E. Shechter and Dr. T. Gulik-Krzywicki of the Centre de Génétique Moléculaire at Gif-sur-Yvette in collaboration with the author [118a].

Using vesicles into which dansylphosphatidylethanolamine (dan-PE) was incorporated, changes in depolarization of fluorescence and fluorescence emission maxima were studied as functions of temperature and compared with changes in X-ray diffraction, transport, and passive permeability. The fluorescent probe was incorporated into the vesicles by means of surface exchange via dan-PE-containing liposomes. Subsequently, the liposomes were separated from the membranes by means of discontinuous sucrose density centrifugation, and it was demonstrated that the fluorescence emission spectrum of dan-PE in the vesicles is markedly different from that of the dan-PE-containing liposomes. Thus, the dan-PE is incorporated into the fabric of the membrane, and the results to be discussed cannot be due to the trivial attachment of dan-PE-containing liposomes to the vesicles. Moreover, it should be emphasized that membrane vesicles containing dan-PE manifest no significant differences in the transport activities studied. The P-enolpyruvate-dependent vectorial phosphorylation of α-MG, as well as the D-lactate-dependent transport of lactose and proline are essentially the same as that of control preparations.

1. X-ray Diffractions Studies

X-ray diffraction measurements carried out with centrifugally "stacked" vesicles at low temperatures reveal a sharp reflection at 4.2 A^{-1} superimposed on a broad band at 4.5 A^{-1}. The 4.2 A^{-1} reflection is characteristic of rigid lipid paraffin chains, while the 4.5 A^{-1} band is characteristic of "liquid-like" lipid paraffin chains [119, 120]. From 0°C to approximately 15°C, the intensity of the 4.2 A^{-1} reflexion remains unchanged; above 15°C, it begins to diminish in intensity; at 30°C, it can no longer be detected. Similarly, a transition in D-lactate oxidation and transport in the vesicles is also observed at approximately 15°C [19, 30]. Therefore, the activity transition may be related to the "melting" of the lipid paraffin chains of the membrane phospholipids. These experiments are generally consistent with studies carried out with fatty acid auxotrophs of E. coli in other laboratories [121-125].

2. Fluorescence Depolarization Studies

Since fluorescence depolarization reflects primarily the mobility of a fluorescent chromophore, variations in this parameter as a function of temperature using dan-PE were studied. Experiments were carried out initially with an artificial lipid-water system in order to demonstrate that fluorescence depolarization of dan-PE in a defined system could detect the "melt" transition discussed above. Thus, dan-PE was incorporated into a lecithin-water system, which had been previously characterized by X-ray diffraction, and it was demonstrated that the transition observed in depolarization of fluorescence when studied as a function of temperature correlated with the X-ray diffraction changes. Subsequently, fluorescence depolarization measurements were made with dan-PE labeled vesicles, and a transition at 15° to 20°C was observed. These studies indicate that the "melting" of lipid paraffin chains of the membrane phospholipids as demonstrated by X-ray diffraction is reflected in the mobility of dan-PE incorporated into the vesicle membrane.

3. Studies of Fluorescence Emission Maximum (λ^{E}_{max})

The fluorescence of dansyl chromophores is sensitive to the polarity of the microenvironment around the probe. The lower the polarity, the greater the displacement of the emission spectra to the blue [126]. Moreover, previous studies have shown that the emission maximum of dan-PE incorporated into protein-lipid-water model systems is dependent on the type of lipid-protein interaction (i.e., hydrophobic versus electrostatic) in the system [116, 117].

The fluorescence of dan-PE incorporated into protein-lipid-water systems displaying pure electrostatic interactions or electrostatic and hydrophobic interactions was studied. One system, made up of lysozyme-phosphatidylinositol-water, exhibits electrostatic interactions at temperatures below 35° to 40° C, and hydrophobic interactions at temperatures above 40°C, as demonstrated by X-ray diffraction [116, 117]. When dan-PE is incorporated into this system, this transition is accompanied by a change in λ^{E}_{max} at approximately 40°C. A second system consisting of lysozyme-cardiolipin exhibits hydrophobic interactions over the temperature range studied [116, 117]. No discontinuity is observed with the latter system; moreover, the change in λ^{E}_{max} of incorporated dan-PE as a function of temperature is similar to that observed after the transition from electrostatic to hydrophobic interactions in the lysozyme-phosphatidylinositol-water system.

With dan-PE labeled membrane vesicles, studies of λ^{E}_{max} as a function of temperature exhibit behavior that is remarkably similar to that of the lysozyme-phosphatidylinositol-water system, that is, two linear functions with different slopes as a function of temperature yielding a sharp discontinuity at a particular temperature. It is important to note that the temperatures at which the discontinuities occur vary from one vesicle preparation to another and correlate fairly well with the temperatures at which these membrane vesicles exhibit leakage transitions [19]. In view of the similarity between these results and those obtained with the artificial model systems, it is tempting to attribute the leakage transition to a temperature-dependent change in the nature of the lipid-protein interactions within the membrane.

VI. CONCLUSIONS AND SPECULATIONS

It has been the intent of this chapter to discuss some recent observations regarding mechanisms of bacterial solute transport and their relevance to studies of membrane structure. In particular, much of the discussion has emphasized respiration-coupled active transport systems as studied in isolated bacterial membrane vesicle preparations. Since a number of these observations are surprising, a few of the more important general conclusions will be reemphasized.

Regarding the P-enolpyruvate-P-transferase system, evidence has been presented that indicates that this system, although of great importance in some bacteria, is not of such widespread significance as was originally thought. A short time ago, it seemed possible that such a mechanism, coupled possibly with specific sugar-P P-hydrolases, could be of singular importance [8, 9, 65]; however, this view is no longer tenable. In some bacterial species, e.g., *Staphylococci*, the P-transferase system is involved in the transport of many sugars; however these observations cannot be generalized. In *E. coli* relatively few, albeit important, sugars are transported via this mechanism. On the other hand, most of the inducible sugar transport systems and almost all of the amino acid transport systems in E. coli are coupled to respiration by means of D-lactic dehydrogenase. Moreover, evidence has been cited that suggests that in many other bacterial species, respiration-coupled active transport as opposed to vectorial phosphorylation is the prevalent mechanism.

Neither the generation nor utilization of high-energy phosphate compounds appears to be involved in respiration-coupled active transport. This conclusion is based upon a large number of experimental observations and appears to be well substantiated, at least in those organisms that have been studied in detail. Alternative means of redox coupling to transport such as proton or potential gradients have been considered and appear to be inconsistent with the data presented. Bhattacharyya et al. [85] demonstrated that the addition of valinomycin to membrane vesicles induces a highly specific active transport system for K^+ or Rb^+. Since the vesicles establish what appeared to be a proton gradient [91], it seemed likely that valinomycin-induced Rb^+ transport might be coupled to a proton gradient and that a detailed study of this system might provide definitive evidence with regard to this question. Surprisingly, the valinomycin-induced transport of Rb^+ behaves, with few exceptions, like the sugar and amino acid transport systems previously studied. Moreover, detailed ion transport studies, as well as studies of pH and ANS fluorescence indicate that proton or potential gradients are not the primary driving force for transport, even for Rb^+ transport induced by valinomycin. Although this possibility still cannot be excluded definitively, it should be obvious that if proton or potential gradients were the driving force for active transport in this system, the proton or ion "carriers" must have the same properties as the solute-specific "carriers" discussed above. That is, the proton- or ion-pumping mechanism must be coupled primarily to a specific dehydrogenase, and the proton or potential gradient that is established must respond to the same perturbants in the same manner as the sugar, amino acid, and valinomycin-induced transport systems (i.e., the generation of the proton or potential gradient must respond to temperature in the same way, and the proton or potential gradient that is established must collapse only in the presence of inhibitors that block electron transfer after the site of energy coupling). In other words, such a mechanism would appear to provide an unnecessary duplication in the system. On the other hand, a proton gradient mechanism would provide a convenient explanation for the action of "uncoupling agents" [76].

Although evidence has been presented that indicates that the "carriers" reflect the redox potential of a specific portion of the respiratory chain between the site of ascorbate-PMS reduction and the cytochrome chain, the conceptual model presented in Fig. 8 is merely the simplest interpretation of the data that are presently available. No evidence has been presented to demonstrate that the "carriers" undergo oxidation-reduction, let alone that they are electron transfer intermediates. Moreover, the mechanism presented provides no immediate explanation for the action of "uncoupling agents" like DNP, CCCP, or azide.

The observation that respiration-coupled transport is coupled to specific dehydrogenases is noteworthy if only because it is completely unexpected. Aside from the surprise element and from the obvious technical importance of this observation, this aspect of the mechanism places certain restrictions upon the means by which transport may be regulated. One can envisage at least three levels at which control could be exerted — at the level of the primary dehydrogenase; at the site of energy coupling; and at the level of the "carrier." Although little is known at the present time about the regulation of these transport systems, an important level of control might be at the primary dehydrogenase. A free-living organism rarely finds itself in an environment containing a single sugar or amino acid. Thus on a teleological basis, it might be expected that a primary mechanism would exist for regulating transport *en bloc*. By controlling the specific dehydrogenase involved with transport, this could be accomplished. Control at other levels of transport might be superimposed upon this basic means of regulation. In this regard, preliminary experiments with *E. coli* membrane vesicles indicate that many glycolytic intermediates, especially 2,3-diphosphoglycerate, inhibit D-lactic dehydrogenase.

Another aspect of the respiration-coupled systems that has only recently begun to be studied is their possible relevance to transport under anaerobic conditions. Obligate anaerobes or facultative anaerobes transport nutrients; moreover, δ-aminolevulinic acid- or heme-requiring mutants of *E. coli* apparently do not manifest transport defects, although they have not been studied in detail from this point of view. Possibly, anaerobically growing cells or cells lacking cytochromes utilize unique transport systems. However, the possibility should be considered that they may use the same general type of transport systems as those used aerobically, with the exception that an alternative electron acceptor is used rather than the cytochrome chain and oxygen. In this regard, experiments carried out in collaboration with Dr. Wilhelmus N. Konings of the University of Groningen in The Netherlands have demonstrated that membrane vesicles prepared from *E. coli* grown anaerobically under the appropriate conditions will transport lactose anaerobically with fumarate or nitrate as electron acceptors [127]. With fumarate as acceptor, α-glycerol-P or D-lactate are effective electron donors; with nitrate, on the other hand, formate is the only electron donor which stimulates transport effectively. It is also noteworthy that despite genetic evidence indicating an involvement of ATPase in anaerobic transport [128, 129], all attempts to drive transport aerobically or anaerobically in shocked cells or membrane vesicles have been uniformly negative.

Rather than terminate this chapter on a purely speculative note, it seems pertinent to discuss some very recent experiments that may have significance for further studies in this area. Dr. Jen-shiang Hong, currently at Brandeis University, has isolated and characterized two classes of mutants from E. coli and S. typhimurium which are altered in respiration-coupled active transport [130]. Mutants defective in D-LDH (dld$^-$) transport lactose or proline normally, as might be expected since other dehydrogenases can drive transport to some extent. Membrane vesicles prepared from these mutants, however, exhibit some unexpected properties. In the presence of ascorbate-PMS, vesicles from the mutants transport proline or lactose at the same rate and to the same extent as vesicles prepared from the wild type. D-Lactate, as expected, does not stimulate transport by the mutant vesicles. L-Lactate and/or succinate (depending upon the carbon source used for the growth of the cells) stimulates transport in mutant vesicles as well as D-lactate in the wild type. Even more striking is the observation that the rate of oxidation of L-lactate or succinate is similar to that observed in vesicles prepared from the wild type. In other words, in the absence of D-LDH, the coupling of L-lactic dehydrogenase or succinic dehydrogenase to transport is enhanced. With vesicles prepared from a mutant defective in both D-LDH and succinic dehydrogenase, only L-lactic dehydrogenase is coupled to transport. D-Lactate-dependent transport and D-lactate oxidation by dld$^-$ vesicles can be completely reconstituted by treating these vesicles with a guanidine extract from wild type membrane vesicles [131].*

ABBREVIATIONS

The abbreviations used in this chapter are D-LDH, D-lactic dehydrogenase; HPr, heat-stable protein; IPTG, isopropyl-β-D-thiogalactopyranoside; TMG, methyl-1-thio-β-D-galactopyranoside; TDG, β-D-galactosyl-1-thio-β-D-galactopyranoside; α-MG, methyl-α-D-glucopyranoside; HOQNO, 2-heptyl-4-hydroxyquinoline-N-oxide; NEM, N-ethylmaleimide; PCMB, p-chloromercuribenzoate; DNP, 2,4-dinitrophenol; CCCP, carbonylcyanide-m-chlorophenylhydrazone; DCI, dichlorophenolindophenol; PMS, phenazine methosulfate; cyclic DBS_3, cyclic dihydroxybenzoylserine triplex; DMO, 5,5-dimethyloxazolidine-2,4-dione; ANS, 1-anilino-8-naphthalene sulfonic acid; dan-PE, dansylphosphatidylethanolamine.

*See Notes Added in Proof, p. 292.

REFERENCES

1. H. R. Kaback, Fed. Proc. Fed. Amer. Soc. Exp. Biol., 19, 130 (1960).
2. H. R. Kaback and A. B. Kostellow, J. Biol. Chem., 243, 1384 (1968).
3. H. R. Kaback and E. R. Stadtman, J. Biol. Chem., 243, 1390 (1968).
4. H. R. Kaback and E. R. Stadtman, Proc. Nat. Acad. Sci. U. S., 55, 920 (1966).
5. H. R. Kaback and T. F. Deuel, Arch. Biochem. Biophys., 132, 118 (1969).
6. M. R. J. Salton, The Bacterial Cell Wall, Elsevier, Amsterdam, 1964.
7. H. R. Kaback, in Methods in Enzymology (W. B. Jakoby, ed.), Academic Press, New York, 1971.
8. H. R. Kaback, in Current Topics in Membranes and Transport (A. Kleinzeller and F. Bronner, eds.), Vol. I, p. 35, Academic Press, New York, 1970.

8a. H. R. Kaback, unpublished observation.

8b. A. S. Gordon and H. R. Kaback, unpublished data.

9. J. Lenard and S. J. Singer, Proc. Nat. Acad. Sci. U. S., 56, 1828 (1966).
10. D. F. H. Wallach and P. J. Zahler, Proc. Nat. Acad. Sci. U. S., 56, 1552 (1966).
11. W. D. Stein, The Movement of Molecules across Cell Membranes, p. 369, Academic Press, New York, New York, 1967.
12. H. R. Kaback, Ann. Rev. Biochem., 39, 561 (1970).
13. H. R. Kaback and L. S. Milner, Proc. Nat. Acad. Sci. U. S., 66, 1008 (1970).
14. H. R. Kaback, E. M. Barnes, Jr., and L. S. Milner, Advances in Microbiology, p. 171, Libreria Internacional, Mexico, 1971.
15. H. R. Kaback, E. M. Barnes, Jr., A. S. Gordon, F. J. Lombardi, and G. K. Kerwar, Molecular Basis of Biological Activity (K. Gaede, B. L. Horecker, and W. J. Whelan, eds.), p. 373, Academic Press, New York.
16. E. M. Barnes, Jr. and H. R. Kaback, Proc. Nat. Acad. Sci. U. S., 66, 1190 (1970).

17. E. M. Barnes, Jr. and H. R. Kaback, J. Biol. Chem., 246, 5518 (1971).

18. G. W. Dietz, Fed. Proc. Fed. Amer. Soc. Exp. Biol., 30, 52 (1971).

19. H. R. Kaback, Biochim. Biophys. Acta, 265, 367 (1972).

20. G. F. Ames, Arch. Biochem. Biophys., 105, 1 (1964).

21. J. Piperno and D. L. Oxender, J. Biol. Chem., 243, 5914 (1968).

22. G. A. Scarborough, M. K. Rumley and E. P. Kennedy, Proc. Nat. Acad. Sci. U. S., 60, 951 (1968).

23. E. Pavlasova and F. M. Harold, J. Bacteriol., 98, 198 (1969).

24. C. F. Fox and E. P. Kennedy, Proc. Nat. Acad. Sci. U. S., 54, 891 (1965).

25. C. F. Fox, J. R. Carter and E. P. Kennedy, Proc. Nat. Acad. Sci. U. S., 57, 698 (1967).

26. W. Kundig, F. D. Kundig, B. E. Anderson and S. Roseman, J. Biol. Chem., 241, 3243 (1966).

27. S. Tanaka, P. G. Fraenkel and E. C. C. Lin, Biochem. Biophys. Res. Commun., 27, 63 (1967).

28. H. R. Kaback, J. Biol. Chem., 243, 3711 (1968).

29. G. K. Kerwar, A. S. Gordon and H. R. Kaback, J. Biol. Chem., 246, 291 (1972).

30. H. R. Kaback and E. M. Barnes, Jr., J. Biol. Chem., 246, 5523 (1971).

31. B. L. Horecker, J. Thomas and J. Monod, J. Biol. Chem., 235, 1580 (1960).

32. H. C. P. Wu, J. Mol. Biol., 24, 213 (1967).

33. A. K. Ganesan and B. Rotman, J. Mol. Biol., 16, 42 (1966).

34. J. Guardiola and M. Iaccarino, J. Bacteriol., 108, 1034 (1971).

34a. M. DeFelice and M. Iaccarino, manuscript in preparation.

34b. G. K. Kerwar, H. R. Kaback and M. Iaccarino, unpublished data.

35. E. M. Tarmy and N. O. Kaplan, J. Biol. Chem., 243, 2579 (1968).

36. E. M. Tarmy and N. O. Kaplan, J. Biol. Chem., 243, 2587 (1968).

37. L. D. Kohn and H. R. Kaback, J. Biol. Chem., in press (1973); M. Futai, Biochem., 12, 2468 (1973).

38. G. B. Cox, N. A. Newton, F. Gibson, A. M. Snoswell and J. A. Hamilton, Biochem. J., 117, 551 (1970).

39. H. Weissbach, E. L. Thomas and H. R. Kaback, Arch. Biochem. Biophys., 147, 249 (1971).

40. W. L. Klein, A. S. Dhams and P. D. Boyer, Fed. Proc. Fed. Amer. Soc. Exp. Biol., 29, 341 (1970).

41. S. A. Short, D. C. White and H. R. Kaback, J. Biol. Chem., 246, 298 (1972).

42. J. D. Butlin, G. B. Cox and F. Gibson, Biochem. J., 124, 75 (1971).

42a. G. Prezioso, J.-s. Hong, G. K. Kerwar, and H. R. Kaback, Arch. Biochem. Biophys., 154, 575 (1973).

43. H. Hirata, A. Asano and A. F. Brodie, Biochem. Biophys. Res. Commun., 44, 368 (1971).

43a. M. Barnes, Jr. and H. R. Kaback, unpublished experiments.

44. A. Kepes and G. N. Cohen, in The Bacteria (I. C. Gunsalis and R. Stanier, eds.), Vol. IV, p. 179, Academic Press, New York, 1962.

45. A. Kepes, in Current Topics in Membranes and Transport (A. Kleinzeller and F. Bronner, eds.), Vol. I, p. 101, Academic Press, New York, 1970.

46. A. Kepes, Biochim. Biophys. Acta, 40, 70 (1960).

47. J. R. Carter, C. F. Fox, and E. P. Kennedy, Proc. Nat. Acad. Sci. U. S., 60, 725 (1968).

48. T. H. D. Jones and E. P. Kennedy, J. Biol. Chem., 244, 5981 (1969).

49. D. Schachter and A. J. Mindlin, J. Biol. Chem., 244, 1808 (1969).

50. J. A. Manno and D. Schachter, J. Biol. Chem., 245, 1217 (1970).

51. A. L. Koch, J. Mol. Biol., 59, 447 (1971).

52. P. T. S. Wong, E. R. Kashket and T. H. Wilson, Proc. Nat. Acad. Sci. U. S., 65, 63 (1970).

53. T. H. Wilson, M. Kusch and E. R. Kashket, Biochem. Biophys. Res. Commun., 40, 1409 (1970).

53a. H. R. Kaback, unpublished experiments.

54. W. N. Konings, E. M. Barnes, Jr. and H. R. Kaback, J. Biol. Chem., 246, 5857 (1971).

55. I. G. O'Brien and F. Gibson, Biochim. Biophys. Acta, 215, 393 (1970).

56. P. D. Bragg, Can. J. Biochem., 48, 777 (1970).

57. A. S. Gordon, F. J. Lombardi and H. R. Kaback, Proc. Nat. Acad. Sci. U. S., 69, 358 (1972).

58. L. A. Heppel, Science, 156, 1451 (1967).

59. A. B. Pardee, Science, 162, 632 (1968).

60. H. C. Neu and L. A. Heppel, J. Biol. Chem., 240, 3685 (1965).

61. W. N. Konings and E. Freese, FEBS Letters, 14, 65 (1971).

62. W. N. Konings and E. Freese, J. Biol. Chem., 247, 2408 (1972).

62a. H. R. Kaback, unpublished data.

62b. E. M. Barnes, Jr., personal communication.

63. M. M. Shemyakin, N. A. Aldanova, E. I. Vinogradova and M. Yu. Feigina, Tetrahedron Lett., (28), 1921 (1963).

64. M. M. Shemyakin, Yu A. Ovichinnikov, V. T. Ivanov, V. K. Antonov, E. I. Vinogradova, A. M. Shkrob, G. G. Malenkow, A. V. Eustatov, I. A. Laine, E. I. Melnik and I. D. Ryabova, J. Membrane Biol., 1, 402 (1969).

65. F. M. Harold, Adv. Microbiol. Physiol., 4, 45 (1970).

66. P. F. J. Henderson, J. D. McGivan and J. B. Chappell, Biochem. J., 111, 521 (1969).

67. P. Mueller and D. O. Rudin, Biochem. Biophys. Res. Commun., 26, 398 (1967).

68. F. M. Harold and J. R. Baarda, J. Bacteriol., 94, 53 (1967).

69. F. M. Harold and J. R. Baarda, J. Bacteriol., 95, 816 (1968).

70. J. B. Chappell and A. R. Crofts, Biochem. J., 95, 393 (1965).

71. D. C. Tosteson, T. E. Andreoli, M. Tiefenberg and P. Cook, J. Gen. Physiol., 51, 3735 (1968).

72. B. C. Pressman, E. J. Harris, W. S. Jagger and T. H. Johnson, Proc. Nat. Acad. Sci. U. S., 58, 1949 (1967).

73. P. Bhattacharyya, W. Epstein and S. Silver, Proc. Nat. Acad. Sci. U. S., 68, 488 (1971).

74. J. P. Reeves, Biochem. Biophys. Res. Commun., 45, 931 (1971).

75. P. Mitchell, Fed. Proc. Fed. Amer. Soc. Exp. Biol., 26, 1370 (1967).

76. P. Mitchell, Biol. Rev., 41, 445 (1966).

77. G. D. Greville, Current Topics in Bioenergetics, 3, 1 (1971).

78. F. M. Harold and J. R. Baarda, J. Bacteriol., 98, 2025 (1968).

79. I. C. West, Biochem. Biophys. Res. Commun., 41, 655 (1970).

80. W. J. Waddel and R. G. Bates, Physiol. Rev., 49, 285 (1969).

81. A. Ottolenghi, Fed. Proc. Fed. Amer. Soc. Exp. Biol., 28, 885 (1969).

82. E. Daniel and G. Weber, Biochem., 5, 1893 (1966).

83. G. M. Edelman and W. O. McClure, Acc. Chem. Res., 1, 65 (1968).

84. B. Rubalcava, D. Martinez de Muñoz and C. Gitler, Biochem., 8, 2742 (1969).

85. B. Chance, G. K. Radda and C. P. Lee, Proc. Nat. Acad. Sci. U. S., 62, 612 (1969).

86. R. B. Freedman and G. K. Radda, FEBS Letters, 3, 150 (1969).

87. J. Vanderkooi and A. Martinosi, Arch. Biochem. Biophys., 133, 153 (1969).

88. J. R. Brokelhurst, R. B. Freedman, D. J. Hancock and G. K. Radda, Biochem. J., 116, 721 (1970).

89. A. Azzi, P. Gherardini and H. Santato, J. Biol. Chem., 246, 2035 (1971).

90. L. Stryer, Science, 162, 526 (1968).

91. W. Kundig, S. Ghosh and S. Roseman, Proc. Nat. Acad. Sci. U.S., 52, 1067 (1964).

92. B. Anderson, N. Weigel, W. Kundig and S. Roseman, J. Biol. Chem., 246, 7023 (1971).

93. W. Kundig and S. Roseman, J. Biol. Chem., 246, 1407 (1971).

94. R. D. Simoni, M. F. Smith and S. Roseman, Biochem. Biophys. Res. Commun., 31, 804 (1968).

95. W. Hengstenberg, W. K. Penberthy, K. L. Hill and M. L. Morse, J. Bacteriol., 96, 2187 (1968).

96. S. Roseman, J. Gen. Physiol., 54, 1385 (1969).

97. E. C. C. Lin, Annu. Rev. Genetics, 4, 225 (1970).

98. H. R. Kaback, in The Molecular Basis of Membrane Transport (D. C. Tosteson, ed.), p. 421, Prentice-Hall, Englewood Cliffs, New Jersey, 1969.

98a. H. R. Kaback, unpublished observations.

99. H. R. Kaback, Proc. Nat. Acad. Sci. U. S., 63, 724 (1969).

99a. O. H. Lowry, J. Carter, J. B. Ward and L. Glaser, J. Biol. Chem., 246, 6511 (1971).

100. W. Kundig and S. Roseman, in Methods in Enzymology, Vol. IX, p. 396, Academic Press, New York, 1966.

101. S. P. Rose and C. F. Fox, Biochem. Biophys. Res. Commun., 45, 376 (1971).

102. J. B. Egan and M. L. Morse, Biochim. Biophys. Acta, 97, 310 (1965).

103. J. B. Egan and M. L. Morse, Biochim. Biophys. Acta, 109, 172 (1965).

104. J. B. Egan and M. L. Morse, Biochim. Biophys. Acta, 112, 63 (1966).

105. W. Hengstenberg, J. B. Egan and M. L. Morse, J. Biol. Chem., 243, 1881 (1968).

106. D. Laue and R. E. MacDonald, Biochim. Biophys. Acta, 165, 410 (1968).

107. S. Tanaka, S. A. Lerner and E. C. C. Lin, J. Bacteriol., 93, 64 (1967).

108. A. H. Romano, S. J. Eberhard, S. L. Kingle, T. D. McDowell, J. Bacteriol., 104, 808 (1970).

109. L. S. Milner and H. R. Kaback, Proc. Nat. Acad. Sci. U. S., 65, 683 (1970).

110. J. Hochstadt-Ozer and E. R. Stadtman, J. Biol. Chem., 246, 5304 (1971).

111. K. Aruma, N. Tsukagoshi and G. Tamura, Biochim. Biophys. Acta, 163, 121 (1968).

111a. J. Ditmer and H. R. Kaback, unpublished observations.

112. W. L. Hubbel and H. M. McConnel, Proc. Nat. Acad. Sci. U. S., 61, 12 (1968).

113. D. Chapman and D. F. H. Wallach, in Biological Membranes (D. Chapman, ed.), p. 125, Academic Press, New York, 1968.

114. D. F. H. Wallach, J. M. Graham and B. R. Fernbach, Arch. Biochem. Biophys., 131, 322 (1969).

115. T. Gulik-Krzywicki, E. Shechter, M. Iwatsubo, J. L. Ranck and V. Luzzati, Biochim. Biophys. Acta, 219, 1 (1970).

116. T. Gulik-Krzywicki, E. Shechter, V. Luzzati and M. Faure, Nature (London), 223, 1116 (1969).

117. E. Shechter, T. Gulik-Krzywicki, R. Azerad and C. Gros, Biochim. Biophys. Acta, 241, 431 (1971).

118. T. Gulik-Krzywicki, E. Shechter, V. Luzzati and M. Faure, Proc. of the International Symposium on the Biochemistry and Biophysics of Mitochondrial Membranes, Academic Press, New York, in press.

118a. E. Schechter, T. Gulik-Krzywicki and H. R. Kaback, Biochim. Biophys. Acta, 274, 466 (1972).

119. V. Luzzati, in Biological Membranes (D. Chapman, ed.), Academic Press, New York, 1968.

120. M. H. F. Wilkens, A. E. Blaurock and D. M. Engelman, Nature (London), 230, 42 (1971).

121. H. R. Schairer and P. Overath, J. Mol. Biol., 44, 209 (1969).

122. C. F. Fox, Proc. Nat. Acad. Sci. U. S., 63, 850 (1969).

123. U. Henning, G. Dennert, K. Rehn and G. Deppe, J. Bacteriol., 98, 784 (1969).

124. M. Esfahani, E. M. Barnes, Jr. and S. Wakil, Proc. Nat. Acad. Sci. U. S., 64, 1057 (1969).

125. M. Esfahani, A. R. Limbrick, S. Knutlon, T. Oka and S. J. Wakil, Proc. Nat. Acad. Sci. U. S., 68, 3180 (1971).

126. R. F. Chen, Arch. Biochem. Biophys., 120, 609 (1967).

127. W. N. Konings and H. R. Kaback, manuscript in preparation.

128. H. U. Schairer and B. A. Haddock, Biochem. Biophys. Res. Commun., 48, 544 (1972).

129. J. D. Butlin, Ph.D. Thesis, The Australian National University, Canberra, Australia, 1973.

130. J.-s. Hong and H. R. Kaback, Proc. Nat. Acad. Sci. U. S., 69, 3336 (1972).

131. J. P. Reeves, J.-s. Hong, and H. R. Kaback, Proc. Nat. Acad. Sci., U. S., in press (1973).

132. J. P. Reeves, E. Shechter, R. Weil, and H. R. Kaback, Proc. Nat. Acad. Sci. U. S., in press (1973).

NOTE ADDED IN PROOF

A second class of mutants is defective in the coupling of electron transfer to active transport. Whole cells, as well as membrane vesicles, prepared from these etc mutants exhibit markedly reduced ability to transport amino acids, despite the ability of the vesicles to oxidize D-lactate, NADH, and succinate. etc Mutants are similar phenotypically to mutants uncoupled for oxidative phosphorylation (uncA) [42], but have normal ATPase activity. Moreover, uncA mutants catalyze respiration-linked transport as well as the wild type [42a]. It is interesting that etc and uncA co-transduce with asn-1 (asparagine synthetase) indicating that they may be linked genes.

Finally, a fluorescent galactoside, 2-(N-dansyl)-aminoethyl β-D-thiogalactoside (dansyl-galactoside), although not actively transported itself, has been shown to be a competitive inhibitor of lactose transport in E. coli membrane vesicles prepared from cells containing the β-galactoside transport system [132]. Fluorescence studies indicate that D-lactate oxidation causes binding of dansyl-galactoside to the β-galactoside carrier protein in such a manner that the dansyl group is transferred to a hydrophobic environment within the membrane. The data strongly suggest that energy is coupled to one of the initial steps in the transport process and that facilitated diffusion cannot be the rate-limiting step for active transport of β-galactosides.

Hopefully, the genetic approach initiated by Dr. Hong, combined with enzymologic and physical techniques currently in use will extend the study of membrane transport to the molecular level.

Chapter 6

COLICINS

S. E. Luria

Department of Biology
Massachusetts Institute of Technology
Cambridge, Massachusetts

I. COLICINS AND OTHER BACTERIOCINS

This is a peculiar time for attempting to present, within the framework of a book on bacterial surfaces, the status of the colicin problem. Within the last year some of the generalizations that were taken as established have proved not to be valid, while new discoveries have cast light on mechanisms that had been only vaguely formulated. It is an exciting but confusing time in the small world of colicin research. Nothing that has happened, however, negates the belief that this research is a promising approach to the functional organization of the bacterial membrane.

The remarkable discovery [1] that certain bacteria produce bacteriocins, that is, macromolecular antibiotics directed against organisms of same or related groups, roused little interest at first among bacterial physiologists, despite the demonstration [2] that one bacteriocin, colicin E1,* caused a rapid arrest of macromolecular syntheses, and that its bactericidal action followed one-hit kinetics, indicating a noncooperative, "one-molecule-to-kill" type of action. Interest in colicin action was rekindled by reports [4, 5] that trypsin could restore viability to bacteria that had taken up certain colicins. This indication that the effective molecules of these colicins may act from the cell surface to produce specific biochemical changes was at the root of Nomura's formulation [6, 7] of a three-element system, including (1) a receptor to which colicin becomes attached, (2) a transmission mechanism that conveys a colicin-initiated stimulus, and (3) a biochemical target that is affected and the damage to which is responsible for cell death. Different bacteriocins proved to have different targets. Specifically, colicins exhibit three main classes of biochemical effects as evidenced at the macromolecular level: (1) inhibition of protein synthesis by E3-type colicins; breakage of DNA by colicin E2 type; and arrest of macromolecular syntheses by the K-E1 colicin group. Some of the bacteriocins of other groups of bacteria have effects similar to those of one or another colicin. The transmission or amplification system has been the object of much speculation [7-9] and should be thought about in terms of current models of membrane structure [9a].

In this chapter, a brief sketch of the background knowledge of the production, structure, and properties of colicins serves as an introduction to sections on their mode of action and on their interaction with the bacterial envelope. The subject of bacteriocins and their induction has been reviewed competently and repeatedly [5, 8, 10-12a]. Those reviews should be consulted for the many aspects of the problem not dealt with in this chapter, such as the genetic and molecular aspects of colicinogeny, and questions of nomenclature and classification. Many of the articles cited in those reviews are not referred to here.

---

*Bacteriocins of coliform bacteria are called colicins. In genetic work a colicin is conventionally identified by the symbol of the strain of origin, for example, E1-ML or E2-P9 [3]. In this chapter, which concerns itself mainly with mode of action, the strain symbol is omitted. This is probably justified since isolates of one colicin from different strains, e.g., E2, have similar effects.

II. PROPERTIES OF COLICINS AND OTHER BACTERIOCINS

Colicins are produced by colicinogenic strains. Colicinogeny is determined by extrachromosomal genetic elements — the Col factor plasmids — which are usually autonomous replicons and which may or may not require the presence of the bacterial DNA polymerase I [13]. The proneness of plasmids to genetic recombination may explain why the Col properties are often found associated within a plasmid with genetic determinants of fertility or other properties [14].

Col factors have been isolated as circular DNA molecules [15, 16]: Col E1, Col E2, and Col E3 monomer DNA consists of circles 4 to 5 x 10^6 daltons in molecular weight. Dimers and trimers are found under abnormal growth conditions. The number of copies per cell is generally small, but may increase to several thousands under conditions of continued plasmic replication without cell growth [16a, 16b]. The DNA molecules of Col E2 and Col E3 appear to differ by a single region of non-homology [16c].

Synthesis and release of colicin by colicinogenic bacteria is not a continuous process — a constant fraction of growing cells goes into colicin production. The fraction can be increased, for many but not all colicins, by inducing agents, such as uv light, nalidixic acid, or thymine deprivation, which have in common the ability to interfere with DNA synthesis, by a mechanism still not understood. The induced cells are non-viable, although they do not lyse, and it is generally assumed that spontaneous colicin production is likewise a lethal synthesis. Induction is probably a result of inactivation or exhaustion of a specific repressor which, for the E group colicins, is somewhat heat sensitive, so that colicin production is induced in most cells by exposure to temperatures of 44° to 47°C [17]. Induced production of these colicins can be triggered by a period of arrested protein synthesis, during which RNA synthesis must be allowed to occur [17]. The presence of a functional recA gene is required for maximal induction of some colicins but not all [18]. Cyclic AMP is required for maximal production of colicin E1 [18a]. That replication of Col factor DNA is a necessary concomitant of colicin production has now been claimed, now questioned [19, 20]. Studies of Col factors with temperature-sensitive replication mechanisms should prove valuable [21]. In addition, the study of colicin release from induced cells would profit from applications of methods used to study the transfer of extracellular or periplasmic enzymes from the cytoplasm, and may provide interesting clues to the underlying mechanisms of membrane activity.

Colicinogenic bacteria are protected against their own colicin, which they adsorb, by a mechanism of immunity. Nonimmune mutants of Col B-carrying bacteria have been found [22], and it seems reasonable to suppose that each Col factor may have a gene coding for a specific immunity substance. Immunity is not temperature-sensitive (ts) in a strain that produces a ts mutant colicin [23]; hence, the immunity substance is probably not a form of the colicin molecule itself. A protein with reasonable claim to be an immunity substance has been found in extracts of bacteria carrying Col E3 (see below). Immunity breaks down when high multiplicities of colicins are added to cells [24], suggesting a stoichiometric combination between colicins and immunity substance. We need to learn how the immunity substances function in the immune cells, how they prevent response to exogenous colicin, and whether they play a role in preventing the production or killing action or both of endogenous colicin. Evidently, immunity to a colicin is a different phenomenon from the immunity of lysogenic bacteria, which involves the association of a repressor with certain operators on the phage DNA [25]. A colicinogenic factor must produce both the immunity protein and the heat-sensitive "repressor" that controls colicin production [17]. Genetic analysis of immunity coupled with biochemical studies should prove very profitable.

The specificity of immunity, while not absolute [18, 26], provides the most satisfactory means of colicin identification. The traditional system of identification and nomenclature, however, has been based on patterns of cross-resistance in bacterial mutants altered in colicin adsorption. A group of colicins has been named, for example, A, or B, or C if a bacterium can in one step acquire resistance to all the colicins of that group because of failure to adsorb them (the so-called loss of receptors). There is no rationale for this classification since it groups together colicins with utterly different modes of action (e.g., E1, E2, E3). Failure to adsorb colicin need not be a loss of receptors; it may be due to masking of receptors by reorganization of some feature of the cell surface. Only rarely, as with E2 and E3, is there evidence for mutual interference at the adsorption level [27], indicative of a true common receptor.

Colicins, when purified, are proteins with a single polypeptide chain [28]. The properties of a few well characterized ones are listed in Table 1 [28a, b, c], together with those of the corresponding colicinogenic factors. One killing unit (KU) of colicin is defined as the amount that when adsorbed gives a survival $S/S_0 = e^{-1}$. This definition is based on the fact that the dependence of S on the amount A of colicin adsorbed is a simple exponential: $S = S_0 e^{-pA}$ (ratio of viable count S after colicin to viable count S_0 without colicin). Deviations from this rule can generally be accounted for by heterogeneity of bacterial cells in adsorption or response, with no evidence of cooperativity among adsorbed colicin molecules [29]. The reported values of the probability p that an adsorbed molecule will kill (equal to ratio of KU's to molecules) vary from 1:10 to 1:1000, although some reports of $p \sim 1$ have appeared [30]. For a given colicin, the p value for different sensitive bacteria may vary by some orders of magnitude.

TABLE 1

Physical Properties of Some Colicins and Colicinogenic Factors

Colicin	Mol. wt.	f/f_o [a]	Isoelectric point, pI	Col factor DNA, mol. wt. [f] ($x10^6$)
E1	56,000 [b]	2.02 [b]	9.05 [b]	4.2
E2	60,000 [c]	(1.41) [e]	7.63, 7.41 [c]	5
E3	60,000 [c]	(1.41) [e]	6.64 [c]	5
K	45,000 [d]	1.70 [d]	5.14, 5.21 [d]	-
	70,000 [g]	-	5.7 [d]	-
Ia	77,000 [e]	1.82 [e]	-	-
Ib	79,600 [e]	1.76 [e]	-	-

[a] f/f_o is the ratio between the friction coefficient of a molecule and the friction coefficient of a sphere of the same mass, and provides a measure of the molecular asymmetry. The values in parentheses were measured by one indirect method only.

[b] Schwartz and Helinski [28].

[c] Hershman and Helinski [67].

[d] Jesaitis [28a].

[e] Konisky and Richards [28b].

[f] Bazaral and Helinsky [15].

[g] Kunugita and Matsuhashi [28c].

The characterization of receptor substances has only recently made some progress. Two proteins that inactivate colicin E3 and K, respectively, have been isolated from the solubilized envelope of E. coli cells sensitive to colicin but not from the resistant mutants [31, 31a]. In both cases these presumed receptor substances were sensitive to trypsin and to plasmic membrane (see also Weltzien and Jesaitis [32]. The E receptors also have receptor properties for the binding of vitamin B_{12} [32a]. A lipoprotein inactivator for a *Proteus* bacteriocin has also been isolated [33]. Except when genetic data support the results from inactivation tests, one should be careful in interpreting inactivators as receptors. For example, colicins B, I, and V are inhibited by enterochelin, an iron chelator that happens to be hyperproduced and excreted by certain colicin B-resistant mutants [34]. Claims that the colicin receptors are located on the cytoplasmic membrane rather than the cell wall have been put forward [35, 36] based on indirect evidence, and appear to be contradicted by the isolations of receptor proteins from the outer wall. However, because the ratio of killing units to colicin molecules is usually low, one cannot exclude the possibility that the *effective* adsorption of colicin may take place on a few accessible receptors located on the cytoplasmic membrane (see also Takagaki et al. [36a]) and that most receptors on the cell wall may be nonfunctional.

Wherever the effective receptors may be located, there is no evidence available as to whether or not the colicins act enzymatically upon the receptor molecules (or vice versa). A bacteriocin of *Bacillus megaterium* has phospholipase A activity [37], but the nature of its receptor is unknown. Whether either colicin or receptor molecules are altered after effective adsorption is also unknown. Cells coated with colicin E2, then treated with trypsin, adsorb a second load of E2 rather poorly [27]. This may be due to colicin fragments remaining on receptors or to trypsin damage to receptors complexed with colicin. There is some evidence that colicin E2 can desorb from cell receptors and be transferred to other cells [38]. Partial recovery of active colicin K from receptors has been reported [38a] but, in view of the evidence that only a small minority of adsorbed colicin molecules do function, such recovery is hard to interpret in terms of colicin action.

III. ACTION OF COLICIN E3

Sensitive cells that have adsorbed this colicin show a progressive inhibition of protein synthesis. The rate and extent of inhibition, as well as the lag before its onset, are correlated with the "multiplicity" of colicin, even though the kinetics of killing are one-hit [7, 8, 26]. This indicates that if killing is due to the arrest of protein synthesis, at low multiplicities synthesis may continue in cells ultimately destined to die. RNA and DNA syntheses are not affected, nor is DNA degraded or ^{32}P metabolism altered for at least 1 hr or even longer. Active transport is also unaffected. Extracts of E3-treated cells are impaired in their ability to support protein synthesis *in vitro* [39]. Fractionation of the extracts revealed that the ribosomal fraction is the one affected. The ribosomes still bound to mRNA, but binding of tRNA was very much reduced. The polysomes extracted from E3-treated cells are unstable and their ribosomes tend to separate into subunits [40]. Further fractionations and experiments of reconstruction of ribosomal subunits from ribosomal proteins and RNA pinpointed the damage, successively, in the 30 S subunit, its 23 S core and finally its 16 S RNA molecule [39, 41-43].*

The 16 S RNA molecule from ribosomes of E3-treated cells is not randomly damaged, but is split into two fragments. The smaller one, identified by fingerprint analysis as coming from the 3'-OH end, is about 50 nucleotides long and probably has a 5'-OH end, leaving a 3'-phosphate end on the larger fragment [42, 43]. Most of the small fragments are found free, but some remain with the ribosomes (which retain a low level of residual activity). The isolated major RNA fragment can form an apparently incomplete assembly _in vitro_ with the protein of 30 S ribosomal subunits.

Workers with colicin E3 reported repeatedly that the colicin was without effect when added directly to protein synthesizing extracts; more specifically, ribosomes treated with partially purified E3 were not rendered inactive in protein synthesizing systems [39]. It came, therefore, as an unexpected shock when Boon [45] reported that partially purified colicin E3 inhibited protein synthesis in an _in vitro_ system using _E. coli_ ribosomes, producing the same damage to the 16 S RNA as it produces _in vivo_! Nomura's group was prompt in reporting similar, independently obtained results [46].

Before turning to the implications of these recent findings, we should consider them in detail since they represent the first intimation of a direct biochemical, possibly enzymatic, action of a colicin. The action of E3 on ribosomes does not require the supernatant fraction. Ribosomes from sensitive cells or resistant mutants are equally sensitive. Cells that are colicinogenic, and therefore immune, yield ribosomes that _in crude extract_ are not attacked by pure E3, but that become sensitive after being _washed_. It can be demonstrated [46, 47] that an inhibitor is present in extracts of immune cells and, not unexpectedly, also in crude colicin preparations. The crude colicin is relatively inactive toward ribosomes _in vitro_ and becomes active after DEAE-Sephadex chromatography, which

---

*An intriguing background for this elegant series of experiments is the report [44] of ribosomal inactivation by a pyocin, that is, a bacteriocin of _Pseudomonas pyocyanea_. This pyocin is not a simple protein, but like some other bacteriocins, it is a phage tail whose effect on the bacteria is to arrest all macromolecular synthesis including protein synthesis. The inactivation of ribosomes by the pyocin may be a secondary effect of some general breakdown, in the same way that ribosomal damage by colicin E2 (see below) is probably an indirect effect of colicin action.

presumably removes the inhibitor. Crude colicin preparations, as might be expected, inhibit the in vitro activity of pure colicin. Attachment of colicin from a crude preparation to cell receptors also presumably removes the inhibitor. There is some evidence that inhibitor and colicin interact in a reversible, stoichiometric relation [47]. Identification of the inhibitor with the "immunity substance" is a logical step, but must await confirmation by genetic studies.

Colicin E3 also attacks ribosomes from bacteria other than E. coli [47a] and possibly also from mouse cells [47b]. It does not appear to attack the free 16 S rRNA molecules [45, 46]. This leaves open a series of choices: Is the colicin a specific nuclease the phosphodiesterase action of which is exerted only on the RNA molecule in situ, or is the colicin a nuclease activated by ribosomal proteins, or does the colicin activate a ribosomal nuclease? New findings by Boon [47] are relevant to these questions without providing the answers. E3 does not cleave the 16 S RNA in isolated 30 S ribosomal units, but does cleave it if 50 S subunits are added (in 30 mM Mg^{2+}). Both subunits must be present for cleavage to occur.

Whatever the final answers will be, the findings with colicin E3 force us to a careful reexamination of our conceptions of the action, not only of E3 itself, but of all colicins. If a colicin acts by a direct interaction with ribosomes it must penetrate beyond the cytoplasmic membrane. We may think of it as either entering the cytoplasm as a free molecule or remaining bound to the membrane, with its active (enzymatic?) site available for attack on ribosomes. The latter alternative, however unappealing intuitively, is not without precedents, at least in the existence of proteins that span the thickness of a membrane and are accessible from either side [48]. The colicin molecules are strongly asymmetrical (see Table 1) with calculated axial ratios between 6 and more than 20, and their longer axis could easily span the thickness of one or more unit membranes.

It may be of interest to point out a possible resemblance of the action of colicin E3 to that of diphtheria toxin, which affects enzymatically one component of the protein-synthesizing machinery of animal cells by a mechanism of which one part of the toxin molecule is needed only for attachment to the cell surface, and the other only for the intracellular action [49].

There is no evidence that colicin E3 causes damage to the transport systems of the membrane, as the colicins of the K-E1 group do. Yet, bacteriocin PF13, which curiously resembles E3 in specifically inhibiting protein synthesis and inactivating ribosomes, has been reported to cause leakage of K^+ and H^+ from the cells [50].

Like a number of other colicins with a variety of modes of action, colicin E3 has been reported to induce in sensitive cells small increases in lysophosphatidyl-ethanolamine (LPE) and in diphosphatidyl glycerol (DPG) at the expense of phosphatidyl ethanolamine (PE) and phosphatidyl glycerol

(PG), respectively [51]. These small changes, of the same order as those observed following treatment with antibiotics and other chemicals, may be the expression of nonspecific damage. Yet, in view of the evidence that E3 gets to the ribosomes, the phospholipid changes should be kept in mind as possible expression of an enzymatic penetration mechanism.

IV. ACTION OF COLICIN E2

The earliest observations by Nomura [6] revealed that colicin E2 specifically damaged DNA in sensitive cells. A number of other isolates of E2 colicin from different bacterial strains appear to act in a similar way [3]. DNA synthesis is slowed down, but long before synthesis has stopped, damage to the DNA can be detected as a progressive release of acid-soluble radioactivity in cells prelabeled with radioactive thymine. The breakdown of DNA to acid-soluble fragments takes place at a rate that increases with multiplicity of colicin per cell, and its extent may reach 60-95% depending on the bacterial strain.

Syntheses of other macromolecules remain normal for a relatively long time, although RNA also undergoes some breakdown after 30 min or longer, suggesting some delayed damage to ribosomes [52]. Very large amounts of colicin E2 also inhibit synthesis of the RNA phage R17 [53] with a substantial delay. Cell division is inhibited [54], but apparently can still occur in cells with damaged DNA [54a].

Possibly because of the damage to DNA, the killing of bacteria by E2 becomes fairly rapidly resistant to reversal by trypsin [5]. The occurrence of irreversible cellular damage can be prevented by previous addition of 2,4-dinitrophenol or of cyanide plus iodoacetate. Chloramphenicol does not prevent or stop E2-initiated DNA breakdown. Colicin E2 has been reported not to exhibit nuclease activity in vitro [7], but the tests may not have been adequate to detect limited endonucleolytic action.

Extensive but relatively unproductive studies by a number of authors threw little light on the mode of action of E2 except for discovering side effects such as the fact that E2 at low multiplicities induces production of phage λ from λ-lysogens [55]. This is probably an aspecific effect of colicin E2 in common with most agents that interfere with DNA synthesis. E2-initiated degradation affects not only the DNA of the bacterial chromosome, but also that of phage, including the DNA supercoils of phage λ [56, 57].

The solubilization of DNA originally had suggested the involvement of one or more exonucleases. The products of DNA degradation, however, were not identified as consisting only or mainly of single nucleotides, and fragmentation to acid-soluble fragments by endonucleolytic action alone could explain the available data.

The rate, extent, and starting time of DNA breakdown to acid-soluble fragments depend on the amount of colicin [54]. There are no published findings as to whether acid solubilization of cellular DNA starts preferentially at any one portion of the bacterial chromosome. Specifically, it is not known whether there is preferential solubilization of those parts of the chromosome, such as the replication origin or the growing sites, which are supposedly membrane-associated. Experiments with E2 on bacterial mutants such as recA or recB, defective in DNA repair and known or suspected to be defective in some nuclease activity, have given uncertain results. DNA solubilization tends to be increased in recA cells and decreased in recB cells. The killing action of E2, however, is similar on the mutants and on the normal parent strains [54, 54b].

The situation started to become clearer when tests were done, not only on acid solubilization of DNA, but on the molecular properties of DNA extracted from E2-treated cells and examined in neutral or alkaline sucrose gradients. The first report by Obinata and Mizuno [58] described the presence of single-strand breaks in the DNA extracted 7 min after addition of E2 to sensitive cells. Neutral gradients also revealed a reduced size of the DNA fragments. Viability could be restored fully by treating bacteria with trypsin up to 4 min after addition of colicin E2. Addition of trypsin up to 5 or 6 min did prevent further breakage of DNA strands. The authors proposed that the first effect of colicin was the production of single-strand breaks in DNA, and that the trypsin-irreversible lethality was due to formation of two-strand breaks.

A more detailed study by Ringrose [59] looked systematically for products of endonuclease action in the labeled DNA of E2-treated cells using a careful size calibration of DNA fragments. The results confirmed the overall picture presented by Obinata and Mizuno [58]. The DNA showed one-strand scissions after 2 min, and two-strand scissions after about 5 min of colicin treatment, several minutes before any acid-solubilization of DNA was detectable. As endonucleolytic degradation occurred the fragments also became less homogeneous in size, the average size of single-strand fragments decreasing from 2.5×10^7 to 10^5 daltons. The data for lower multiplicities of colicin show evidence that 1-strand breaks accumulate faster than 2-strand breaks. Trypsin added before the 2-strand breaks appear causes repair of already broken strands to occur, a repair clearly visible in sucrose gradient by return of fragments to the original size. Presumably, the repair is carried out by the bacterial polynucleotide ligase [60]. Lethality, indicated by trypsin-irreversible loss of colony-forming ability, sets in between 5 and 10 min, well correlated with the onset of two-strand breaks. (Trypsin reversion of E2 bactericidal action lasted longer in other experiments [38].) On the basis of these observations Ringrose postulated that E2-determined attack on DNA includes (1) a phase of single-strand endonucleolysis; (2) a phase of 2-strand nuclease action; (3) a phase of exonucleolysis of the fragments. It is not clear, however, that phases 2 and 3 need be sequential in that order. To explain the relative uniformity of fragments, Ringrose suggested that single- and possibly also double-strand breaks may occur preferentially or exclusively at certain DNA sites or nucleotide sequences, as is known to be the case in the action of restriction nucleases [61].

Experiments utilizing a more sensitive method that follows colicin action not on the bacterial chromosome, but on circular λ-DNA molecules, which shift from the supercoiled to the open-circle form (one-strand break) to linear molecules of full or partial length [56, 56a] have given results that confirm Ringrose's picture. The observation that energy metabolism is needed for E2-treated bacteria to lose the capacity to be rescued by trypsin can be interpreted as either an energy requirement for a first step of colicin action or a need for energy for the action of a critical nuclease. It is known, for example, that some restriction endonucleases have specific requirements for ATP [61], possibly involved in migration of the enzyme molecules from a recognition site to the substrate site for endonucleolysis [61a].

The nature of the nucleases responsible for E2 action has been considered by a number of authors. Endonuclease I (= endo I) has received special attention, although mutants with supposedly greatly reduced activity of this enzyme are killed by E2 [58]. Endo I is a periplasmic enzyme [62] that is normally found combined with inhibitory tRNA and is fully activated by RNase. The inhibited form produces mostly single-strand breaks in DNA [63], while the uninhibited form produces only double-strand breaks [64]. Obinata and Mizuno [58] found that RNase added to sonicated extracts of E2-treated cells activated endo I more rapidly than in extracts of normal cells and suggested that E2-treated cells might contain less of the inhibitory RNA.

Almendinger and Hager [65] have conducted a series of experiments whose results tend to incriminate endo I in at least some of the effects of colicin E2 by correlating both bacterial killing and DNA degradation by E2 with the level and activity of endo I in the cells. First, release of endo I (and other proteins) by osmotic shock prior to E2 treatment increased survival severalfold and sharply reduced acid-solubilization of DNA by E2. Second, cells treated with E2 and then exposed to osmotic shock released less and less endo I as the colicin concentration used became higher and higher; release of other enzymes was not affected. The authors interpreted this as indication of the transfer of a major fraction of endo I from the periplasmic to the endoplasmic space, and pointed out that a need for such entry of endo I would explain why spheroplasts, which have no endo I left, adsorb E2 without undergoing DNA degradation [26].

As a more direct test for a role of endo I in E2 action, Almendinger and Hager compared bacteria with normal and mutant endo I. Young cells of the "endo I-less" mutant strain were more resistant to E2 than normal cells; in older cells the difference disappeared. These differences paralleled changes in amount of endo I activity as measured by hydrolysis of poly dAT, a substrate that revealed substantial amounts of endo I activity in a supposedly endo-I-less mutant strain. The correlation between sensitivity to E2 and the level of endo I activity was not as good when the substrate used was sonicated E. coli DNA, which should be closer to the in vivo substrate than poly dAT. The authors pointed out that endo I can make both one- and two-strand breaks in DNA and may therefore be responsible for both stages 1 and 2 of E2 action as defined by Ringrose [59]. As an extension of their interpretation, Almendinger and Hager suggested that colicin-initiated transfer of periplasmic enzymes to the endoplasm might be a more general mechanism of colicin action than just for E2 (see, however, the situation for colicin E3).

Almendinger and Hager's findings are susceptible to alternative interpretations. Osmotic shock may decrease the effectiveness of E2 by reducing contact between receptors and the cytoplasmic membrane [66]. In addition, infection with E2 may cause alterations of the cell wall that reduce differentially the release of enzymes by osmotic shock. Mutants truly lacking endo I should help in deciding the role of endo I in colicin E2 action. If this enzyme is in fact involved, it remains to be explained why in the earliest stage of E2 action, when DNA breaks are already present, the cells can be rescued following trypsin treatment [58, 59]. One explanation could be that endo I, although transferred and active on DNA, is still on the cytoplasmic membrane — removal of colicin may reverse the transfer process. Alternatively, the single-strand breaks might be produced by another nuclease — either another periplasmic enzyme or the colicin E2 itself. Tests on E2 nuclease activity in vitro may not have been exhaustive enough to detect localized one-strand scissions. The serological cross-reactivity between colicin E2 and colicin E3 [67], which may have endonucleolytic activity on RNA, does not discourage speculation on a possible enzymatic activity of colicin E2.* A destabilizing effect of colicin E2 on DNA in vitro has been reported [67a]. Whether it plays any role in the in vivo action of E2 is uncertain.

Trypsin reversal still poses a difficult problem even if only the first stage of DNA damage is reversible, presumably by ligase action. Use of ligase-negative bacterial mutants could help answer part of the question.† The action of nuclease(s) located on the cytoplasmic membrane might be more readily reversible by removal of colicin than the activation of nucleases released into the inner endoplasm. Alternatively some nucleases, including the colicin itself, might be unstable intracellularly and DNA degradation may depend on a continuous entry of enzyme maintained by the colicin, an unlikely

*See p. 440, note (1).
†See p. 440, note (2).

but not impossible alternative. It has been reported [68] that colicin E2 added to a preparation of E. coli spheroplast membranes in the presence of ATP (or ADP, or pyrophosphate) causes most of the DNA to detach from the fast sedimenting membrane fraction. The release of DNA from membranes occurs only above 20°C° The released DNA is undegraded; some protein and RNA are also released (but no lipids), suggesting a change in membrane conformation rather than a digestive action. The detachment of DNA might conceivably be related to the entry of nuclease(s) postulated by Almendinger and Hager [65]. One would like to know whether the change in endo I location that has been reported is observed also in bacteria treated with colicin E2 at the lower temperatures.

Yet, adsorbed colicin E2 can produce physiological effects apart from its action on DNA and other macromolecules. This is shown by a remarkable E. coli mutant [68a] the cells of which stop dividing without undergoing DNA degradation upon adsorbing E2 at 30°C. The cells synthesize cell materials, including DNA, and show elongation. Trypsin treatment even after 1 hr restores cell division. At higher temperatures there is DNA degradation and trypsin-irreversible killing. It is evident that cell division per se is a target for colicin E2 action, probably mediated by a reversible action on the cytoplasmic membrane. This mutant, incidentally, responds normally to the other colicins tested.

An additional feature of colicin E2 action on normal bacteria is that the breakdown of bacterial DNA to acid-soluble fragments can be prevented by infection with phage T4, even after DNA breakdown has already started [69]. This interference is readily observed by using T4 mutants that have lost by mutation their own ability to degrade bacterial DNA. The data might be explained by postulating a phage-directed inhibition of bacterial nucleases, similar to the one observed in irradiated bacteria [70]. The mechanism of this inhibition remains obscure, however, and may involve the reversal of activation of nucleases originally present in or outside the membrane. It may be related to the changes present in the mutant bacteria that respond to E2 by arrest of cell division only [68a]. Earlier reports that colicin E2 causes some breakdown of T4 DNA when added to infected bacteria [6] can probably be reconciled with the inhibitory effect of T4 [69] on the basis of different experimental conditions.

Among other bacteriocins a vibriocin [71] resembles colicin E2 in causing degradation of DNA and in other respects. Unlike E2, however, it also causes leakage of K^+ ions and of nucleotides from the cells. Moreover its action is inhibited by chloramphenicol, suggesting that some protein must be synthesized after bacteriocin uptake. The fact that the vibriocin resembles on the one hand colicin E2 in its selective action on DNA and, on the other hand, may cause damage to the membrane is of interest in connection with the general problem of primary colicin action. More on that in the concluding section.

V. ACTION OF COLICINS K AND E1

These colicins are discussed together because they share many features of a pattern of action on cellular processes that is also shared by colicins Ia and Ib [71a] and other less-well-studied colicins. There are, however, in the ways K and E1 act some differences that should be accounted for in any overall interpretations.

All colicins of this group cause a rapid arrest of the synthesis of DNA, RNA, proteins, and polysaccharides [2, 6]. This arrest is observed whether syntheses are measured by chemical tests or by incorporation of isotopic precursors. No general damage to cellular structures is evident—the DNA remains intact, enzymes do not leak out, nor do nucleotides. Synthesis of phospholipids is reduced by only 70 to 80% [26]. Respiration, at least with certain substrates, continues at normal rates. In colicin-treated aerobic cells, the levels of ATP are reduced to less than half, usually about 20% of normal [72] without any leakage of ATP or ADP from the cells.

A reasonable interpretation appeared at first to be that macromolecular syntheses are blocked because of the shortage of ATP. In fact, a mutant with a temperature-sensitive adenylate kinase also stops making proteins and nucleic acids upon transfer to a high temperature, at which the ATP levels become about 20% of normal [73]. It was suggested that AMP:ATP ratios, rather than ATP levels, could be the controlling factor [72]. An important observation by F. Levinthal (unpublished) was that colicin E1 allowed macromolecular syntheses to continue at least for some time in *E. coli* cells growing anaerobically on glucose, and that syntheses stopped when the colicin-treated cells were shifted to O_2. This observation suggested a block in oxidative phosphorylation as the cause of the reduction of ATP levels and raised the possibility that colicins E1 and K acted as uncouplers.

Studies on uptake and exit of various substances into and from colicin-treated cells have revealed additional features and provided a different interpretation of the action of the colicins of this group. In certain respects, E1 and K do affect active transport systems as the uncouplers do. Like uncouplers, they prevent the energy-dependent accumulation of amino acids [14] and of thiomethyl-β-D-galactoside (= TMG) and cause their exit from preloaded cells [74]; they have only a slight effect on the rate of hydrolysis of o-nitrophenyl-β-D-galactoside by intact cells, an indication that the permeability barrier of the cell is not damaged. The utilization of glucose and the accumulation of α-methyl glucoside, both driven by the phosphoenolpyruvate-dependent phosphotransferase system [75], are not inhibited [74]. In fact, glucose is utilized by K- or E1-treated cells at about the same rate as by normal cells; but a substantial part of its carbon atoms reappear promptly in the medium [72] as phosphorylated intermediates (glucose-6-phosphate, fructose diphosphate, dihydroxyacetone phosphate, and 3-phosphoglycerate), which are not excreted by normal cells even in the presence of uncouplers such as carbonyl cyanide m-chlorophenylhydrazone (CCCP) or 2,4-dinitrophenol (2,4-DNP). Thus, a major action of these colicins is to cause a selective inability to accumulate or retain certain substrates and metabolites.

Studies on ion transport are further evidence that colicins K and E1 do not function like the chemical uncouplers. Both colicins cause efflux of accumulated K^+ ions from E. coli cells [26, 76-79]. After addition of colicin K to cells preloaded with $^{42}K^+$, efflux starts after a lag that decreases as the amount of colicin and the temperature increase. The rate of K^+ escape increases with the same two variables; below 10°C there is no exit [78]. CCCP does not cause K^+ efflux. The difference between colicin E1 or K and the chemical uncoupler was elegantly shown by using cells suspended in an unbuffered acid medium [77]; addition of colicin did not permit entry of H^+ but allowed K^+ to escape; CCCP alone did not allow H^+ to enter; but CCCP added after colicin allowed entry of H^+, presumably because the colicin permitted the counterflow of K^+. Formally the colicin acted as valinomycin does on mitochondria or gram-positive bacteria [80].

Experiments with $^{28}Mg^{2+}$ [81] have confirmed the difference between uncouplers and colicins and have revealed an interesting distinction between colicins K and E1. At low Mg^{2+} concentrations E1, like the uncoupler p-trifluoromethoxycarbonylcyanide phenylhydrazone (=FCCP), inhibits the exchange of Mg^{2+} between the medium and the cells (it is not clear whether a similar inhibition occurs for the exchange of K^+, as distinct from exit). Colicin K affects Mg^{2+} exchange only slightly, but strongly increases the permeability to Mg^{2+} at high Mg^{2+} concentrations and, therefore, causes the actual exit of preaccumulated Mg^{2+}, with modalities similar to those encountered with K^+ ions [78]. In a mutant strain [82] that has lost a Mg^{2+} transport system responsible for permitting uptake of Co^{2+}, colicin K causes entry of Co^{2+}. Colicin E1 has a slighter effect on permeability to Mg^{2+} or Co^{2+}. The effects of colicin K or E1 on entry and exit of various ions can be prevented by pretreatment of the cells with CCCP (or FCCP or KCN) before addition of colicin [81, 83]. FCCP added after the efflux of K^+ or Mg^{2+} caused by colicin K has started is without effect, whereas in the absence of colicin it can stop the exchange of external and internal Mg^{2+}.

We can now draw a more precise pattern. Colicins E1 and K alter the functional membrane in such a way that accumulation of various substances is inhibited. In most cases the inhibition may be caused by an uncoupling action, which may also be responsible for the inhibition of oxidative phosphorylation that is responsible for the low ATP levels. A likely target, the damage of which by colicin E1 or K can explain the uncoupling action, is the system described by Kaback [84, 85], which couples the transport of amino acids, sugars, and also the valinomycin-induced transport of some ions to the transport of electrons from D-lactate or other electron donors to oxygen. This system is active in the membrane vesicles the transport activities of which are inhibited by E1 [86] and K [36].

In addition, both colicins E1 and K promote the exit of phosphorylated intermediates of glycolysis and of various cations, colicin K being especially effective in promoting exit and entry of Mg^{2+} and Co^{2+} ions. To explain these alterations we may envisage the creation by the colicins of specific "channels" for passage of the ions and possibly of other substances. These channels might represent either specific intermolecular pathways in the membrane or molecular carriers (analogous to the macrocyclic antibiotics) created by the action of the colicin, possibly at the sites of the corresponding intake systems. Chemical uncouplers or low temperatures may prevent the initial action of the colicins by preventing some of the reactions leading to formation of these channels.

It is clear that the failure of colicin-treated cells to accumulate substrates such as amino acids or galactosides is not an indirect effect of the lowered ATP levels. Colicins K and E1 inhibit accumulation also in membrane vesicles [36, 86, 86a] that do not make ATP [84]. On the other hand, K^+ exit after colicin treatment also occurs [77] under conditions when ATP levels are not lowered by colicin, that is, in the presence of N, N-dicyclohexylcarbodiimide, an inhibitor of membrane-bound ATPase [87]. More important still, under these conditions the syntheses of proteins and nucleic acids are still inhibited by colicin E1 despite the high ATP levels [77].

If ATP shortage is not the cause of the arrest in biosynthesis, then what is? Some relevant data have been obtained in experiments [88] in which the incorporation of labeled leucine into protein was followed together with its level in the intracellular pool. When the labeled leucine was added a few minutes before colicin E1, so that the pool was preloaded, incorporation into protein continued for a longer time at a decreasing rate that paralleled the decrease of the size of the leucine pool. These experiments suggest that one cause for the arrest of protein synthesis is unavailability of amino acids owing to failure to be retained in the cell.

A number of aspects of the action of colicins K and E1 remain unexplained. For example, the *in vivo* function of pyruvic dehydrogenase seems to be blocked since pyruvate is excreted during glucose catabolism by colicin-treated *E. coli* bacteria [72]. Some regulatory mechanism may be involved. Similarly, the anaerobiosis question remains open. Protection by strict anaerobiosis against colicins E1 or K is not due simply to the type of energy metabolism carried out by the cells, because cells grown under incomplete anaerobiosis metabolize glucose in the typical fermentative pattern and yet are inhibited by colicin E1 [72]. The protection afforded by anoxia may indicate that the damage to membrane functions requires either that electron transport is actively taking place, or that molecular oxygen be present, or both. In anaerobic cells, active transport might be driven endogenously by ATP derived from substrate-level phosphorylation without involvement of electron transport [88a, 88b]. Such a dual way of driving a membrane-associated reaction in E. coli would not be unique [89]. It now appears that in E. coli, even under aerobic conditions, some amino acids are transported by a mechanism activated directly by ATP rather than by an activated state generated by electron transport [89a].

Feingold [77] has hypothesized that the bacterial killing by E1 or K might be an oxidative process, possibly of unsaturated fatty acids, resembling the killing of bacteria by phagocytes. This brings us to the question of the primary action of the colicins of this group upon the cellular envelope.

Assuming that the significant events affect the cell membrane, the primary change produced by the colicins may be either enzymatic or conformational. A conformational change, because of local shifts in steric interactions, may then be followed by enzymatic changes, for example, by activation of membrane enzymes. The idea of a conformational change that could spread by propagation on a two-dimensional membrane [80] has been suggested as one mechanism for amplification of colicin action [7-9a]. Studies have been reported [91a] of the effect of colicin E1 on the fluorescent probes 8-anilino-1-naphthalenesulfonate (=ANS) and N-phenyl-1-naphthylamine (=NPN) in the presence of sensitive bacteria. After a short lag the fluorescent signal caused by the cell-dye association increased twofold, the rate of increase being dependent on colicin multiplicity. Increased uptake did not seem sufficient to explain the change in signal, which was accompanied by a blue shift in the emission spectrum and was interpreted as indicative of a conformational change in the envelope exposing dye-binding sites to a changed environment.

The question of the primary actions of colicins K and E1 also concerns the reversibility, specifically by trypsin but also by dodecyl sulfate [38a]. The reversibility of colicin K effects [4] was one of the mainstays of Nomura's formulation of colicin action [6, 7]. Experiments on reversibility give results that are not always in full agreement. True reversibility requires not only recovery of colony-forming ability under conditions in which no actual damage has occurred, but recovery of function and viability in cells that have already manifested functional damage. Such true reversibility was indeed observed in the original experiments with colicin K on E. coli B cells [4], but some questions have been raised recently as to the general applicability of these findings [79, 79a, 92]. Plate and Luria [92b] found that cells of E. coli K-12 treated with colicin K lose their trypsin rescuability, as measured by viable counts, at an exponential rate directly proportional to the amount of colicin adsorbed. Similar results were obtained with colicin E1. A careful comparison of viability rescue and residual levels of transport or of protein synthesis at various times clearly indicated that only those cells that had manifested no functional damage could be rescued by trypsin. Thus each adsorbed colicin molecule that can damage a cell and kill it has a constant probability per unit time of becoming an actual killer. The potential killers might be those colicin molecules adsorbed at special sites of interaction between wall and cytoplasmic membrane [92d].

These findings indicate that the action of these colicins goes through an early phase (stage I) in which colicin attachment had had no physiological effects, followed by a phase (stage II) in which physiological effects are detectable and rescuability by trypsin is lost or greatly reduced. The rate of the transition from stage I to stage II is strongly dependent on temperature [92a], explaining why at low temperature colicin K-treated cells remain rescuable for very long times [78]. The Arrhenius plot plot for the stage I — stage II transition is discontinuous and resembles the biphasic Arrhenius plots for several membrane-associated systems in E. coli [92c]. In an unsaturated fatty acid auxotroph of E. coli the temperature of the discontinuity varies with the type of unsaturated fatty acid used to supplement growth [92a]. Thus the action of colicin K appears to be influenced by changes in the packing characteristics, or "fluidity," of the lipid components of the E. coli envelope. Similar features in the dependence of the action of colicin El on temperature have been observed with fluorescent probes [92e].*

VI. GENETIC STUDIES

Many biochemical problems have been clarified, at least in part, by genetic studies. Colicin action is an exception. It is fair to state that the genetic approach has contributed little toward an understanding of the way colicins act. Resistance to colicin action acquired by mutation and resulting in loss of ability to adsorb one or more colicins has been correlated with the absence in cellular extracts of colicin-inactivating proteins, presumably the receptors [31]. A search for mutants that have receptors but fail to respond to colicins has been carried out in many laboratories in the hope of discovering "target mutants," that is, mutants altered in the key biochemical process responsible for the action of a certain class of colicins. Instead, what has been found is a variety of "tolerant" mutants [93-96a], whose lesions generally are not related to the biochemical action of colicins, but probably alter the envelope so that the colicins, though adsorbed, cannot reach their targets or cannot alter them. Most tolerant mutants are insensitive to sets of colicins with different mode of action. Only for colicin E2 is there some evidence that one group of temperature-dependent mutants, called CetC [97], may have a lesion related to the action of the colicin since these mutants are UV-sensitive and have a defective DNA repair system. When extracts of the envelope of these

*See p. 440, note (3).

mutants are submitted to SDS gel electrophoresis one protein peak, about 44,000 MW, is found to be increased above the level in the parent strain [98].

Most tolerant mutants exhibit serious defects in their surface layer. For example, Tol C mutants, tolerant only to colicin E1 and sensitive to K and other colicins of similar action [93, 99], are very permeable to certain dyes [96], resembling mutants isolated as acridine sensitive (but colicin sensitive; see Sugino [100] and Nakamura [101]). Tol A and Tol B mutants, besides being resistant to a series of colicins with different modes of action, are hypersensitive to deoxycholate (as are many other envelope mutants) and undergo spontaneous lysis at a frequency rising with temperature of incubation [96]. In all the cases analyzed the Tol phenotype is recessive to Tol^+ in diploid cells [95, 96, 102]. Tolerance to some colicins has also been found in mutants isolated in connection with some unrelated aspect of envelope physiology [103].

Given the task of a colicin to carry or send biochemical signals from surface receptors to some deeper layer of cellular organization, it is not surprising that a variety of mutational changes can interfere with these (unknown) transmission processes (see Burman and Nordström [103] for a recent discussion) or that these mutations should alter some proteins of the cellular envelope [98, 98a, 104], especially since some of the mutations are of a kind suppressible by amber or ochre suppressors [105]. The difficulty is to attribute to such proteins any role in the action of colicins unless this action can be tested, as for colicin E3, in an in vitro subcellular system.

Some classes of mutations to tolerance provide some insight, however, into the functional relation of the colicins to the cell envelope. One class consists of mutants that are sensitive at 30° C, but when transferred to 40° C become insensitive to E2 and E3 and grow poorly [95]. The conversions from sensitivity to resistance and back upon shifts of temperature also occur in the presence of chloramphenicol, suggesting that some component is reversibly shifted between two alternative states. Mutants of a second type, when shifted to 40° C, stop dividing and become insensitive to E2 and E3; they still adsorb these colicins, but are not damaged and can be rendered viable by trypsin treatment. In these mutants colicin action appears to be dependent on a protein also essential for continued cell growth.

A third set of temperature-dependent tolerant mutations has been found in Tol A and Tol C genes [105]. Here tolerance or sensitivity require growth at 41° or 30° C, respectively, and cannot be acquired by temperature shift in the absence of protein synthesis. It appears that only those colicin molecules that are adsorbed to cell material made at 30°C can kill the cells. The parts of the envelope made at the higher temperature do adsorb colicin but appear to be irreversibly "disconnected" from the mechanism for signal transmission — possibly the transfer of colicin itself. Cells grown at 41 °C and transferred to 30 °C have for several generations a mosaic of sensitive and insensitive regions, revealed by their progressively increasing sensitivity.

Methods for selecting "target" mutants, that is, mutants that do not respond to a colicin because of failure to undergo the ultimate damage, have been tried but have not yet been productive. One might expect that some ribosomal mutations would produce resistance to E3. Changes in some membrane component might render cells refractory to K, E1, and other colicins with the same mode of action. If the action of E2 should prove to be exerted through endonuclease I [65], mutants altered in that enzyme may exhibit interesting relations to this colicin.

VII. CONCLUDING REMARKS

There must be some order in the apparent madness of a set of agents, produced by a common process, and resembling each other in their bactericidal effect, and which nevertheless act in such disparate ways at the biochemical level. At the present stage of our knowledge we must search for order from two sources — analysis and analogy.

An analysis of the mode of action of the various colicins suggests that, beyond trivial elements such as combination with specific receptors, an important similarity may be the requirement for energy metabolism for an initial action. The requirement is well documented for colicins K, E2, and E3, which are prevented from acting by the presence of uncouplers or inhibitors [5, 26, 81, 83]; it is present for E1, which like E2 and K does not act irreversibly at low temperatures. If we assume that the first "meaningful" action of a colicin is on the cytoplasmic membrane, we can visualize an initial energy-requiring action on the membrane. This action may then lead to the various specific lesions: entry of E3 into the endoplasm; inhibition of cell division and entry of nucleases (including possibly the colicin itself) in cells that adsorb E2; and damage to the membrane by K or E1. The energy-requiring step in the initial reaction is unlikely to concern the transfer of colicin from an external receptor to the membrane. It is more likely an interaction that requires the membrane to be in a coupled condition in order to be acted upon in a way that leads to creation of "channels" for the colicin itself, or for periplasmic enzymes, or for a variety of other substances (or to a permanent uncoupling of electron transport). The protection afforded by anaerobiosis against colicins K and E1 might be the result of the state of the membrane in anaerobic cells.

There are no data available yet on the action of colicins E2 or E3 in strict anaerobiosis.

In this view the key step in the primary action of the colicin is an alteration of the membrane. We have no way as yet to assert whether this primary alteration is enzymatic or conformational. When an enzymatic effect is detected, it clearly can account for the "amplification" effect that can make a single colicin molecule responsible for cell death. But in examining the membrane phenomena, which may be the critical primary action of colicins, we cannot exclude amplification mechanisms based on spreading conformational disturbances [90], as well as on actual diffusion of molecules in a more or less homogeneously "fluid" membrane [9a]. The period during which the killing action of colicin can be prevented by trypsin may be one in which reversible conformational alterations have taken place.

Analogies are less helpful the less one knows about the phenomena one is trying to analogize. The analogy with phages is not very useful because phages introduce a nucleic acid genome that generates a variety of proteins in the bacterial endoplasm. An analogy that has seemed attractive [12] is that of T-even phage "ghosts," i.e., phage particles deprived of nucleic acids. The effect of ghosts resembles in some respects that of colicins E1 and K, but it differs in fundamental respects such as the loss of ATP from ghost-treated cells to the medium and the inhibition by ghosts of glucoside transport mediated by the phosphotransferase system [106]. Thus, ghosts have a much less selective effect on membranes. It seems likely that the resemblance is purely coincidental. After all, it would be surprising if several agents that damage the membrane without disrupting it did not have some effect in common. The very interesting work on phage ghost action [107] is probably of little direct relevance to colicins. This remark applies also to those supposed bacteriocins that have turned out to be phage tails.

In speculating about the biological nature and role of colicins, colicinogeny, and sensitivity to colicins, the two most popular themes have been the relationships to phages and to mating systems [3, 12]. Phages have seemed attractive, not only because of the cell-killing capacity of some virulent strains, but also because prophages resemble (some) Col factors in being controlled by a repressor subject to induction. This analogy may be trivial: many bacterial regulatory systems, including temperate phage repression, have in common all sort of properties, for example, sensitivity to cyclic AMP [108]. Colicin repressors may just be a subset of bacterial repressors, some being heat-labile, others heat-stable, some uv-sensitive, others insensitive.

The same can be said for the analogy with mating systems. Specific surface receptors are probably involved in both cases (and also with phages), but this basis for analogy is almost meaningless, as would be the assertion that light and sound are related phenomena because they are both perceived through specific receptors. Association of Col factors with fertility factors [109, 110] can be explained by well-known recombination patterns and, in fact, new associations and dissociations do arise in the course of laboratory manipulations [111]. A reasonable argument in the case made for a relation to mating systems is that colicins usually attack only, or at least preferentially, bacteria related to the producer strain. Such poisons to be used only within the family not unreasonably suggest marriage.

But if we look at what colicins actually do, it is easier to think of another relationship: that of bacterial proteins, either enzymatically active or not, which have evolved from ancestors that were functionally integrated in the bacterial cell, and which probably had specific functions in the cytoplasmic membrane, the periplasmic space, or the cell wall. Shift of the corresponding structural and regulatory genes to episomes, and thereby to other bacteria, may have rendered these proteins incapable of functioning in their original setting. Natural selection would then put a premium on those that, by possessing or gaining secretory and bactericidal properties, gave to the harboring strain a selective advantage in the wide open world — that is, for the enteric bacteria, the gut of warm-blooded animals. This narrow view has one sole advantage: it is potentially testable by searching for cellular proteins that possess colicin-like activity. If such were found, a study of their production and regulation might be illuminating.

ACKNOWLEDGMENTS

The work by the author and his collaborators has been supported by grants from the National Science Foundation (GB-5304X) and The National Institutes of Health (AI-03038).

REFERENCES

1. A. Gratia, C. R. Soc. Biol., 93, 1040 (1925).

2. F. Jacob, L. Siminovitch, and E. L. Wollman, Ann. Inst. Pasteur, 83, 295 (1952).

3. P. Reeves, Bacteriol. Rev., 29, 24 (1965).

4. M. Nomura and M. Nakamura, Biochem. Biophys. Res. Commun., 7, 306 (1962).

5. B. L. Reynolds and P. R. Reeves, Biochem. Biophys. Res. Commun., 11, 140 (1963).

6. M. Nomura, Cold Spring Harbor Symp. Quant. Biol., 28, 315 (1963).

7. M. Nomura, Proc. Nat. Acad. Sci., U. S., 52, 1514 (1964).

8. S. E. Luria, Ann. Inst. Pasteur, 107 (Suppl.), 67 (1964).

9. J. P. Changeux and J. Thiery, J. Theoret. Biol., 17, 315 (1967).

9a. S. J. Singer and G. L. Nicolson, Science, 175, 720 (1972).

10. P. Fredericq, Ann. Rev. Microbiol., 11, 7 (1957).

11. Y. Hamon, Ann. Inst. Pasteur, 107, 18 (1964).

12. M. Nomura, Ann. Rev. Microbiol., 21, 257 (1967).

12a. P. Reeves, The Bacteriocins, Springer-Verlag, New York, 1972.

13. D. T. Kingsbury and D. R. Helinski, Biochem. Biophys. Res. Commun., 41, 1538 (1970).

14. P. Fredericq, Zentr. Bakt. Parasitenk., Abt. I (Orig.), 196, 142 (1965).

15. M. Bazaral and D. R. Helinski, J. Mol. Biol., 36, 185 (1968).

16. D. B. Clewell and D. R. Helinski, Proc. Nat. Acad. Sci., U. S., 62, 1159 (1969).

16a. D. B. Clewell, J. Bacteriol., 110, 667 (1972).

16b. D. G. Blair, D. J. Sherratt, D. B. Clewell, and D. R. Helinski, Proc. Nat. Acad. Sci. U. S., 69, 2518 (1972).

16c. J. Inselburg, Nature (London), 241, 234 (1973).

17. C. K. Kennedy, J. Bacteriol., 108, 10 (1971).

18. D. R. Helinski and H. R. Hershman, J. Bacteriol., 94, 700 (1967).

18a. A. Nakazawa and T. Tamada, Biochem. Biophys. Res. Commun., 46, 1004 (1972).

19. M. Kohiyama and M. Nomura, Zentr. Bakt. Parasitenk., Abt. I (Orig.), 196, 211 (1965).

20. C. Haussman and R. C. Clowes, J. Bacteriol., 107, 900 (1971).

21. W. Goebel, Eur. J. Biochem., 15, 311 (1970).

22. V. R. Lemoine and R. Nagel de Zwaig, Acta Cient. Venezolana, 22, 44 (1971).

23. C. K. Kennedy and E. Sodergren, J. Bacteriol., 112, 736 (1972).

24. P. Fredericq, Symp. Soc. Exptl. Biol., 12, 104 (1958).

25. M. Ptashne, in The Bacteriophage Lambda (A. D. Hershey, ed.), Cold Spring Harbor Laboratory, 1971.

26. M. Nomura and A. Maeda, Zentr. Bakt. Parasitenk., Abt. I (Orig.), 196, 216 (1965).

27. A. Maeda and M. Nomura, J. Bacteriol., 91, 685 (1966).

28. S. A. Schwartz and D. R. Helinski, J. Biol. Chem., 246, 6318 (1971).

28a. M. A. Jesaitis, J. Exp. Med., 131, 1016 (1970).

28b. J. Konisky and F. M. Richards, J. Biol. Chem., 245, 2972 (1970).

28c. K. Kunugita and M. Matsuhashi, J. Bacteriol., 104, 1017 (1970).

29. R. Shannon and A. Hedges, J. Bacteriol., 93, 1353 (1967).

30. E. Mitusi and D. Mizuno, J. Bacteriol., 100, 1136 (1969).

31. S. F. Sabet and C. A. Schnaitman, J. Bacteriol., 108, 422 (1971).

31a. S. F. Sabet and C. A. Schnaitman, J. Biol. Chem., 248, 1797 (1973).

32. H. U. Weltzien and M. A. Jesaitis, J. Exp. Med., 133, 534 (1971).

32a. C. A. Schnaitman, personal communication.

33. J. A. Smit, L. A. Stocken, and H. C. De Klerk, J. Gen. Microbiol., 65, 249 (1971).

34. S. K. Guterman, Biochem. Biophys. Res. Commun., 44, 1149 (1971).

35. J. Šmarda and U. Taubeneck, J. Gen. Microbiol., 52, 161 (1968).

36. P. Bhattacharyya, L. Wendt, E. Whitney, and S. Silver, Science, 168, 998 (1970).

36a. Y. Takagaki, K. Kunugita and M. Matsuhashi, J. Bacteriol., 113, 42 (1973).

37. M. Ozaki, Y. Higashi, H. Saito, T. An, and T. Amano, Biken's J., 9, 201 (1966).

38. B. L. Reynolds and P. R. Reeves, J. Bacteriol., 100, 301 (1969).

38a. D. Cavard, J. Marotel-Schirmann, and M. E. Barbu, C. R. Acad. Sci., 273, 1167 (1971).

39. J. Konisky and M. Nomura, J. Mol. Biol., 26, 181 (1967).

40. B. W. Senior, J. Kwasniak, and I. B. Holland, J. Mol. Biol., 53, 205 (1970).

41. M. Nomura and P. Traub, in Organizational Biosynthesis (H. J. Vogel, J. O. Lampen, and V. Bryan, eds.), pp. 459-476, Academic Press, New York, 1967.

42. C. M. Bowman, J. E. Dahlberg, T. Ikemura, J. Konisky, and M. Nomura, Proc. Nat. Acad. Sci., U. S., 68, 964 (1971).

43. B. W. Senior and I. B. Holland, Proc. Nat. Acad. Sci., U. S., 68, 959 (1971).

44. Y. Kaziro, M. Tanaka, and N. Shimazono, Biochem. Biophys. Res. Commun., 17, 624 (1964).

45. T. Boon, Proc. Natl. Acad. Sci., U. S., 68, 2421 (1971).

46. C. M. Bowman, J. Sidikaro, and M. Nomura, Nature New Biol., 234, 133 (1971).

47. T. Boon, Proc. Nat. Acad. Sci. U. S., 69, 549 (1972).

47a. J. Sidikaro and M. Nomura, FEBS Letters, 29, 15 (1973).

47b. F. Turnowsky, J. Drews, F. Eich and G. Högenauer, Biochem. Biophys. Res. Commun., 52, 327 (1973).

48. M. S. Bretcher, Nature New Biol., 231, 225 (1971).

49. T. Uchida, D. M. Gill, and A. M. Pappenheimer, Jr., Nature New Biol., 233, 8 (1971).

50. F. K. De Graaf, E. A. Spanjaerdt Speckman, and A. H. Stouthamer, Ant. v. Leeuwenhoek, 25, 287 (1969).

51. D. Cavard, C. Rampini, E. Barbu, and J. Polonovski, Bull. Soc. Chim. Biol., 50, 1455 (1968).

52. K. Nose and D. Mizuno, J. Biochem., 64, 1 (1968).

53. R. K. Fujimura, J. Mol. Biol., 17, 75 (1966).

54. E. M. Holland and I. B. Holland, J. Gen. Microbiol., 64, 223 (1970).

54a. T. Beppu and K. Arima, J. Biochem., 70, 263, 1971.

54b. L. S. Saxe, unpublished data.

55. H. Endo, T. Kamiya, and M. Ishizawa, Biochem. Biophys. Res. Commun., 7, 306, 1963.

56. L. S. Saxe, and S. E. Luria, Bacteriol. Proc., 1971, p. 50.

56a. L. Saxe, Ph.D. Thesis, Mass. Inst. of Technology (1973).

57. R. R. Hull and P. R. Reeves, J. Virol., 8, 355 (1971).

58. M. Obinata and D. Mizuno, Biochim. Biophys. Acta, 199, 330 (1970).

59. P. Ringrose, Biochim. Biophys. Acta, 213, 320 (1970).

60. M. Gellert, Proc. Nat. Acad. Sci., U. S., 57, 148 (1967).

61. T. J. Kelly, Jr., and H. O. Smith, J. Mol. Biol., 51, 393 (1970).

61a. K. Horiuchi and N. D. Zinder, Proc. Nat. Acad. Sci., U. S., 69, 3220 (1972).

62. N. G. Nossal, and L. A. Heppel, J. Biol. Chem., 241, 3055 (1966).

63. W. Goebel and D. R. Helinski, Biochemistry, 9, 4793 (1970).

64. F. W. Studier, J. Mol. Biol., 11, 373 (1965).

65. R. Almendinger, and L. P. Hager, Nature New Biol., 235, 199 (1972).

66. T. Beppu, and K. Arima, J. Bacteriol., 93, 80 (1967).

67. H. R. Hershman and D. R. Helinski, J. Biol. Chem., 242, 5360 (1967).

67a. P. S. Ringrose, FEBS Letters, 23, 241 (1972).

68. T. Beppu and K. Arima, Biochim. Biophys. Acta, 262, 453 (1972).

68a. T. Beppu, K. Kawabata, and K. Arima, J. Bacteriol., 11, 485 (1972).

69. R. L. Swift and J. S. Wiberg, J. Virol., 8, 303 (1971).

70. J. D. Chapman, J. Swez, and E. C. Pollard, Nature (London), 218, 690 (1968).

71. A. Jayawardene, and H. Farkas-Himsley, J. Bacteriol., 102, 382 (1970).

71a. R. Levinsohn, J. Konisky, and M. Nomura, J. Bacteriol., 96, 811 (1967).

72. K. L. Fields, and S. E. Luria, J. Bacteriol., 97, 64 (1969).

73. D. Cousin, Ann. Inst. Pasteur, 113, 309 (1967).

74. K. L. Fields, and S. E. Luria, J. Bacteriol., 97, 57 (1969).

75. W. Kundig, S. Ghosh, and S. Roseman, Proc. Nat. Acad. Sci., U. S., 52, 1067 (1964).

76. H. Hirata, S. Fukui, and S. Ishikawa, J. Biochem., 65, 843 (1969).

77. D. S. Feingold, J. Membr. Biol., 3, 372 (1970).

78. L. Wendt, J. Bacteriol., 104, 1236 (1970).

79. J.-P. Dandeu, A. Billault, and E. Barber, C. R. Acad. Sci., 269, 2044 (1969).

80. F. M. Harold, and J. R. Baarda, J. Bacteriol., 96, 2025 (1968).

81. J. E. Lusk and D. L. Nelson, J. Bacteriol., 112, 148 (1972).

82. D. L. Nelson and E. P. Kennedy, Proc. Nat. Acad. Sci., U. S., 69, 1091 (1972).

83. H. Hirata, S. Fukui, and S. Ishikawa, 1972, in press.

84. H. R. Kaback, Ann. Rev. Biochem., 39, 561 (1970).

85. H. R. Kaback, Biochim. Biophys. Acta, 265, 367 (1972).

86. J. P. Kabat and S. E. Luria, Bacteriol. Proc. 1970, p. 62.

86a. J. P. Kabat, Ph.D. Thesis, Mass. Inst. Technol. (1971).

87. F. M. Harold, J. R. Baarda, C. Baron, and A. Abrams, J. Biol. Chem., 294, 2261 (1969).

88. J. E. Lusk, unpublished observations.

88a. W. L. Klein and P. D. Boyer, J. Biol. Chem., 247, 7257 (1972).

88b. H. U. Schairer and B. A. Haddock, Biochem. Biophys. Res. Commun., 48, 544 (1972).

89. R. Fisher and D. R. Sanadi, Biochim. Biophys. Acta, 245, 34 (1971).

89a. E. A. Berger, Proc. Nat. Acad. Sci., U. S., 70, 1514 (1973).

90. J. P. Changeux, J. Thiery, Y. Tung, and C. Kittel, Proc. Nat. Acad. Sci., U. S., 57, 335 (1967).

91. W. A. Cramer and S. K. Phillips, J. Bacteriol., 104, 819 (1970).

91a. S. K. Phillips and W. A. Cramer, Biochemistry, 12, 1170 (1973).

92. J. Marotel-Schirmann, D. Cavard, L. Sandler, and E. Barbu, C. R. Acad. Sci., 270, 230 (1970).

92a. C. A. Plate, Antimicrobial Agents and Chemotherapy, in press;

92b. C. A. Plate, and S. E. Luria, Proc. Nat. Acad. Sci., U. S., 69:2030-2034 (1972).

92c. J. E. Cronan, Jr., and P. R. Vagelos, Biochim. Biophys. Acta, 265, 25 (1972).

92d. M. E. Bayer, J. Virol., 2, 346 (1968).

92e. W. A. Cramer, S. K. Phillips, and T. W. Keenan, Biochemistry, 12, 1178 (1973).

93. R. C. Clowes, Zblatt f. Bakt., 196, 152 (1965).

94. C. Hill and I. B. Holland, J. Bacteriol., 94, 677 (1967).

95. M. Nomura and C. Witten, J. Bacteriol., 94, 1093 (1967).

96. R. Nagel de Zwaig and S. E. Luria, J. Bacteriol., 94, 1112 (1967).

96a. A. Bernstein, B. Rolfe, and K. Onodera, J. Bacteriol., 112, 74 (1972).

97. I. B. Holland, E. J. Threlfal, V. Darby, E. M. Holland, and A. C. R. Samson, J. Gen. Microbiol., 62, 371 (1970).

98. I. B. Holland, A. C. R. Samson, E. M. Holland, and B. W. Senior, in "Growth Control in Cell Cultures," Ciba Found. Symp, 1971, p. 221.

98a. I. B. Holland and S. Tuckett, J. Supramol. Struct., 1, 77 (1972).

99. E. N. Whitney, Genetics, 63, 30 (1971).

100. Y. Sugino, Genet. Res. Cambridge, 7, 1 (1966).

101. H. Nakamura, J. Bacteriol., 96, 987 (1968).

102. E. J. Threlfal and I. B. Holland, J. Gen. Microbiol., 62, 383 (1970).

103. L. G. Burman and K. Nordström, J. Bacteriol., 106, 1 (1971).

104. B. Rolfe and K. Onodera, Biochem. Biophys. Res. Commun., 44, 767 (1971).

105. R. Nagel de Zwaig and S. E. Luria, J. Bacteriol., 94, 1112 (1967).

106. H. H. Winkler and D. H. Duckworth, J. Bacteriol., 107, 259 (1971).

107. D. H. Duckworth, Bacteriol. Rev., 34, 344 (1970).

108. J.-S. Hong, G. R. Smith, and B. N. Ames, Proc. Nat. Acad. Sci., U. S., 68, 2258 (1971).

109. H. Ozeki, B. A. D. Stocker, and S. M. Smith, J. Gen. Microbiol., 28, 671 (1962).

110. P. Fredericq, Ergebn. Mikrob. Immunol. Exp. Therap., 37, 114 (1963).

111. R. Nagel de Zwaig, Genetics, 54, 381 (1966).

Chapter 7

CELL SURFACE STRUCTURES AND THE ABSORPTION OF DNA MOLECULES DURING GENETIC TRANSFORMATION IN BACTERIA

Alexander Tomasz

The Rockefeller University
New York, New York

I. INTRODUCTION

This chapter is concerned with early stages in genetic transformation in pneumococci. However, it does not intend to be a complete and critical survey of all the facts. Recent reviews of this type are available for the experts [1, 2]. The purpose of the present treatise is to develop genetic transformation as a phenomenon of cell physiology in which bacteria can bind and absorb DNA molecules from their environment. Studies on the mechanism of this phenomenon can contribute in a unique way to our understanding of the functioning and structure of the bacterial surface and have already provided interesting information concerning such general problems as cell-to-cell communication, transport of macromolecules through the cell boundaries, functionally specific surface alterations, and replication of macromolecular surface components.

II. GENETIC TRANSFORMATION: A BIOLOGICAL CONTEXT

Some bacteria (about a dozen) species may, under certain growth conditions, show the peculiar capacity of physically binding and absorbing extracellular DNA molecules and transporting them to the cellular interior where these molecules (if genetically homologous with the recipient) may undergo recombination with the resident chromosome. It seems that this phenomenon was quickly recognized and used as a unique genetic technique in which the genetic material (DNA) is accessible to experimental manipulation. On the other hand, transformation as a biological phenomenon, comparable to conjugation or transduction, has not been considered with equal seriousness until quite recently. The main reasons for this seem twofold. One was the simple fact — true until relatively recently — that the source of donor material in a physiological sense (i.e., something analogous to a coli cell with male polarity) did not seem to exist; while it was known that bacteria develop a peculiar condition in which they were "receptive" or competent (analogous to the female coli cell) to absorb DNA molecules from the environment, the source of these molecules was invariably the investigator. There was further reason for paying relatively little attention to competence as a physiological state — some early observations implied that bacterial "competence" may be a pathological condition appearing only in the early stationary phase of growth, i.e., at a time when cultures are likely to accumulate "unhealthy" cells and one can find various manifestations of damage to cellular permeability barriers (leakage of various cell components). The amply documented pathological condition of competent B. subtilis cells seemed to confirm these thoughts.

Today it is clear that, for the case of pneumococci at least, both of these views are incorrect. First, it was shown that the coincidence of competence and the stationary phase of growth is fortuitous: it only occurs in certain complex media and pneumococci in the competent state are healthy, logarithmically growing bacteria [3]. Second, a "biological" source of donor material does exist. It was shown in Hotchkiss's laboratory by Ottolenghi that when two properly labeled (by genetic markers) pneumococcal mutants were grown together in a common growth medium, extensive and bilateral exchange of genetic markers could be observed without any experimental intervention [4]. Such an exchange of genes was also observed in a more natural habitat of these bacteria — in mixed infection of mice with two pneumococcal strains [5]. The emergence of these recombinant cells could be prevented by adding DNase to the medium, and the process giving rise to the recombinants was presumably an endogenous transformation. Cells of both strains must have participated in a two-phase phenomenon, some cells "liberating" genetically active DNA (in a DNase-sensitive form) into the medium, while another group of cells developed the "receptive" or competent condition, absorbed the DNA and underwent genetic recombination. Whereas DNA release has since been demonstrated in all major transformation systems, the mechanism of exit of the DNA molecules from the donor cells is not clear yet. In some cases they may arise from disintegrating cells (uptake of such DNA molecules should properly be considered "genetic cannibalism"); there is some evidence, on the other hand, which suggests that DNA may be actively excreted into the medium [6]. In any case, it seems clear that the two phenomena of DNA release and competence together form a potential recombination system analogous to transduction or sexual recombination in enteric bacteria. One further analogy between the processes of transduction, conjugation, and transformation emerged from studies on the mechanism of induction of competence. It seems that in all three processes the introduction of donor DNA into the recipient cell is preceded by an interaction between recipient surface and extracellular factors of nongenetic nature. In the virus-mediated process viral tail-tip proteins, and possibly other parts of the viral coat, interact with viral receptors on the cell surface [7]; in conjugation, pili and other male-specific factors react with periodate-sensitive sites on the female cell envelope [8]; in transformation prior to the uptake of DNA molecules the surface of the recipient cells has to be made "competent" by interaction with molecules of a specific extracellular protein, the Activator [9], which is produced endogenously by pneumococcal cultures as they reach a critical cell concentration range [3]. In virus-mediated DNA uptake and in conjugation these surface interactions can be hardly separated from the injection of donor DNA, i.e., recipient and donor functions appear in a concerted fashion. By contrast, in transformation, the recombinational mechanism seem fragmented in time and space, e.g., DNA may appear in the medium even in the absence of competent recipient cells and the recipient (competence) state may also appear in the absence of donor material. On the other hand, even in this imperfect gene-exchange system there appear on occasion suggestions of "attempts" at coordination between recipient and donor states; maximal competence during growth of pneumocci may coincide with the time of maximum concentration of DNA in the medium [4]. In turn, maximal competence of the culture is achieved through the inductive effect of the "activator substance." One may also mention the observation that pneumococci in the competent phase have a tendency to clump [10]. One may speculate that if such clumps contained both donor and recipient (competent) cell types, then the proximity of the cells may be analogous to a primitive conjugal condition.

III. CHARACTERISTICS OF THE "COMPETENT PHASE" OF BACTERIAL POPULATIONS

In any discussion of competence it is important to distinguish two aspects — one is the mechanism responsible for the abrupt, synchronous, and very rapid appearance of the competent condition in the culture. I shall briefly discuss these population-level phenomena separately from the second aspect of competence, which is concerned with the specific biochemical features of competent bacteria.

Competent cells appear in cultures of transformable species of bacteria under almost any growth condition with varying frequencies, which may be as low as 10^{-5} to 10^{-6}. Of course, for the biochemical analysis of the nature of the competent state one needs higher frequencies of competent cells. Experience has shown that for each transformable species there is a characteristic set of nutritional or physiological requirements that favor high frequency of competence usually without affecting bacterial growth. For example, in H. influenzae the presence of inosine and lactate in the medium is needed for optimum levels of competence [11]; in B. subtilis Mn^{2+} and Fe^{2+} ions inhibit competence, and therefore, in media containing these divalent ions, chelators may appear as stimulants of competence [12]. The mechanism of action of these factors is complex and poorly understood. Figure 1 illustrates the influence of such factors on the development of the competent phase in three pneumococcal cultures,

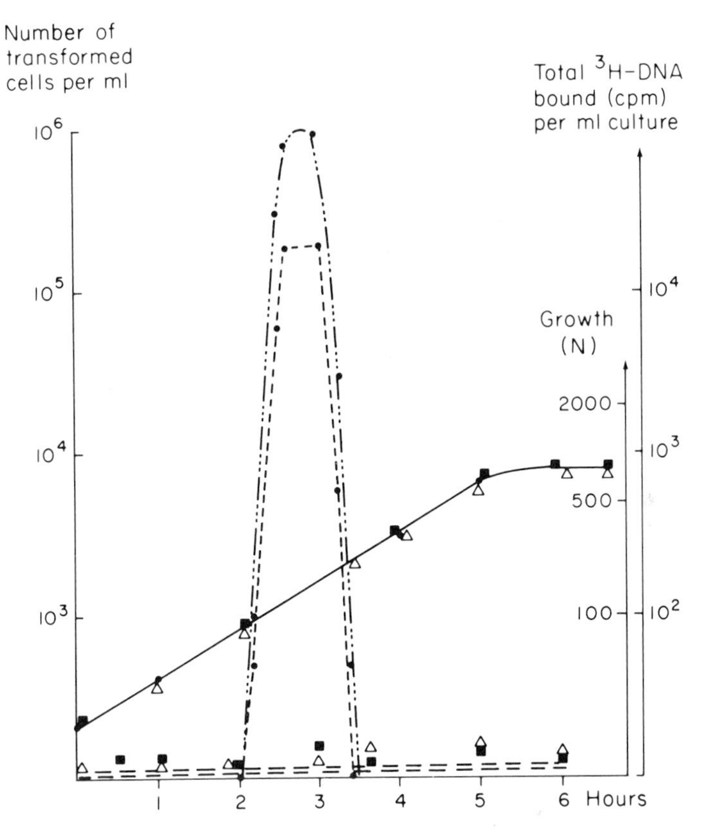

FIG. 1. The influence of growth conditions on the DNA-binding capacity of pneumococci. Three pneumococcal strains were grown in growth media at pH 8 (●), pH 6.8 (Δ) and pH 8 with traces of subtilisin (0.1 μg/ml, ■). Growth was monitored by nephelometry (solid line; N: nephelos value). At frequent intervals the capacity of the cultures was tested for binding of radioactive pneumococcal DNA (carrying the streptomycin resistance marker) and for transformation to streptomycin resistance. (---). Cell-bound radioactivity; (-.-) number of transformed cells.

one growing at pH 8, the second at pH 6.8, and the third one at pH 8, but in a medium containing trace amounts of subtilisin. Competence only appears under the first condition.

This figure also shows the unique temporal course of appearance of the competent phase, a property that, formally at least, is shared by all transformation systems; competent cells appear fairly synchronously in the cell population from low or nondetectable levels; they increase in number with a rate very much faster than the average division time. For instance, in pneumococci the number of competent cells may double in each 20-40 sec, while the average cell division rate is 60 min. After having reached a peak, the absolute number of competent cells may fall precipitously to precompetence level, may stay steady for some time, or the absolute number of competent cells may decline only slowly with only the frequency of competent cells decreasing (since they are diluted out by the continued growth of the culture as a whole without the appearance of new competent bacteria). Besides fulfilling the special environmental conditions optimal for competence, a good synchrony in the appearance of competence is often provoked by additional physiological intervention of a more drastic sort, such as a stop in the aeration [13] of the culture or nutritional step-down [14]. The principle of design for these methods was suggested by the observation that in undisturbed test-tube cultures, competent cells are least frequent during balanced growth, while higher frequencies are detected during times when the cell population is likely to be undergoing self-induced physiological shifts such as in the early log phase (streptococci, *B. licheniformis*, *B. subtilis*) late log phase (pneumococci and streptococci and *Hemophilus*) or early stationary phase of growth (*B. subtilis*). Most procedures used to provoke competence attempt to imitate this unbalanced growth by exposing the bacteria to sudden environmental changes and in the *B. subtilis* or *H. influenzae* transformation systems the synchrony of competence is undoubtedly "selected for" by these provoking methods. In an interesting contrast to *B. subtilis* and *H. influenzae* in pneumococcal [3] and streptococcal cultures [15], the synchronous appearance of competence occurs in logarithmically growing cultures without any obvious experimental provocation. In all of these cases the synchronous onset of the competent phase in a large number of cells suggests the operation of some kind of an endogenous induction process. *A priori* one should consider inducers of at least four types. (1) Induction could be the result of some aspect of the complex physiological changes initiated by the "provoking" procedure. (2) In the specific case of *B. subtilis* induction could be related to the onset of sporulation [16]. (3) The induction could occur as a response to a change in the chemical composition of the nutritional environment caused by the bacterial metabolism itself (excretion of metabolic end products or selective depletion of the medium of some components). (4) A specific competence inducer substance may be produced during growth. This latter seems to be the case in pneumococci [9] and streptococci [17].

A. The Competent Phase in Pneumococcal Cultures

1. Genetic Basis

As in all other transformation systems, in pneumococci too, there are clonal isolates or mutants known, that show only abnormally low transformation frequencies because of a lowered capacity to bind DNA. The behavior of such a genetically incompetent ("com$^-$") mutant during culture growth is compared to that of the "wild type" culture in Fig. 2.

2. Nonconstitutive Nature and Control of Expression

Both Figs. 1 and 2 also demonstrate that even in genetically competent strains the ability to bind radioactive DNA is not a constant property of the cells but rather it is limited to a transient period within a culture cycle. As was already mentioned earlier, in pneumococci the unique temporal course in the onset, development and decay of competent culture phase is induced by a specific extracellular macromolecule, "the Activator" or "competence factor" (CF) produced by the cells themselves. The relevant findings upon which this conclusion is based in pneumococcus are as follows.

1. By experimenting with mixed cultures of properly marked, and thus distinguishable, bacteria it was shown that competent cells can communicate their competent state to the incompetent cells in the medium through a protein-like macromolecule, which seems to be a specific component of the cells' surface [9]. Addition of even a relatively few competent bacteria (10^4/ml) to a majority of yet incompetent cells (10^6/ml) can trigger development of competence as well as production of more CF in the majority cells. The action of CF and its autocatalytic reproduction can explain both the synchronous appearance and the explosive spread of competence state [3]. In contrast to the pneumococcal and streptococcal activators, the competence stimulating factor described in _B. subtilis_ [18] does not seem to participate in the cell to cell propagation of the competent state [19].

2. As previously mentioned, pneumococcal cultures growing from low inocula develop their competent state only after a delay period during which the cells are normally growing. It can be shown that at any time during this incompetent period cells can be made maximally competent without delay by the addition of purified CF [9]. This experiment shows that the factor limiting the development of competence is the availability of endogenous CF. Thus, the endogenous production of initial quantities of CF must occur in a process that is dependent on cell concentration prior to the appearance of competent state [3].

3. After having reached a peak the absolute number of competent cells either declines slowly or precipitously depending on the growth conditions. In the latter case culture filtrates contain a factor (inhibitor) that seems to inactivate CF, and which is rapidly produced by the bacteria [9]. Under conditions when the number of competent cells declines more slowly no inhibitor can be detected in the filtrates and CF remains intact and active in the presence of such cells. However, it is impossible to

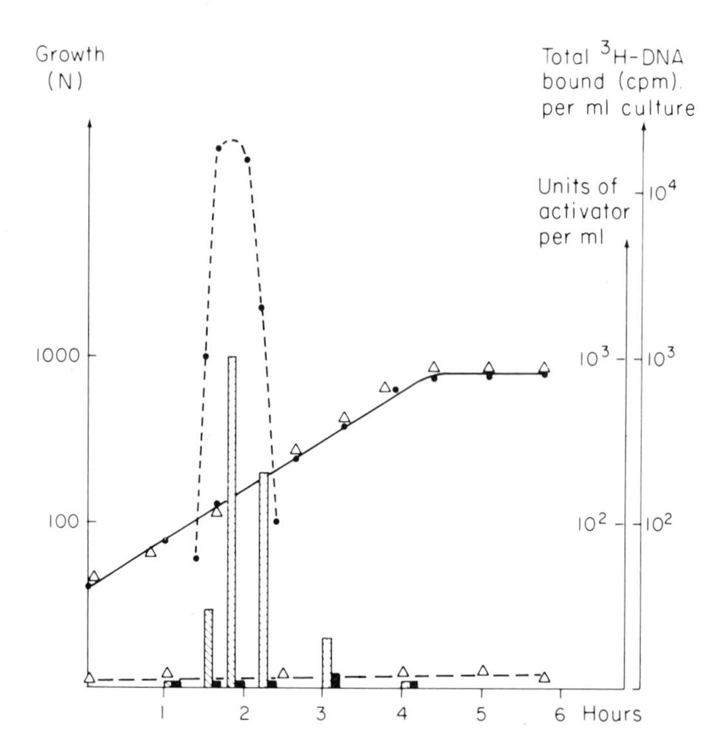

FIG. 2. DNA-binding capacity and Activator production in wild-type and mutant (genetically incompetent) cells. (●, △) Experimental points for the wild-type and mutant cells, respectively. (—) Growth (nephelometric-N-units); (---) cell-bound radioactive DNA (cpm/ml culture); (cross hatched bars, wild type; solid lines, mutant) represent Activator concentration (U/ml culture).

stop the decay of competence in such bacteria by adding fresh CF or by washing the cells and transferring them to fresh medium. It seems that after a "round of competence" decay of the competent state must follow and during this period cells temporarily lose their reactivity to CF. For the sake of easier discussion it has been proposed that "sites" for the CF exist on the bacterial surface ("CF-site") [3]. After reacting with CF these sites seem to decay and then regenerate again in a growth requiring process. Clearly, then, there are at least two different kinds of incompetent pneumococci — bacteria growing at low cell concentration, at low pH or in the presence of proteolytic enzymes are incompetent because the action (or production or both) of CF is inhibited. Such bacteria retain their reactivity to CF (i.e., have CF-sites). A second type of incompetence is characteristic of cells which have undergone a "round of competence." These bacteria have temporarily lost their CF-sites and thus cannot be made competent by CF.

These observations suggest that the appearance of the competent state in pneumococcal cultures is "controlled" in the physiological sense of the word by a specific macromolecular agent acting on the level of cell population and, in an abstract sense, this represents a transient "differentiation" in a bacterial population [20]. The relevant parts of this system of controls can be graphically summarized as in Fig. 3.

It may be of interest to mention that the abrupt appearance of the competent physiological state at a certain stage of culture growth shows analogies to a group of phenomena in microbial physiology sometimes referred to as "modulations" [21]. It is known that bacterial cultures are capable of undergoing rapid and drastic changes in several physiological properties of the majority of cells. Numerous examples for such massive nongenetic variability are known: such as qualitative and quantitative changes in surface antigens, in virulence and in cell shape, sporulation, cyst formation, production of polypeptide antibiotics, etc. The nature of the signals invoking these changes is poorly understood.

3. Frequency of Competent Cells

Under optimal conditions, at the peak of a culture's competent phase, the majority or all the pneumococci present share this physiological condition, at least to the extent that they all are capable of binding DNA in a nuclease-resistant form. This fact was unequivocally demonstrated by radioautography of competent cells exposed to both homologous and heterologous DNA labeled with tritiated thymidine of high specific radioactivity [10]. In contrast, the same method only showed a maximum of about 16% of the cells binding DNA in competent-phase *B. subtilis* cultures.

4. Competence as a Nonpathological Property

The ability of pneumococci in the competent phase to react with DNA does not seem to have any deleterious effect on the major parameters of growth; such cultures grow logarithmically with the normal generation times and a high constant frequency of competence can be stabilized in such cultures for prolonged periods by a continuous dilution device [3]. If competent bacteria were nongrowing or slowly growing cells, they would be expected to be diluted out under these conditions.

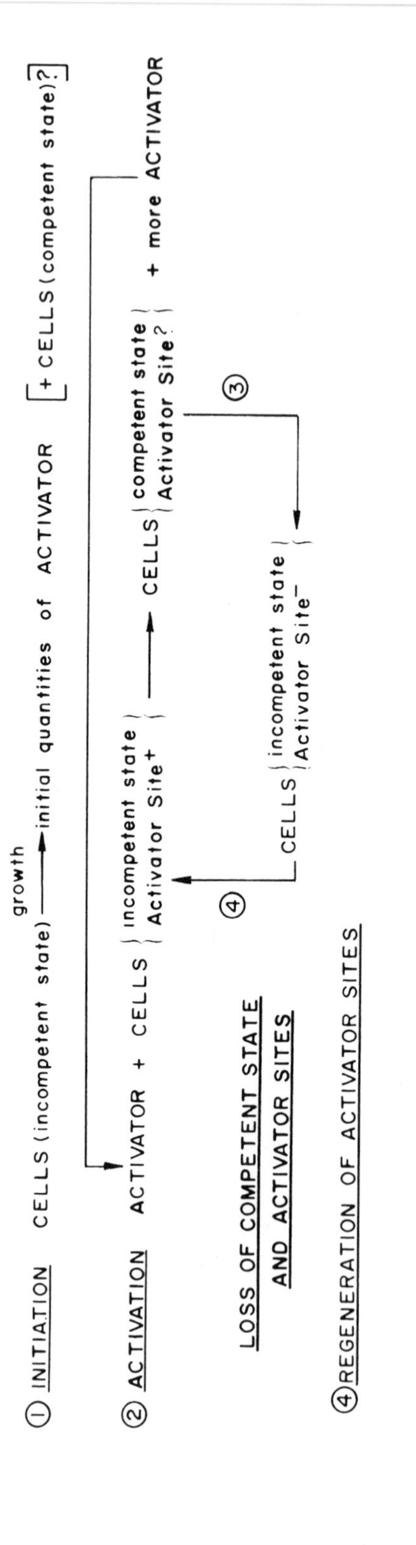

FIG. 3. Control of the competent phase in pneumococcal cultures.

IV. BIOCHEMICAL CHARACTERISTICS OF COMPETENT PNEUMOCOCCI

What are the peculiar biochemical features of pneumococci in the competent phase? Are there any detectable biochemical changes that may be correlated specifically with the cell's ability to bind and transport polydeoxynucleotides? In what follows I briefly describe the several specific macromolecular "factors" identified so far as participants in the cell surface alterations that appear to be triggered by the interaction between bacteria and molecules of the activator substance. The recognition of these macromolecules would have been very difficult or even impossible without the availability of highly purified activator [22]. The most frequently used design in our experiments was as follows. Pneumococci were grown in synthetic media at low pH or in the presence of traces of subtilisin, i.e., under conditions that prevent spontaneous expression of competence. Such physiologically incompetent bacteria were then washed and transferred to fresh growth medium that permits expression of competence (pH 8 and without proteolytic enzymes). The culture was then divided in two parts; one received purified activator (0.1-0.2 μg protein/10 ml); the other did not. By properly adjusting the concentration of cells and activator and incubation time (10-20 min) one can obtain very high levels of competence in the activator-treated culture and no competence at all in the control. (Eventually, after longer periods of incubation, the competent phase will also appear in the control tube. However, this spontaneous expression of competence requires that the bacteria first accumulate sufficient quantities of endogenous activator.) It seems reasonable to assume that such "twin" pairs of cultures only differ from one another in those physiological and biochemical properties that are related to competence.

A. The Competence Antigen

Formalin-killed vaccines of competent pneumococci can invoke the formation of specific antibodies in the rabbit [23, 24]. Antisera prepared in this way can inhibit binding of radioactive DNA to competent pneumococci. Antisera prepared against incompetent pneumococci do not contain such activity. Apparently, the immunological system of the rabbit can recognize a new antigenic determinant which is unmasked or produced on the surface of pneumococci in the competent state. The chemical nature of this antigen is unknown. However, it seems that this determinant represents only a small fraction (several percentiles) of the total cell surface, since over 95% of the immunoglobulins in sera prepared against competent and incompetent cells are identical. Furthermore, although the anti-competent cell sera are specific to pneumococci (i.e., they do not inhibit other transformation systems), there is one notable exception to this: Transformation of Streptococcus viridans D (a strain taxonomically related to pneumococcus) is also inhibited by the anti-pneumococcal sera [24]. The major surface antigens of pneumococci and these streptococci are quite different and yet these bacteria apparently share a minor antigenic determinant specific for DNA-binding. Competence-specific antigens were also detected in the streptococci [25] and Hemophylus influenzae [26].

B. Activator Substance or Competence Factor

Some properties of the pneumococcal Activator are listed in Table 1. Figure 4 shows the chromatographic behavior on DEAE- and CM-Sephadex and on gel filters. An exact cellular localization of activator is difficult because of the unusual avidity with which this substance adsorbs to all kinds of surfaces [22]. Nevertheless, the substance appears to be surface component, since (a) whole bacteria (live or acetone powder) can act as "activators"; (b) treatment of competent cells with high concentrations of trypsin seems to destroy cell-bound Activator (without affecting cell viability) [27]; (c) the conditions used to solubilize Activator (brief heating of competent cells in physiological salt solution, pH 8 and 3 mM mercaptoethanol) seems to be selective in removing surface components. Studies with the Receptor for the Activator substance (see below) indicate that at least 70% of the Activator may be bound to the plasma membrane.

The "reaction" between incompetent cells and Activator can be described as follows.

1. Incompetent cell + Activator (soluble) → competent cell (bound Activator) + more soluble Activator

2. Competent cell (bound Activator) + DNA → transformed cell

Process 1 shows saturation kinetics with the initial rate being proportional to the concentrations of both Activator (slope = 3.5-4) and cells (slope = 1)[3]. Under our routine conditions of assay there is no detectable time-lag preceding the appearance of competent cells measured either by the genetic assay or by the binding of radioactive DNA. Figure 5 shows that preincubation of DNA with purified Activator has no effect on the biological activity of these molecules. Similar results were reported in the case of streptococcal Activator [28]. A previously reported nuclease-like activity of crude pneumococcal Activator preparations may be an impurity [29]. Of course, this does not mean that nucleases may not have a role in transformation. On the other hand, these findings unequivocally show that induction of the competent state is the result of a reaction between cells and Activator, and not the consequence of some sort of

TABLE 1

Some Properties of the Activator

Approximate molecular weight	10,000
Net charge at pH 7-8	Positive
Inactivated by:	Trypsin, subtilisin, pronase (1 μg/ml, 10 min) HIO_4 (10^{-2} M, 0 °C, 10 min); heating at 100 °C, 10 min.
Resistant to:	RNase (Pancreatic), DNase I, lysozyme, snake venom, phospholipase C (10 μg/ml, 30 min.)
Inactivation on surfaces	Glass, nitrocellulose, cellulose fiber, polypropylene, dialysis membranes (Visking)

structural modification of the transforming DNA by molecules by the Activator. Whether the Activator is used up in the process of activation according to some stoichiometry or is acting only as a catalyst is not known, mainly because of the autocatalytic production of Activator during conversion of the cells to competence. Until now we have not been able to separate these two processes. The autocatalytic production of Activator during interaction with the cells also makes the straightforward interpretation of the kinetics of the process impossible.

C. Specificity of the Activation Process

Treatment of pneumococci with Activator induces a number of specific effects that are not detectable in the control bacteria (Table 2).

Table 3 shows that pneumococcal Activator only induces the competent state in the homologous organism and in streptococcus viridans D, a bacterium taxonomically related to pneumococci which can also undergo low-efficiency transformation by pneumococcal DNA. Furthermore, no Activator could be detected in six genetically incompetent strains [30] nor in pneumococci that were physiologically incompetent, such as cells grown at low pH or in the presence of subtilisin.

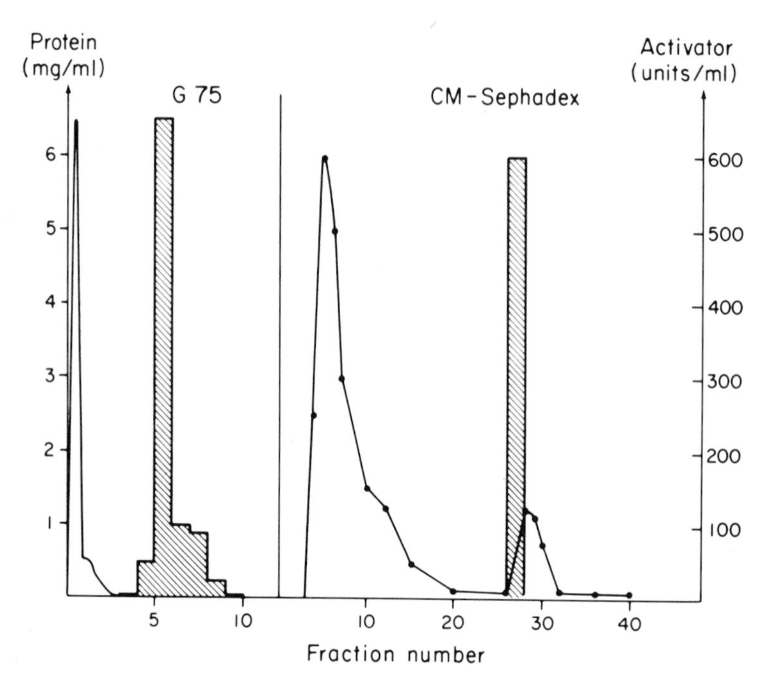

FIG. 4. Fractionation of Activator on gel-filters and cation-exchange columns. Solid lines indicate protein (OD_{280}); cross-hatched bars, the biological activity (in arbitrary units).

A large number of experiments were performed with the purpose of detecting possible changes in several general properties of bacteria treated with Activator. Specifically, we carefully examined the rates of cellular polymer syntheses (protein, RNA, DNA, teichoic acid), uptake of amino acids, "leakage" of intracellular markers (B-galactosidase, nucleotides, phosphorous and amino acids), the rates of potassium uptake and efflux [31]. All of these experiments yielded negative results, i.e., we found no differences between the properties of cells undergoing massive conversion to competence and the control cells that received no Activator and thus remained incompetent. This lack of effect on the general permeability properties of cells and on the rates of cellular polymer syntheses is consistent with a highly specific action of the Activator.

D. Receptor for the Activator

The recent work of Dr. Ziegler in my laboratory led to the discovery of a specific Receptor for the Activator [32]. Protoplast membrane preparations of pneumococci — but not of B. subtilis or group A streptococci—

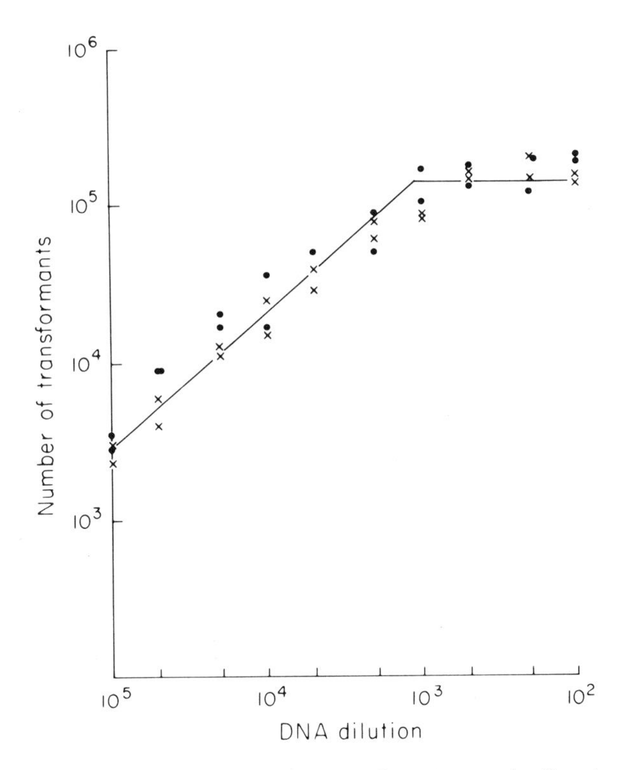

FIG. 5. Biological titration of transforming DNA after treatment with Activator. (x) DNA pretreated with 100 U/ml Activator at 30° C for 30 min; (●) controls.

were found to physically bind and reversibly inactivate soluble Activator. The bound (and "inactive") Activator could be quantitatively recovered by brief heating of the membrane preparation in alkaline salt solution containing mercaptoethanol. A scheme of the assay for membrane-bound Receptor is shown in Fig. 6. The Receptor could be solubilized from the membrane preparations by sonication or detergent treatment, and about 500-fold purification could be achieved by combined gel-filtration and DEAE-Sephadex chromatography. Some properties of the Receptor are shown in Table 4. The 500-fold purified receptor has no effect on the competence of cells, had no exonuclease, autolysin, or protease activity (the latter measured by solubilization of radioactive amino acid-labeled pneumococcal proteins), and contains no detectable choline (specific marker for teichoic acids in this bacterium). Besides inactivating the Activator substance, purified Receptor also caused about 50% drop in the transforming activity of DNA. Whether or not this latter activity represents an impurity or has

TABLE 2

Specific Effects of the Activator on Pneumococci

Experiment	Incubation procedure	Genetic transformation (per ml)	Radioautography (%)	DNA binding[a] Total cell-bound (cpm/ml)	DNA binding[a] Nuclease-resistant cell-bound (cpm/ml)	Agglutinin[a]	Excess Activator[a] (U/ml)	Clumping
I	Cells (10^7/ml) +Activator (100 U/ml) 30 °C, 30 min	9 9×10^5	83	14,000	12,000	++++	1,700	+
II	Cells (10^7/ml) +Activator (43 U/ml) 30 °C, 30 min	6×10^5	80	11,000	10,000	++++	1,670	+
III	Cells (10^7/ml) No Activator 30 °C, 30 min	< 10	< 2	50	45	none	none	no

[a] DNA binding was evaluated by two assays: (1) radioautography, to determine the frequency of cells which bound radioactive DNA, and (2) an assay based on centrifugation of the bacteria and the cell-associated DNA. Agglutination [35] and the amount of Activator [3] were determined by published procedures.

TABLE 3

Species Specificity in the Action of Activator

Strain	Effect of pneumococcal Activator
D. pneumoniae	Induction of competence
S. viridans D	Stimulation of competence
Streptococcus group H (Challis)	No effect
B. subtilis 168	No effect
H. influenzae	No effect
E. coli K12 spheroplasts (transfection with f 1 DNA)	No effect
E. coli K12 spheroplasts (helper phage + DNA)	No effect

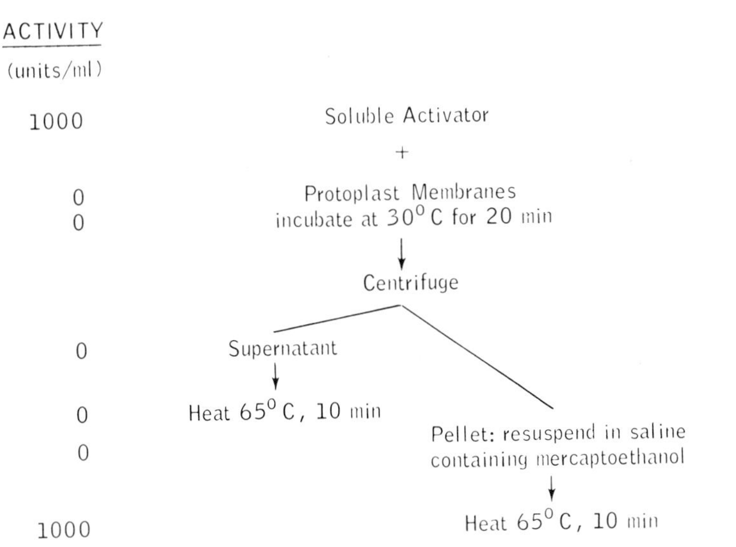

FIG. 6. Outline of the method used to determine membrane-bound receptor activity.

TABLE 4

Properties of the Receptor

	Membrane-bound receptor
Activity	Binds and reversibly inactivates soluble Activator. Reversibility: 100%
Localization	In protoplast membranes
Sensitivity	Inactivated by heating at 65° C, 10 min
Resistance	Trypsin (100 γ/ml, 30 min), lysozyme, RNase, phospholipase C
Solubilization	Sonication, detergent (deoxycholate, SDS)
	Soluble receptor
Activity	Inactivates Activator; reversibility: maximum 20%
Properties	Molecular weight ~ 50,000 (excluded on G 75, retained on G 100. At pH 7-8: not retained by CM-Sephadex, retained by DEAE-Sephadex; eluted by 0.5 M salt

something to do with the function of the Receptor, is presently being investigated.

Combination with the membrane-bound Receptor may indeed be the first step in the interaction of cells and Activator. Upon treating live, incompetent bacteria with Activator one can show that the initially soluble Activator quickly becomes cell-bound and at least 70% of the input activity can be recovered in the protoplast membrane [32].

E. Agglutinin

Treating pneumococci with saturating concentrations of Activator only induces competence (i.e., DNA-binding) if normal protein and/or RNA synthesis is allowed to proceed concomitantly [33]. A variety of inhibitors or inhibitory conditions affecting these syntheses can completely block the appearance of competence even in the presence of excess Activator. It can be shown that the presence of these protein-synthesis inhibitors does not interfere with binding of the Activator to its correct cellular Receptor, since cells "loaded" with Activator in the presence of chloramphenicol can subsequently (after removal of the drug) express normal competence even in the presence of subtilisin or at low pH, i.e., under conditions that would typically prevent activation by extracellular activator. These findings suggest that Activator molecules can become attached to the inhibited cells and upon removal of the inhibitor and excess soluble activator, the same bacteria that adsorbed the Activator molecules can complete a subsequent step in competence development, a step that requires the synthesis of new protein [33]. At least one such protein can be detected in activated pneumococci by applying Pakula's agglutination test [34] ; cells treated with Activator while maintaining normal protein synthesis agglutinate in medium of low ionic strength and low pH [35]. The same cells, before treatment with Activator do not agglutinate. The substance responsible for agglutination is trypsin-sensitive and is, presumably, a protein. It seems to be deposited somewhere between the outer and inner cell envelope, since both competent cells and protoplasts prepared from them can agglutinate. Inhibitors of protein synthesis completely prevent the appearance of agglutinin. No agglutinin formation can be detected in a strain of genetically incompetent pneumococci (RA7), which nevertheless is capable of binding and reversibly inactivating Activator (and thus seems to have Receptor) (Table 5).

F. Role of the Cell Envelope Growth Zone in DNA Binding and Uptake

Very recently we made an interesting observation that indicates that the cell envelope growth zone has an important role in the process of DNA binding and uptake. We treated pneumococci with Activator in a medium allowing syntheses of protein, RNA and DNA but preventing, specifically, the incorporation of choline molecules into the pneumococcal surface polysaccharide. The result was a complete inhibition of activation [36]. The polysaccharide contains covalently bound choline and makes up about 50% of the mass of the cell's outer envelope [37]. Since choline is located exclusively in these macromolecules and since the bacteria require choline for growth [38], one can selectively inhibit the synthesis of these macromolecules by omitting choline from the medium. After the removal of choline from the medium the bacteria will undergo a residual growth

TABLE 5

Effect of Protein Synthesis Inhibitors on Activation to Competence [a]

	Transformants	Binding of Activator to cells	Agglutination	DNA-H^3 binding (total cpm/ml)
Cells (10^7/ml) + Activator (100 U/ml) 30° C, 30 min	1.0×10^6	Yes	++++	8500
with CAP (100 γ/ml)	$< 10^2$	Yes	None	42
Rifamycin (0.1 γ/ml)	$< 10^2$	-	None	40
Tryptazan (100 γ/ml)	$< 10^2$	-	None	34
without leucine	$< 10^3$	Yes	None	60

CELLS + ACTIVATOR

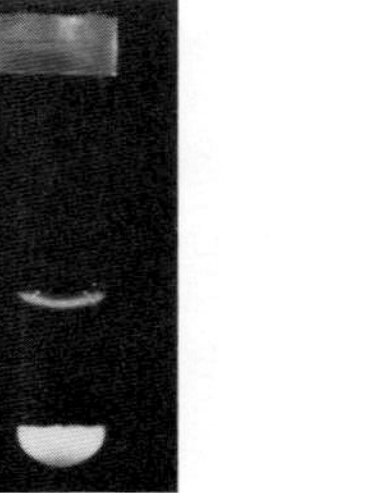

CELLS + ACTIVATOR + CAP

a Binding of Activator: (-) indicates not tested.

for about one mass doubling accompanied by normal rates of protein, RNA, and DNA syntheses. Figure 7 shows loss and recovery of cellular DNA-binding capacity during the incorporation of two pulses of choline into the cell envelope polysaccharide. It can be seen that the cessation and resumption of choline incorporation is extremely closely followed in time by a drop and recovery, respectively, in DNA-binding capacity. Table 6 shows that cells are able to bind Activator and form agglutinin in the absence of choline, and yet are unable to bind DNA molecules either reversibly or irreversibly.

The speed of the response in cellular competence to the cessation or resumption of choline incorporation calls one's attention to the sites at the cell surface at which nascent choline-containing material is incorporated into the cell envelope. Earlier work in our laboratory has demonstrated that in pneumococci the choline-containing surface-component is incorporated into the surface at a limited number of sites, and the incorporated surface polysaccharides are conserved during growth and are passed on to the progeny cells in a symmetrical fashion, suggesting the presence of a single growth zone around the middle region of the cells [39]. Recently we reexamined this point by a direct electron microscopic method and confirmed the localization of the cell envelope growth zone at the cellular equator [36].

In this experiment pneumococci were grown in a medium containing ethanolamine in place of choline [40]. Under these conditions pneumococci utilize ethanolamine and incorporate it into the same surface-polysaccharide which normally, i.e., in choline-containing media, contains choline. Such a biosynthetic substitution results in a number of physiological changes in the cells, one of which is the resistance of the ethanolamine-containing cell envelope to the lytic action of the homologous (pneumococcal) autolytic enzyme. We exploited this fact in our experiment to localize the cell envelope growth zone, in the following manner: pneumococci grown in ethanolamine-medium received a short (5 min) pulse of ^{3}H-choline at such concentration that the cells started immediately incorporating choline (in preference to ethanolamine) into the surface polysaccharide. The pulse was terminated by briefly heating the bacteria to 65 °C (2 to 5 min) and the suspension was next treated with purified pneumococcal autolysin until 80% of the radioactive choline was released from the cells. (Control experiments showed that under these conditions there is no release of ethanolamine-containing material from the bacteria.) The enzyme-treated suspension was then fixed by glutaraldehyde and osmium tetroxide, dehydrated, embedded in epon; thin sections were examined by electron microscopy for the sites of enzymatic cell-wall damage. The only visible sites of wall digestion were invariably located at the exact middle zone of cocci, at the cellular equator. Because under the experimental conditions the autolytic enzyme could only damage choline-containing segments of the cell envelope

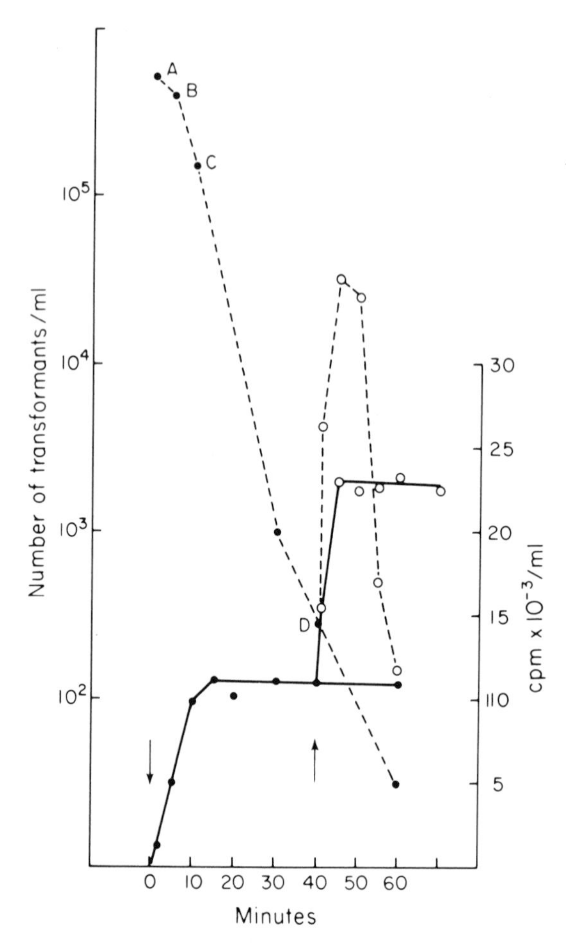

FIG. 7. Decay and regeneration of cellular response to the Activator during utilization of cellular choline pools. Physiologically incompetent bacteria were transferred (arrow) to a medium containing a low concentration of choline (0.1 μg, 0.5 μCi/ml). The incorporation of choline into macromolecular (cold TCA precipitable) material and cellular response to Activator were tested at frequent intervals. A second pulse of choline was administered to the cells at the time indicated by the second arrow. (—) Choline incorporation; (– – –) response to the Activator (number of transformants). At the times indicated by the letter A, B, C, and D samples of the activator-treated cells were exposed to ^{3}H-DNA and the amount of cell-bound radioactive DNA was determined (see Table VII).

TABLE 6

Effect of Choline Deprivation on Cellular Competence

Length of removal (min)[a]	Binding of Activator	Agglutination	DNA binding (cpm/ml)	
			Total	Nuclease-resistant
5	Normal	++++	10,827	7,331
5	Normal	++++	2,800	2,000
10	Normal	++++	1,509	1,038
20	Normal	++++	79	48
30	Normal	++++	53	48

[a] Cells (10^7/ml) + Activator (100 U/ml) in full growth medium; choline removed (before addition of Activator).

the sites of wall damage in the choline-pulsed bacteria represent the sites of nascent wall synthesis, i.e., the envelope growth points (Fig. 8). The combination of these findings clearly establishes the importance of the zone of cell division for DNA-binding and uptake. It should be mentioned that in two other transformable bacteria there is evidence also implying a similar topographic area of the cell surface. Cell-bound radioactive DNA has been localized to the tips and centers of competent B. subtilis cells [10], and in transformable streptococci localized damage to the outer envelope was detected in cells treated with streptococcal competence factor [41].

V. DNA: BINDING AND UPTAKE

The complex surface alterations described up to now: the appearance of a new antigenic determinant, binding of activator molecules to the plasma membrane and deposition of agglutinin finally led to a modified surface now capable of DNA binding and absorption.

With homologous DNA the absorption is very rapidly followed within 5 to 15 min by the physical (and genetic) integration of these macromolecules into the longitudinal continuity of the resident chromosome [42]. This latter process is an extremely efficient one. A single absorbed molecule of DNA can cause genetic change and the physical integration of long stretches of the donor nucleotide sequences occurs with great rapidity and at the correct allelic sites of the resident chromosome. The intracellular fate of the absorbed molecules shows considerable variation, besides genetic integration a portion of the polynucleotides seems to be degraded by cellular nucleases and other catabolic enzymes to 5'mononucleotides [43], nucleosides and bases and can either be reutilized or excreted into the growth medium. Foreign polynucleotides may undergo specific endonucleolytic degradation by specific "restriction enzymes" [44]; viral polynucleotides may initiate an infectious process [45].

Our concern in the present discussion will bypass these intracellular events and will be focused on the early stages of transformation: the association of DNA with cell surface structures and its transport into the cellular interior.

Very little is known about the mechanism by which DNA molecules initially associate with the cell surface. It appears that, similarly to phage adsorption, the "trapping" of DNA molecules by some as yet unidentified surface elements occurs with a high probability following random collisions of cells and molecules [46]. The sensitivity of radioactive DNA-binding to pH, ionic conditions, and divalent ion concentration

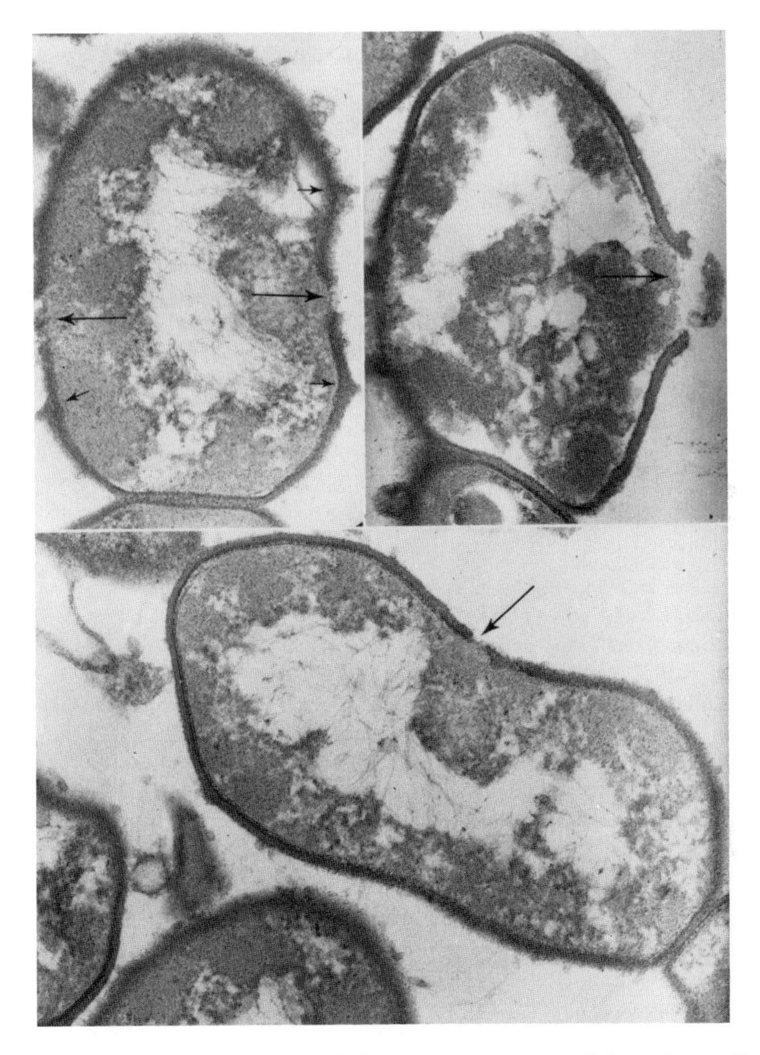

FIG. 8. Selective removal of the nascent regions of the outer cell envelope.

in the H. influenzae transformation system was interpreted as evidence for association by ionic bonds [47]. In pneumococci, as well as in other transformation systems, "uptake" or "absorption" of DNA is usually defined as the cellular process that makes the cell-bound radioactive DNA resistant to the action of pancreatic DNase added from without. Transformable bacteria can "bind" radioactive DNA in excess of the amount that is resistant to such a DNase treatment ("reversibly" bound DNA) [48]. In practice, however, it is not easy to get an exact measure of the reversibly bound portion of the molecules. The main reason for this difficulty is proably that DNA molecules enter the cells in a longitudinal fashion [49-51], and "reversibly" bound DNA probably represents molecules bound but not yet absorbed, as well as the ends of molecules, in the process of uptake. A further difficulty is that the exact cytological location of the "irreversibly" bound (i.e., DNase-resistant) DNA is not clear; it may represent truly intracellular material (i.e., material inside the cytoplasmic membrane) or molecules protected from extracellular nucleases in some compartment within the periplasmic space. Inhibitors of energy metabolism seem to cause a preferential block in the appearance of irreversibly bound DNA [52], a finding consistent with the existence of an energy-requiring "transfer" process which is preceded by an energy independent, initial "binding" step.

An electron microscopic study of the transformation process (using a modification of Kleinschmidt's technique) revealed DNA-cell associations [53] the morphology of which was also consistent with a longitudinal entry of molecules (Fig. 9).

Pneumococcal cells were grown to develop various degrees of competence, were concentrated by filtration to a density of about 1-5 x 10^8 cells/ml, and were resuspended in a fresh medium containing 20 μg/ml bovine serum albumin and 0.001 M $CaCl_2$. To one-half of the cells, 0.02-0.05 μg pneumococcal DNA carrying the streptomycin marker was added, and the suspension was incubated at 30° C for 10 min. The other half of the cells were incubated at 30° C for 10 min without added DNA, as control. At the end of the incubation, the suspensions were chilled and an aliquot of the cells was assayed for a viable cell count and also for transformability. To the rest of the suspensions 40-50 μg methylated albumin was added, and the suspensions were delivered to the surface of a 0.1 M ammonium acetate solution in a Langmuir trough.

The protein, which spread on the surface of the salt solution formed a monolayer holding the bacterial cells and DNA molecules in the surface. It was compressed to a film at a pressure of 0.6-0.8 dyne/cm. Areas of this compressed film were picked up on copper grids coated by Formvar and carbon. The grids were dried in alcohol and isopentane and shadowed with uranium and were subsequently examined in the RCA II electron microscope.

Ten to twenty areas on each grid were searched first at low magnification to locate the bacterial cells, then the surface and the area around each cell were examined at higher magnification sufficient to recognize strands of DNA attached to bacteria. The DNA molecules were identified solely on the basis of their characteristic morphology in the electron microscope, a feature well established by the work of several laboratories.

FIG. 9. DNA molecule attached to competent pneumococci during genetic transformation.

In order to eliminate accidental overlap of DNA strands and cells, a relatively low concentration of DNA was used — about one molecule per 10-25 μ^2 area of the film. The amount of film forming protein was chosen so that the total surface occupied by the cells in the Langmuir trough was less than 1/1000 of the total surface of the film.

In preparations of highly competent cells, one could observe bacteria in association with DNA molecules in the electron microscope. In an experiment, in which the frequency of transformation to streptomycin resistance was 0.8-1% of the viable units, of 360 cells examined in the electronmicroscope, 12 cells were found with DNA strands attached to their surface. Of the same number of cells examined in the control preparation (i.e., competent cells to which no DNA was added) none of the cells had DNA strands attached to them.

In another experiment using cells with 0.1% transformability, of 100 cells examined, 14 had strands attached to their surface, while of the 90 control cells without added DNA, none had strands attached.

When DNA was incubated with cells of very low competence, i.e., populations that only yielded 10^{-5} transformants per viable bacteria — no strands were found in 400 cells examined.

The length of the free ends of cell-attached molecules which we observed by this electronmicroscopic method varied from about 0.5 μ to as long as 9 μ. It is not possible to say whether this represents a lack of synchrony in the initiation of cell attachment, different rates of penetration, or size heterogeneity in the population of DNA molecules. (Lack of synchrony in the conjugal transfer of DNA is known to occur [8]. The entry of a 9-μ long piece of DNA would correspond to the introduction of about 10^6 daltons of DNA capable of coding for the synthesis of 9-12 average size proteins.

Unfortunately, at the moment we have no clear-cut methods to dissect the process of DNA entry into an unequivocal sequence of temporal and cytological steps. Nevertheless, the process can be interrupted by a variety of means (inhibitors [47], dilution [49], mechanical shear [51], suboptimal temperature [54], excess competing DNA [49], antibodies against cellular sites [24] or against single-stranded DNA [55]) at apparently different stages. The number and sequential order of these is not clear as yet. However, several things can be said with fair certainty.

1. DNA molecules are bound in two forms — a DNase-sensitive and a DNase-resistant form and it is likely that these two represent the temporally earlier and later stages of entry, respectively.

2. DNA binding and uptake can be clearly separated from the eventual genetic integration of these molecules, since heterologous DNA is also taken up. Table 7 shows that the maximal amount of DNA bound per average competent pneumococcus at saturating DNA concentrations is practically identical for homologous (pneumococcal: 27 molecules) and heterologous (adenovirus 2 DNA: 22 molecules). The table also shows that the cellular binding and transport system operating in transformation requires the double-stranded polydeoxynucleotide structure, but simple polymers such as pdAdT, pdAT, or pdGdC can all associate with the same cellular sites that are used by the homologous DNA molecules. A peculiar feature of the competent state is that the affinity of polydeoxynucleotides for cellular binding and uptake drop rapidly below the molecular weight range of several hundred thousand daltons [56].

3. The immunological data discussed earlier suggest that only a relatively small fraction of the total bacterial surface takes part in DNA uptake.

4. A remarkable feature of the phenomenon is that while the surface of bacteria is commonly impermeable to charged molecules (such as free nucleotides) and molecules even of the size of sucrose, in transformation DNA molecules with the molecular weight of 10^5 daltons, as well as entire

TABLE 7

Binding of Polydeoxynucleotides by Competent Pneumococci

Polydeoxynucleotide	Association with cell
Pneumococcal DNA (labeled with ^{3}H-thymidine)	18,974 cpm/ml (=0.046 γ)
Adenovirus II DNA (labeled with ^{3}H-thymidine)	913 cpm/ml (=0.038 γ)
Pneumococcal DNA, heat-denatured	120 cpm/ml (< 0.0004 γ)
pdAdT	Yes[a]
pdAT	Yes[a]
pdGdC	Yes[a]
prAprU	No[a]

[a] Association with the competent cells was determined by assaying the ability of the polynucleotides to competitively inhibit transformation and cellular binding of radioactive transforming pneumococcal DNA.

bacteriophage genomes (10^8 daltons) (i.e., DNA segments corresponding to stretched-out lengths ranging from few tenths of a micron to several hundred microns) can traverse the cellular boundaries with a considerable rate. It is estimated that 1-2 μ long DNA molecules can penetrate the bacterial surface in a matter of 1-5 min, or under certain experimental conditions, in a few seconds [57], i.e., in the same order of magnitude as in conjugational DNA transfer [8]. These facts suggest very strongly the operation of a highly specific macromolecular transport system for polydeoxynucleotides. This view is all the more justified since in transformation the "reactants" are bare DNA molecules and cells, and thus the entire mechanism for DNA uptake must reside in the recipient bacterium. It seems that even with heterologous polynucleotides the final stage of the transport process is _within_ the boundaries of the plasma membrane since the uptake of viral DNA (transfection) can initiate a normal infectious process.

VI. SUMMARY

Although the chemical nature of DNA receptors and the mechanism of DNA-transport across the cell surface membranes is not known, several surface components can be identified that seem to be specifically involved with these early steps in genetic transformation. These components are the Activator, the membrane-bound Receptors for the Activator, the Competence Antigen, and the Agglutinin. In addition, the need for a functioning cell envelope growth zone can be recognized. The appearance of these factors on the cell surface seems to follow a temporal order, which is depicted in Fig. 10.

The surface modification seems to be initiated by the Activator. While the Activator is clearly produced by the bacterial culture, the nature of the signal that provokes the formation of initial quantities of this factor is a mystery at present. In the model drawn in Fig. 10 molecules of the Activator penetrate the outer cell envelope and attach to preexisting receptors bound to the plasma membrane. This binding is followed by a protein-synthesis requiring step, one recognizable product of which is the Agglutinin. One can see at least two mechanisms for this protein-synthesis-requiring step — the product may be a specific new membrane-component involved either with DNA uptake or with stabilization of a local membrane configuration produced by the binding of the activator. Alternatively, one can also visualize the products of this essential protein synthesis as new autolysin molecules needed to "unmask" some membrane-bound DNA transport site. There is some evidence in the literature for a high functional "turnover" of these enzymes in bacteria, and the work of Shockman and his colleagues indicates that in *Streptococcus faecalis* the physiologically relevant activity of these enzymes is restricted to the newly made cell wall located at the cellular equator [58]. The timing of the appearance of the competence antigen is not clear. In fact, it is even possible that the agglutinin and the competence antigen may be the same surface component recognized by two different techniques. The active involvement of the cell envelope growth zone in DNA binding may be interpreted in several ways. One attractive possibility is that DNA binding and uptake may occur in this region of the surface and that the structural integrity, or possibly a precise juxtaposition of several structural elements in this area of the surface, requires an uninterrupted surface replication.

The picture that emerges from this discussion is that we succeeded in identifying several specific biochemical components of a complex and fascinating activity of the pneumococcal surface — the binding and transporting of extracellular DNA molecules as a prelude to genetic transformation. Of course, the surface of bacteria and eukaryotic cells as well contains a large number of specific receptors and systems of receptors,

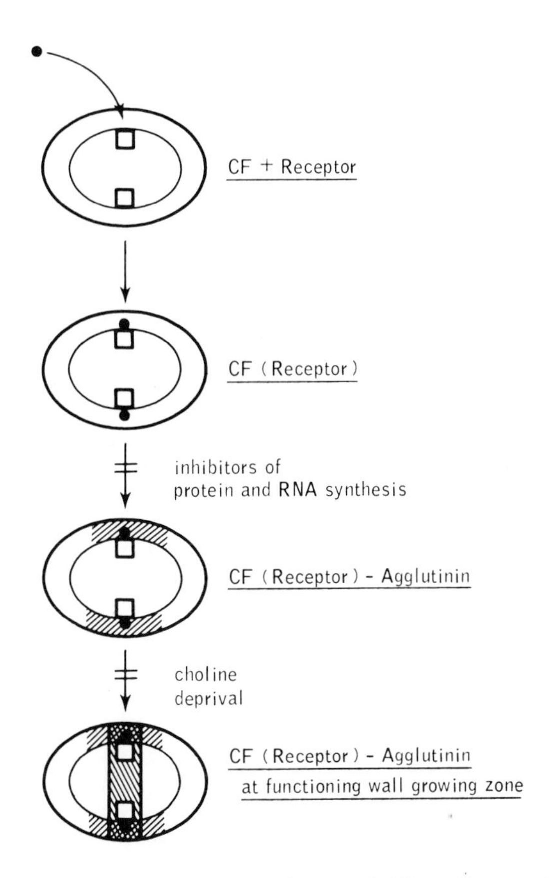

FIG. 10. Model for the stages in the acquisition of competent state in pneumococcus.

which enable cells to recognize small molecules (e.g., transport carriers), macromolecules (e.g., colicin receptors), viruses, and cells; one of the present major activities in the field of cell-surface research is to attempt to isolate and biochemically characterize these receptors [59]. Studies on the mechanism of DNA binding and uptake are among these activities.

Such studies have another general context: they are important for investigations concerning association between DNA and cell-surface structures in other systems. Some form of attachment of the bacterial chromosome to the cell surface has been proposed on theoretical grounds to explain the hypothetical triggering of DNA synthesis in male bacteria upon contact with female cells, and also to provide a mechanism for chromosome segregation during division [60]. There have been numerous experimental demonstrations, by cytological and biochemical techniques, of physical association between particulate cellular matter and portions of the chromosomal material, both in bacteria and in eukaryotic cells [61-65]. In addition, some form of interaction between polydeoxynucleotides and cell-surface elements must occur in all phenomena in which DNA molecules traverse cellular boundaries in either direction (i.e., both during entry as well as during exit from the cells) as occurs in conjugation or viral invasion. In a broader sense, one may also mention here the other phenomena of macromolecular export (e.g., exoenzymes) and import (e.g., uptake of colicines [66]. These examples then provide the general, broad context within which studies of the mechanism of DNA binding and uptake may provide important analogies and models of more general use. For studying interactions between cell-surface components and DNA, transformation seems to have the advantage of the cells reacting with DNA molecules, rather than with entire virions or whole donor cells. Furthermore, among all the experimentally demonstrated DNA-surface association it is only in genetic transformation that the binding of DNA to the cell surface has a clear physiological significance. This fact may facilitate the chemical identification of the DNA-binding surface components.

REFERENCES

1. A. Tomasz, Ann. Rev. Genet., 3, 217 (1969).
2. R. D. Hotchkiss and M. Gabor, Ann. Rev. Genet., 4, 193 (1970).
3. A. Tomasz, J. Bacteriol., 91, 1050 (1966).
4. E. Ottolenghi and R. D. Hotchkiss, J. Exp. Med., 116, 491 (1962).
5. E. Ottolenghi and C. M. McLeod, Proc. Nat. Acad. Sci. U. S., 50, 417 (1965).
6. E. Ephrati-Elizur, Genet. Res., 11, 83 (1968).
7. J. H. Wilson, R. B. Luftig and W. B. Wood, J. Mol. Biol., 51, 423 (1970).
8. R. Curtiss, III, Ann. Rev. Microbiol., 23, 69 (1969).
9. A. Tomasz and R. D. Hotchkiss, Proc. Nat. Acad. Sci. U. S., 57, 480 (1964).
10. G. Javor and A. Tomasz, Proc. Nat. Acad. Sci., U. S., 60, 1216 (1968).
11. J. M. Ranhand and R. M. Herriott, Biochem. Biophys. Res. Commun., 22, 591 (1966).

12. F. E. Young and J. Spizizen, J. Bacteriol., 86, 392 (1963).

13. S. H. Goodgal and R. M. Herriott, J. Gen. Physiol., 44, 1201 (1961).

14. J. Spizizen, B. E. Reilly and A. H. Evans, Ann. Rev. Microbiol., 20, 371 (1966).

15. R. Pakula, M. Piechowska, E. Bankowska and W. Walczak, Acta Microbiol. Polon., 11, 205 (1962).

16. F. E. Young, Nature (London), 207, 104 (1965).

17. R. Pakula and W. Walczak, J. Gen. Microbiol., 31, 125 (1963).

18. A. Akrigg and S. R. Ayad, Biochem. J., 117, 397 (1970).

19. M. R. Goldsmith, S. W. Havas, R. I. Ma, and N. R. Kallenbach, J. Bacteriol., 102, 774 (1970).

20. A. Tomasz, Nature (London), 208, 155 (1965).

21. J. P. Duguid and J. F. Wilkinson, in Microbial Reaction to Environment XIth Symp. Soc. Gen. Microbiol., 69, (1961).

22. A. Tomasz and J. L. Mosser, Proc. Nat. Acad. Sci. U. S., 55, 58 (1966).

23. G. Nava, A. Galis, and S. M. Beiser, Nature (London), 197, 903 (1963).

24. A. Tomasz and S. M. Beiser, J. Bacteriol., 90, 1226 (1965).

25. R. Pakula, J. Bacteriol., 94, 75 (1967).

26. H. T. Spencer and R. M. Herriott, J. Bacteriol., 90, 911 (1965).

27. A. Tomasz, Bacteriol. Proc., p. 52 (1967).

28. J. Nalecz, H. Osowiecki, and W. T. Dobrzanski, Bull. Acad. Pol. Sci. XVIII, 18, 687 (1970).

29. M. Kohoutova, H. Brana and I. Holubova, Biochem. Biophys. Res. Commun., 30, 124 (1968).

30. A. Tomasz, Genetics, 52, 480 (1965).

31. A. Tomasz, in Symposium on Cellular Uptake of Informative Molecules (L. Ledoux, ed.), North-Holland Publ., 1971.

32. R. Ziegler and A. Tomasz, Biochem. Biophys. Res. Commun., 41, 1342 (1970).

33. A. Tomasz, J. Bacteriol., 101, 860 (1970).

34. R. Pakula, P. Ray and L. R. Spencer, Can. J. Microbiol., 16, 345 (1970).

35. A. Tomasz and E. Zanati, J. Bacteriol., 105, 1213 (1971).

36. A. Tomasz, E. Zanati and R. Ziegler, Proc. Nat. Acad. Sci. U.S., 68, 1848 (1971).

37. J. L. Mosser and A. Tomasz, J. Biol. Chem., 245, 287 (1970).

38. A. Tomasz, Science, 157, 694 (1967).

39. E. B. Briles and A. Tomasz, J. Cell. Biol., 47, 786 (1970).

40. A. Tomasz, Proc. Nat. Acad. Sci. U. S., 59, 86 (1968).

41. J. M. Ranhand, C. G. Leonard and R. G. Cole, J. Bacteriol., 106, 257 (1971).

42. M. S. Fox and M. K. Allen, Proc. Nat. Acad. Sci. U. S., 52, 412 (1964).

43. S. Lacks, J. Mol. Biol., 5, 119 (1962).

44. H. O. Smith and K. W. Wilcox, J. Mol. Biol., 51, 379 (1970).

45. D. M. Green, J. Mol. Biol., 10, 438 (1964).

46. A. Garen, Biochim. Biophys. Acta, 14, 163 (1954).

47. B. J. Barnhart and R. M. Herriott, Biochim. Biophys. Acta, 76, 25 (1963).

48. L. S. Lerman and L. J. Tolmach, Biochim. Biophys. Acta, 26, 68 (1957).

49. M. Gabor and R. D. Hotchkiss, Proc. Nat. Acad. Sci. U. S., 56, 1441 (1966).

50. N. Strauss, J. Bacteriol., 89, 288 (1965).

51. D. Dubnav and C. Cirigliano, J. Mol. Biol., 64, 31 (1972).

52. A. Tomasz, in Biological Membranes (C. F. Fox, ed.), Academic Press, New York, 1972.

53. A. Tomasz and W. Stoeckenius, XIth Congr. Inter. Genet., Hague (1963).

54. A. G. Richardson and F. Leach, Biochim. Biophys. Acta, 174, 276 (1969).

55. W. Braun, O. Plescia, M. Kohoutova and J. Grellner, Proc. Symp. Intern. Proc., Prague (1965).

56. A. Cato, Jr., and W. R. Guild, J. Mol. Biol., 37, 157 (1968).

57. J. H. Stuy and D. Stern, J. Gen. Microbiol., 35, 391 (1965).

58. G. D. Shockman, H. M. Pooley and J. S. Thompson, J. Bacteriol., 94, 1525 (1967).

59. A. Tomasz, Nature (London), 234, 389 (1971).

60. F. Cuzin, G. Buttin and F. Jacob, J. Cell Physiol., 70, 77 (1967).

61. A. T. Ganesan and J. Lederberg, Biochem. Biophys. Res. Commun., 18, 824 (1965).

62. D. W. Smith and P. C. Hanawalt, Biochim. Biophys. Acta, 149, 519 (1967).

63. A. Ryter, Bacteriol. Rev., 32, 39 (1968).

64. G. Y. Tremblay, M. J. Daniels and M. Schechter, J. Mol. Biol., 40, 65 (1969).

65. F. W. Schull, J. A. Fralick, L. P. Stratton and W. D. Fisher, J. Bacteriol., 106, 626 (1971).

66. T. Boon, Proc. Nat. Acad. Sci. U. S., 69, 549 (1972).

PART III

MORPHOGENESIS AND REPRODUCTION

Chapter 8

BACTERIAL DIVISION AND THE CELL ENVELOPE

Arthur B. Pardee, Po Chi Wu, and David R. Zusman

Departments of Biochemical Sciences and Biology
Princeton University
Princeton, New Jersey

I. OUTLINE OF BACTERIAL DIVISION

We will start with a very brief outline of bacterial division in order to lay a framework for the more detailed discussion to follow. The bacterial division cycle consists of a series of events that occur during the interval in which one cell doubles all of its parts and separates into two daughter cells, each essentially identical to the mother cell. We can start the description of this cycle at any point. The most obvious event, cell division, actually is the terminal one, and is not the best point to take as the start of the cycle. Rather, according to current ideas [1], we should begin with certain biochemical events that occur between cell divisions. Their function is to initiate bidirectional DNA synthesis [1a], but the molecular nature of these events is unknown. In the next step, which is chromosome replication, the DNA chromosome behaves as a unit of replication, a replicon, with a definite locus of initiation and a definite terminal locus [2]. Replication from start to end requires considerable time; in Escherichia coli it requires at least 40 min. The next stage of the cycle occurs when DNA replication is complete. The end of DNA replication appears to allow septation, and the septation events also require a definite time interval (at least 15 min for E. coli). Finally, after septation is complete and one cell has become two, some time is required before daughter cells separate. This separation time is short for E. coli (a few minutes), but is several hours for Bacillus subtilis [3]. During all of these stages the rest of the cell contents — ribosomes, enzymes, wall and membrane, etc. — are being synthesized, and these components are partitioned more or less equally between the daughter cells. Each daughter cell has also received one-half of the DNA, by a nonchance mechanism that might depend on attachment of DNA to the bacterial membrane [2].

The entire round of replication, from formation of the DNA initiation apparatus to final cell separation, takes some time — at least 1 hr for

E. coli. Thus, the cycle can be longer or shorter than the interval between cell divisions. Division times of E. coli depend on growth conditions. They can be as long as several hours or as short as 20 min. Thus, new cycles can start as often as every 20 min; as a consequence several cycle sequences can be going on in the same cell, each at a different stage. Then we observe DNA undergoing dichotomous replication [4] with two replication points well advanced along the chromosome and another two pairs of replication points that have started recently. We can also observe new septum sites starting before the first is completed.

Bacterial growth can be thought of as composed of linear sequences that can overlap one another in time. As a result, we observe the most evident events occurring periodically, and this periodicity gives the appearance of a cycle.

We would like to be able to describe the cell precisely at each stage of the division cycle. The description will be of three sorts — genes that are active, chemical substances that are present and are being made, and the morphology of the cell — the relative locations of the cell parts. These descriptions, linked in time, would allow us to describe the cell cycle as a genetic, chemical, morphological sequence of events. We appreciate that the sequence is dependent on growth rate, i.e., on nutrients provided by the environment as well as on physical and chemical inhibitors and stimulators of growth. Bacteria have different sizes, DNA contents, compositions, etc., when grown at different rates. We can observe dichotomous behavior at rapid growth rates, or nonseptation after inhibitors of many sorts are applied. We thus have to answer the questions about the interrelations and regulations that coordinate the cycle under diverse conditions.

Current knowledge of both of these problems at all three levels — genetic, biochemical, and morphological — is fairly rudimentary. We can draw an analogy with putting a jigsaw puzzle together. At this time, most of the pieces (facts) have not yet been identified or examined. We have built inward a little from the edges (from other scientific disciplines). A few workers have grouped similar-appearing pieces, and have even fitted a few of them together. The combination of these facts and some bold hypotheses has allowed a guess as to the general appearance of the picture. But much remains to be discovered before the puzzle can be sorted and connected, and before a clear, complete picture can emerge.

Progress during the past decade has been considerable [5]. Approaches being applied include nutritional variations, inhibitions, variations with conditional lethal mutants of division, synchronous cultures, and single cell methods (for example radioautography). Effects are observed on the synthesis of small molecules and macromolecules as well as morphological changes, using both light and electron microscopes to see cell separation as well as nuclear and membrane morphologies. We can, in particular, mention as outstanding the nutritional biochemical studies of Maaløe and his associates [6]; the DNA membrane concepts and experiments particularly those based on mutants, of Jacob and his colleagues [2, 7]; and synchrony studies of Helmstetter and his associates [1], which have led to the broad coordinating generalization of sequential linked stages. In addition, there are the genetic-biochemical studies on DNA replication of Lark's group [8], the morphological plus biochemical studies of Donachie and Begg [9] on cell growth and septation, and the morphological-biochemical studies on separation and surface growth by Higgins and Shockman [10]. Reviews by these workers provide a large part of the information on the subject.

Although we discuss important work done with other organisms, this chapter stresses the studies done with E. coli and B. subtilis. The references cited are those we consider to be most worth the reader's time, in view of their germinal importance or their coverage of the literature; no attempt is made to be either totally comprehensive or historically exact. This is in order to keep a vast literature within bounds. We particularly stress work that relates to the cell membrane, which is the subject of this book. Literature through 1971 was surveyed. We offer an apology to those whose work, often excellent, has not been directly cited here.

II. THE HELMSTETTER-COOPER MODEL

The sequence of events proposed by Helmstetter and Cooper is so fundamental for present concepts of bacterial division that we now present it in some detail. The work on which this model rests was done with E. coli B/r, and so the conclusions strictly apply only to this organism. They very likely will have to be modified for other organisms. Yet this model is extremely valuable for comparison with new findings, much as β-galactosidase induction is the standard to which other work on enzyme formation is compared.

Experiments that attempt to describe the major events of the cell cycle have been performed for many years with the use of synchronized cells [11]. Since the methods for obtaining synchrony usually involved alignment of the cell cycle by blocking some event of the cycle, meaningful conclusions have been difficult to obtain. Helmstetter and Cooper [12] devised a very clever and gentle method of obtaining synchrony by selectively eluting newly divided cells from a randomly growing culture adsorbed to an inverted Millipore membrane filter. In this way, little metabolic stress is experienced by the cells. Thus, the key events of the cycle including DNA initiation, DNA completion, and cell division can then be timed accurately under various growth conditions.

Particularly valuable, since the cells are perturbed the least, is an inverse synchrony procedure in which all of the cells are briefly labeled, for example with radioactive thymidine to label DNA, and then the thymidine originally taken up is measured at invervals in cells as they come off the filter, i.e., just after they have divided. The first cells eluted are those closest to division at the time of labeling; succeeding fractions consist of cells that were further from division. When this kind of experiment is done under the simplest conditions, a jump of about twofold in the rate of DNA synthesis at a definite time in the cycle is observed (see Fig. 1). The results are consistent with a constant rate of DNA synthesis at each chromosomal replication fork. After the chromosome has doubled, the rate becomes twice as great, since there are now twice as many points of replication per cell. This is consistent with Maaløe's postulate [6] that the rate of macromolecular synthesis in a cell depends on frequencies of initiation rather than on rates of replication; at each replication point synthesis is proceeding at a constant rate. We will see that this is true only when growth is rapid; when nutrients limit growth, they can also limit the rate of DNA replication. The duration of DNA synthesis in different media reflects this rate.

The Helmstetter-Cooper model [1] separates the division cycle into three parts that occur in sequence — preparation for DNA initiation I, DNA replication C, and cell division D (Fig. 2). Divisions are timed according to the rule that a new cycle starts as soon as the preceding I period is completed. That is, completion of I not only starts DNA synthesis, but it also starts a new round of I synthesis. This is easily conceived of if one imagines that the initiator substance is made continually, and when it reaches a critical quantity it is used to initiate DNA; the I substance made thereafter immediately starts preparations for the next DNA round. As a consequence, cycles are initiated every I min, a time interval dependent on nutrition. C + D also vary with nutrition, but are never observed to be more rapid than 42 and 21 min, respectively. Since a division occurs at a constant time (I + C + D) after each initiation, divisions also occur every I min (Fig. 2). The nature of this initiator substance is not known, but protein synthesis is required for DNA initiation.

There are two main variations on this theme. The first is seen when the bacteria are in poorer media that decrease growth rates to division times longer than 63 min. Then C and D depend on the nutritional supply; experimentally, C becomes about two-thirds of the division time and D about one-third (Fig. 2B). The result is a division time of C + D = I. DNA initiation and cell division occur at the same time and DNA completion occurs two-thirds of the way through the cycle. This point has been disputed by Kubitschek and Freedman [13] who studied slowly growing *E. coli* in glucose-limited chemostat cultures.

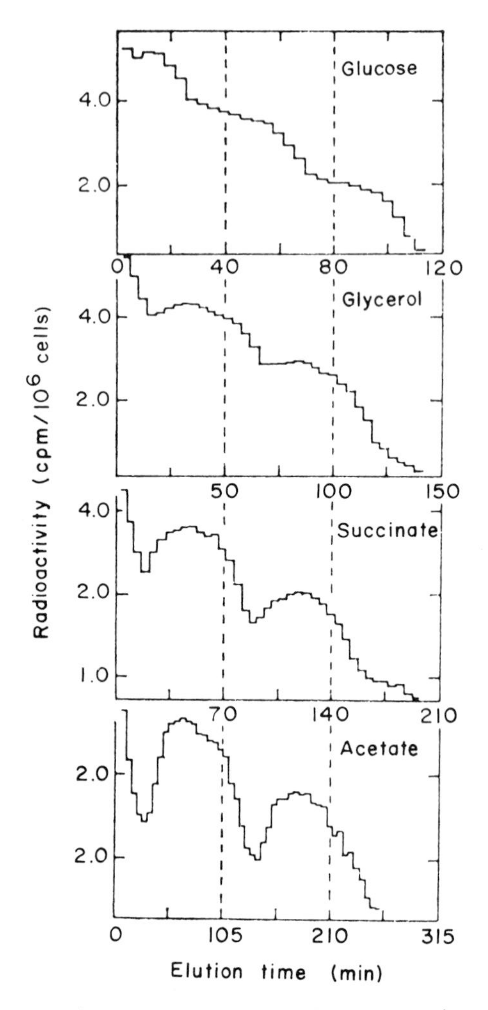

FIG. 1. DNA synthesis in synchronized E. coli B/r; Radioactivity per cell was measured in the effluent from membrane-bound cultures, which had been pulse-labeled with ^{14}C-thymidine. Cells growing in glucose, glycerol, succinate, and acetate medium were exposed to ^{14}C-thymidine at 0.1 μCi/ml for 1 min, 0.1 μCi/ml for 2 min, 0.1 μCi/ml for 2 min, and 0.15 μCi/ml for 2 min, respectively. Immediately after labeling the cultures were bound to membrane filters and washed and eluted with conditioned medium containing the same carbon source. Reprinted from reference 1 by courtesy of Academic Press, New York.

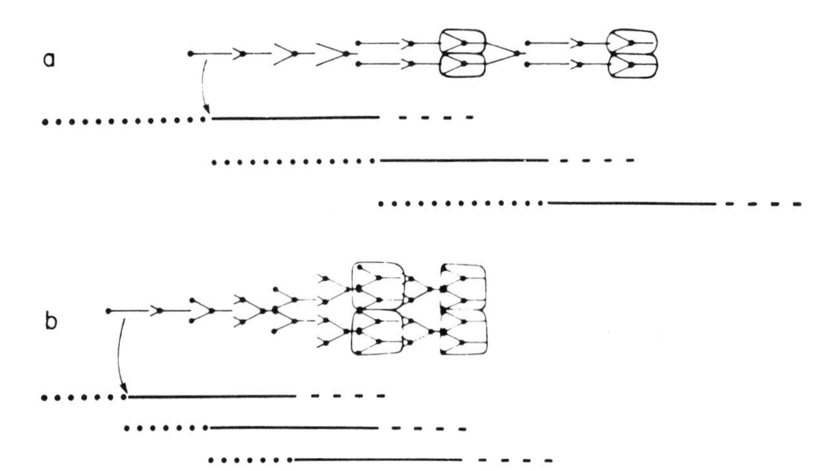

FIG. 2. (A, B) Construction of chromosome replication and division patterns for E. coli B/r. The production sequence is I, the time for accumulation of a fixed level of initiator per chromosome origin (.....); C, the time for a round of chromosome replication (_____); and D, the time between the end of a round and division (-----). Initiator accumulates continuously leading to a repetitive pattern of overlapping sequences. The chromosome configurations are shown at the top with a filled circle indicating a replication point, and enclosed chromosomes indicating a division. After each division, only one daughter is followed, but in the overall development of a culture from a single cell, the number of cells formed at successive divisions would be 2, 4, etc. For simplicity, the chromosomes have been shown as linear structures with the origin at the left and the terminus at the right. In (A) the time relationships between the phases are I = C and D = 1/2C; (B) I = D and D = 1/2C. Reprinted from reference 103 by courtesy of Federation of American Societies for Experimental Biology.

Events become more complicated when the cells are in very rich medium, capable of decreasing initiation time I to less than C (42 min). The initiation of the following DNA round occurs before the present round is finished, and one observes what is called dichotomous replication [14]. Several replication points then move down a single chromosome at the same time. The intervals C and D are still 42 and 21 min, respectively; that is, they are occurring at maximal rates. Division occurs I + 63 min after the start of each cycle, and since cycles start at intervals of I min, division also occurs every I min. The number of replication points, the number of chromosomes per cell, and the amount of DNA at each stage of the cycle can be calculated from these simple rules. This is best appreciated by working through a few examples using different I intervals [1].

One test of the model is suddenly to change the nutrition of an exponentially growing culture, or to inhibit its DNA or protein synthesis and observe the changes of replication pattern [1]. For example, if the medium is made richer (a shift-up experiment) one does not see any change in the rate of division until an interval C + D has elapsed. This is because the effect of the richer medium is only to shorten I; C and D are not affected and remain at 63 min. The cells dividing at 63 min after the shift are ones that had just completed I at the time of the shift (Fig. 3A).

The Helmstetter-Cooper model is of primary importance in our thinking about bacterial division. It links the principal events of the cycle (mass increase, DNA replication, septation, and cell separation) into a consistent sequence. These events are bound together by minimal, necessary requirements that limit the progression from one stage to the other. But the model does not show that these are the only requirements at each stage. There are probably multiple pathways that converge at each stage as shown so elegantly for replication of phage T4 [15]. For example, events other than DNA completion are required for septum formation, but these are normally completed early enough so as not to be limiting. Nor does the model define most of the events at a biochemical level. It presents us with a new set of questions, as any good theory should. For example, what is the chemical nature of the DNA initiator or the chemical-structured basis of the septum initiation process?

III. CHROMOSOME REPLICATION

Some aspects of DNA replication are briefly summarized here, with particular emphasis on the evidence that links DNA with bacterial division and with the membrane. Although most studies have been done with *E. coli* and *B. subtilis*, there is no reason to believe that replication is fundamentally different in other microorganisms.

All of the genes of *E. coli* are carried on one molecule of DNA, although some strains also carry additional, smaller ones (episomes). The DNA is double stranded, and it replicates semiconservatively according to the Watson-Crick model. The chromosome is $1200 \pm 100\ \mu$ long, and has a molecular weight of 2.5×10^9 daltons in its simple, nonreplicated form. Note that growing cells are synthesizing DNA and hence on the average their DNA has a molecular weight about 1.4 times the above value [6, 16]. The DNA is almost 1000 times as long as the cell that contains it; it must be compactly organized in the cell with an average distance of only about 40 Å between strands. This geometry raises questions as to how the DNA can be replicated and transcribed at all times in the cycle, and how the daughter strands finally can be separated. These problems are intensified because the DNA structure is circular, as shown by both genetic and electron microscopic studies. That the DNA can rotate about its long axis during replication is hard to conceive, particularly during dichotomous replication. No really satisfactory answers to this problem have been suggested, although Wang [17] very recently found an enzyme that continually appears to break and rejoin the strands.

A. Membrane-DNA Attachment

A physical connection of DNA to the membrane of *E. coli* was noted as early as 1961 by Goldstein and Brown [18], and newly formed DNA in particular was found associated with the membrane fraction. Jacob et al. [2, 19] and Ryter et al. [7] stressed the significance of this association for replication and equipartitioning of DNA. Evidence for the association is obtained from both morphological studies using the electron microscope and from biochemical fractionations.

Many electron microscopic studies show apparent attachment of chromosomal material to the cell membrane. Often the connection appears to be at a special membranous body named the mesosome. Mesosomes are particularly prominent in bacilli (Fig. 4); they are harder to see in *E. coli*. Morphologically, their appearance as curled up inward growths of the membrane, depends on the treatment to which the cells have been subjected. They can be extruded under hypotonic conditions. A great many biochemical properties have been attributed to mesosomes [20]. These include oxidative-phosphorylation and enzymatic activities [21], septum formation [10], and DNA replication [19].

The role of mesosomes in DNA replication is now questioned by Ryter [22] on the grounds that the number of DNA-mesosome attachment sites in *B. subtilis* are fewer than would be expected if DNA replicates at the mesosome. Van Iterson and Groen [23] have observed many fibrils, possibly DNA, connected to mesosomes that had been extruded from *B. subtilis*. Pontefract et al. [24] report *E. coli* to contain numerous mesosomes to which DNA appears connected at polar positions. Thus, at present the role of the mesosome as a site for DNA segregation must be considered as possible but hardly established.

Thin sections of *E. coli* normally show the nuclear material as translucent areas (in electron micrographs) distributed diffusely throughout much of the cell. Condensation of the nuclear material into compact bodies has been observed after alterations in the ionic environment [25], phage

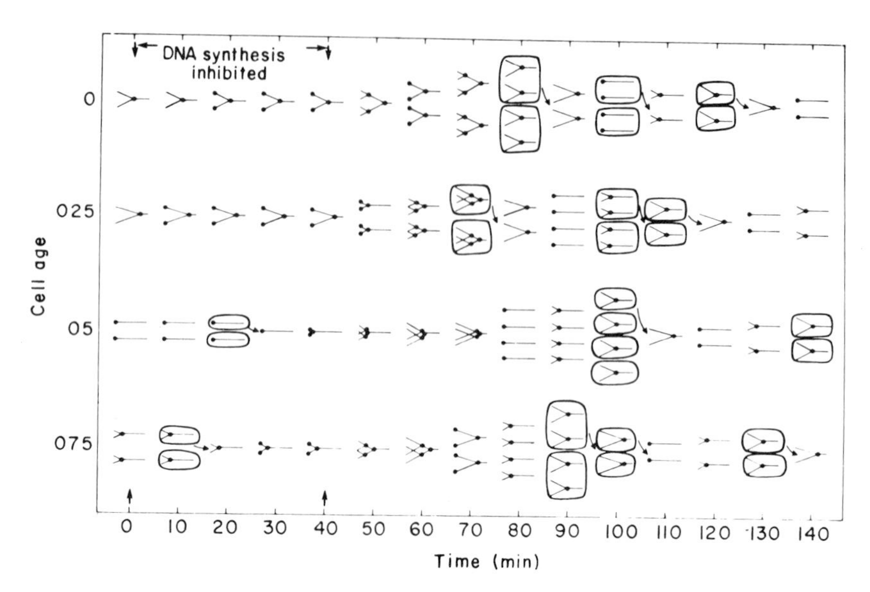

FIG. 3. (A) Theoretical response of individual cells to a nutritional shift-up. Four representative cells from an exponentially growing culture with I = 40, C = 40, and D = 20 min are shifted after 40 min to I = 20, C = 40, and D = 20 min. The ordinate shows the ages of the cells containing the chromosome configurations shown in the left-hand column. After a division, only one of the two cells is followed as indicated by the arrows. Reprinted from reference 1 by courtesy of Academic Press.

(B) Theoretical response of a culture to a nutritional shift-up. Summation of the effect of the shift-up (I = 40 to I = 20 min) over an exponentially growing culture in terms of cell number (_____), total mass (.....), and total DNA (---). Reprinted from reference 1 by courtesy of Academic Press.

infection [26], treatment with a bacteriocin [27], or the addition of chloramphenicol [28, 29]. The chloramphenicol-induced nuclear condensation is dependent on energy metabolism (inhibited by sodium azide) but does not require continued DNA synthesis [29]. Starvation of an oleate-requiring mutant for oleate resulted in the gradual cessation of DNA synthesis followed by nuclear condensation [30]. The condensed nuclear bodies were observed to be attached to the cell membrane in each ultrathin section examined; however, attachment was rarely seen in cells treated with chloramphenicol [29]. The oleate-starved cells were shown to be defective in new membrane synthesis. Condensation of the DNA might result from a failure to initiate new rounds of chromosome replication that could depend

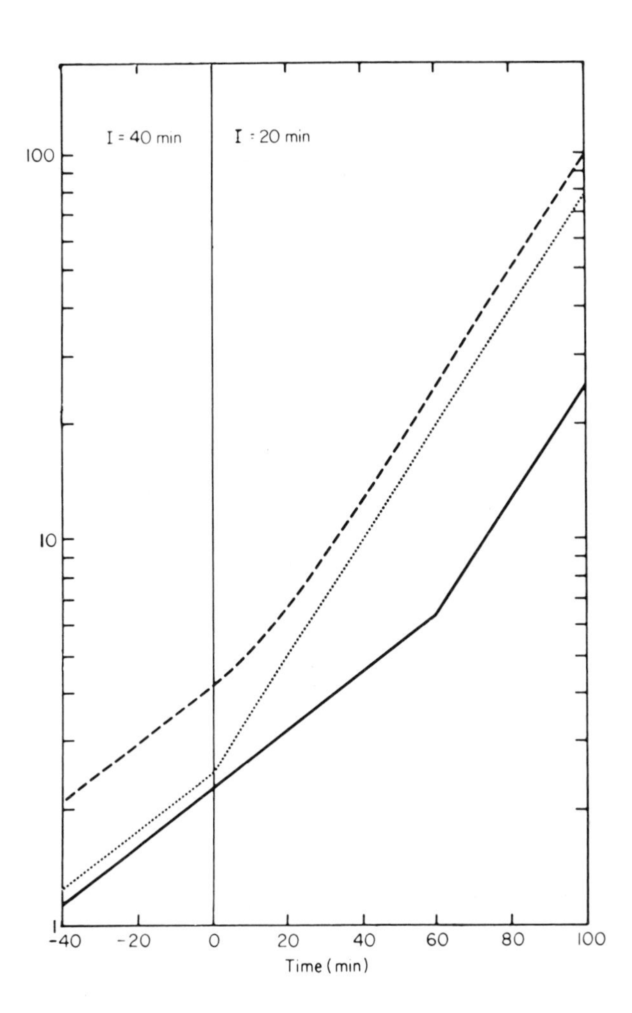

on new membrane synthesis. Organization of the nuclear material might be a step necessary for chromosome segregation into daughter cells. Dividing bacilli are often observed to have condensed nuclei at the center of the cell (Fig. 4; reference 22).

Crude bacterial membrane fractions contain DNA, as shown by many workers [31-35]. Schactle et al. [36] found that membranes isolated from E. coli grown with the arginine analog canavanine had DNA bound at multiple sites. The DNA could be released either by detergents or by proteolysis with pronase. DNA at both the origin-terminus and at the replicating point appear to be preferentially associated with the membrane fraction, although the other parts of the DNA are also found to a two- to

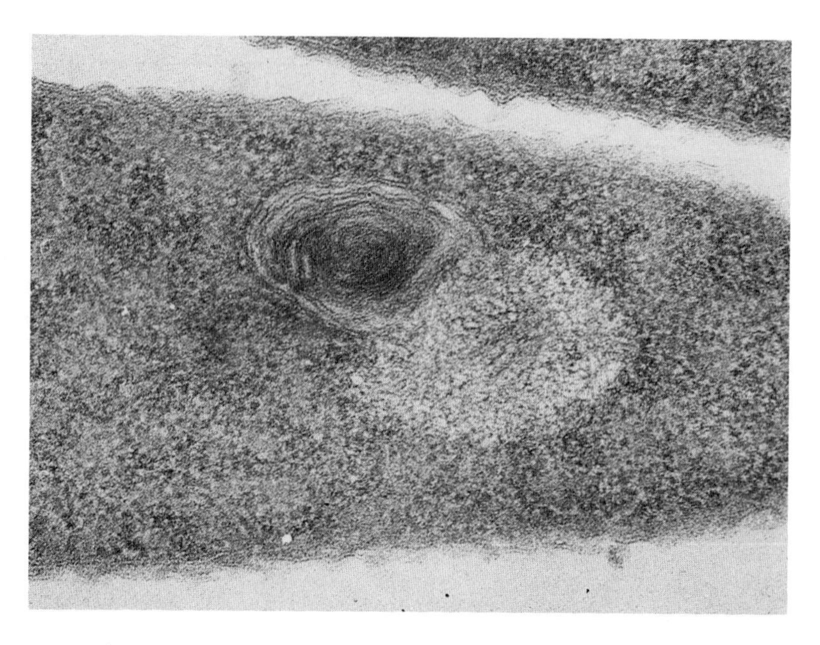

FIG. 4. Section of a well-developed mesosome in a thermolabile mutant of E. coli grown at 40°C for 2 hr. Magnification: X130,000.

fivefold lesser extent [37, 33]. Fractionation of the crude membrane preparations to identify specific DNA attachment components has just been started. Unfortunately, detergents that are commonly used for membrane solubilization probably also dissociate the DNA from the membrane [38]. Some new methods that look promising have recently been reported. Tremblay et al. [39] showed that membranes can adsorb to Mg-laurylsarcosinate crystals; DNA bound to the membranes is also separated onto these crystals.

Attachment of DNA to membrane leads to the prediction that there should be cosegregation into daughter bacteria of the attached DNA strands and membrane site. However, numerous experiments on this question have given conflicting results (see Section V). It will be valuable to correlate formation and properties of DNA-membrane attachment proteins with the properties of membrane proteins modified by mutation.

DNA is bound to membranes in several simpler systems [40]. Episomal DNA or DNA fragments can be transferred into DNA-less cells that are produced by an E. coli minicell mutant. This DNA becomes attached to the membrane even though the DNA does not replicate; DNA injected into minicells by an Hfr donor also does not replicate [41]. The DNA of several phages is associated with the bacterial membrane:

øX174 [42], T4 [43], and lambda [44]. The DNA of ø X174 and lambda replicates on the membrane.

A mutant of E. coli that carries lambda can be induced to produce the phage upon elevating the temperature. The same mutant cured of the phage exhibits defective division (filament formation) at the higher temperature, indicating a role of some bacterial locus in control of both phage replication and bacterial membrane (septum) formation [45].

B. DNA Initiation and Replication

Most of our information about DNA initiation comes from experiments in which the cycle is temporarily inhibited, either with chemical inhibitors such as chloramphenicol, or by starvation of an auxotroph for an essential nutrient [46, 8]. The most suggestive observation is that protein synthesis is required [6]. If an auxotroph is starved for an essential amino acid, DNA that is in the process of replication is completed, but initiation of the next round does not occur. Thus after the auxotroph has been starved for about 90 min at 37° C, all chromosomes appear to be completely replicated. They then all should be ready to start a new round shortly after the nutrients are restored. Unfortunately, not all cells commence replication at exactly the same time, probably because of heterogeneity in cell age; each cell contains only those proteins that were synthesized prior to amino acid starvation. Some synchrony of division has actually been observed following long (3-hr) amino acid starvation [47]. But care must be taken in interpreting the observations made with these cells since they are no longer in balanced growth.

Lark has pursued the question of how many proteins have to be made and when they are produced during the division cycle [8]. Amino acid starvation was used to align the chromosome of E. coli 15T⁻ [48]. Upon readdition of amino acids, the ability of inhibitors to block the reinitiation of DNA replication was followed. A low concentration of chloramphenicol (25 μg/ml, which can block the bulk of protein synthesis) stopped DNA initiation if added longer than about 30 min before initiation was due to occur. That is, a step sensitive to low concentrations of chloramphenicol must be completed at about minus 30 min. If replication was allowed to proceed past this time, 25 μg/ml was ineffective. However, 150 μg/ml chloramphenicol was still able to block DNA initiation, until about 15 min before initiation. Thus, at least two proteins with different sensitivities of synthesis to chloramphenicol seem to be involved, and their synthesis is completed at different times. Phenethyl alcohol, an inhibitor that profoundly affects membrane structure, blocks DNA initiation at the same time as does the higher concentration of chloramphenicol [8]. Chloramphenicol is reported to be less effective in blocking synthesis of membrane proteins than of cytoplasmic proteins in Micrococcus lysodeicticus [49], but this was not the case for E. coli K12 [50].

Lark and Renger [48] also propose that a third step is required for DNA initiation, on the basis that reinitiation following amino acid starvation and restoration has a 15-min lag; hence some process must take place before the completed chromosome can even start to be replicated. This third process does not require protein or amino acid synthesis. Furthermore, none of these events requires DNA synthesis. RNA synthesis might however be involved in DNA initiation, as suggested by Brutlag et al. [51].

Experiments identical in principle to the above have been performed by Ward and Glaser [52], by using E. coli B/r that have been synchronized by elution from Millipore filters instead of by amino acid starvation. The filtration method is preferable since it avoids metabolic imbalances such as those that follow amino acid starvation. The steps in E. coli B/r appear to be more sensitive to chloramphenicol (2 and 30 μg/ml) as compared to E. coli $15T^-$ (25 and 150 μg/ml). More interestingly, the step sensitive to the lower concentration of chloramphenicol was finished at the time of the prior initiation. The second, more resistant step occurred 21 min before the new initiation, independent of the growth rate for cells with division times between 27 and 49 min. Ward and Glaser did not find evidence for the third requirement. Both groups thus find two periods at which protein synthesis is required, although the timing is different perhaps because different strains were used or because of the synchrony methods.

Cooper and Weusthoff [52a] have recently challenged experiments in which chloramphenicol is used to study the initiation of DNA synthesis. They observed that the amount of residual DNA synthesized in the presence of chloramphenicol yields a smooth curve over a wide range of chloramphenicol concentrations and not the expected plateau at an intermediate concentration. They conclude that no particular step can be defined by this method.

We do not know the roles of the proteins that have to be made in order for DNA synthesis to commence, but it is tempting to equate them to the initiator that must accumulate to some critical amount according to the Helmstetter-Cooper model. This idea is consistent with numerous attempts to relate cell mass to the DNA replication cycle, if we assume that mass and quantity of I are proportional. Donachie and Hobbs [53] have calculated that the mass per DNA initiation site is constant at the time of initiation, over a wide range of growth rates. Donachie [54] observed that DNA synthesis is accelerated after thymidine starvation has ended and the DNA/mass ratio rises and reaches the normal value. The delay before resumption of cell division increases as a function of the length of the period of thymine starvation. Donachie proposes that initiator of DNA synthesis accumulates linearly during the starvation period, and that extra rounds are initiated when thymine is added. Since one finds it hard to see how a bacterium could estimate its own mass, a plausible explanation is that the initiator protein is made constitutively in proportion to total protein (approximately equal to the cell mass), so the mass would have a constant value when this protein reaches its critical quantity.

Lark and co-workers [8] have made extensive studies on DNA synthesis following inhibition of DNA synthesis. They find that the rate of synthesis after reinitiation is less than the expected value, and conclude that the new initiations occur at the usual origin, but only half of the strands are replicated. This last conclusion is questioned by other workers.

In short, all of these experiments show, by creating imbalances between DNA and other substances (proteins) necessary for initiation, that conditions for new initiations can be achieved without DNA synthesis or chromosome completion.

Mutants with modified DNA initiation properties (dnaA) have been isolated in both E. coli [55] and B. subtilis [56]. These mutants cannot initiate new replication forks at elevated temperatures although they appear to complete replication in progress. In contrast, a mutant has been found that makes extra chromosome origins at the high temperature [57]. A genetic locus for the dnaA mutants has been identified; however, the total number of genes is not known [58]. As with most mutant studies in this area, the results are preliminary, and need to be correlated with biochemical data.

The replication of an episome provides examples of regulated DNA initiation. On the one hand, an episome maintains a constant one to one ratio with the chromosome; hence replication of the two pieces of DNA must be coordinated. On the other hand, each episome appears to be a replicon, in the sense that it has genes that control its own replication [2].

The interdependence of episome and chromosome replication is indicated not only by the constant ratio during many generations of growth, but by segregation patterns when replication of the episome is blocked. Episome replication can be selectively prevented either by putting a temperature-sensitive replication-negative mutation into the episome (evidence that the episome carries elements essential to its own replication) [59], or by using an inhibitor of episome replication, Acridine Orange [60]. If the total cell DNA is radioactively labeled and then growth is allowed to proceed for several generations in nonradioactive medium under conditions of blocked episome replication, the labeled chromosomal DNA is conserved but segregated into only a fraction of the daughter cells. Since the number of episomes does not increase beyond the original number, they are found in only a few of these final cells. The striking observation is that the episomes remain associated with cells that retain the original (radioactive) DNA strands. Thus, there must be a cosegregation mechanism. This cosegregation is possibly accomplished through attachment of the two DNA's to nearby sites on the membrane, although this has not been proved directly.

Nishimura et al. [61] used an E. coli mutant T46 (temperature-sensitive for chromosome initiation) to investigate the relationship between chromosomal and episomal replication. In this mutant, at the restrictive temperature, neither chromosome nor episome can replicate. However, in many temperature-resistant revertants, DNA did replicate. Analysis of these colonies revealed that the episome had integrated into the chromosome preferentially at specific loci. Thus, the regulatory system of the episome can provide for replication of the entire chromosome at the non-permissive temperature. The two systems seem to operate similarly, but they cannot complement one another when they are on different pieces of DNA.

Episome replication has a negative control aspect. One cell cannot commonly support two episomes [62]. Thus, when another episome is injected into an episome carrying cell the second one cannot replicate; similarly an episome cannot replicate in an Hfr cell that carries a chromosomally integrated episome. Only exceptional mutants can carry two episomes; in these mutants, function of a replication-defective episome can be complemented by a normal one [2].

Episome exclusion can be overcome in mutant DNA-ts43 when it is grown at the permissive temperature. This same mutant shows an immediate cessation of DNA synthesis following a shift of the culture from 30° to 42°C [63]. Palchoudhury and Iyer [64] suggest that the basis for the temperature sensitivity of DNA synthesis is a membrane alteration in the mutant. This would explain the pleiotropic effects of the mutation, even at the permissive temperature, i.e., (1) prevention of stable attachment of M13 and R17 phages; (2) removal of the exclusion that one episome normally exerts on others; and (3) increased compatibility of multiple episomes.

Bacteriophages that contain DNA provide model systems for the study of DNA regulatory mechanisms. As an example, phage φX174 has to produce a special protein (the synthesis of which is chloramphenicol-resistant) in order that its DNA can be replicated in the host cell, E. coli [65]. The protein seems to be membrane associated, but its role does not appear to be to attach the DNA to the membrane.

Regulation of DNA synthesis is at the level of initiation rather than replication. DNA replication, once initiated, proceeds at a fixed rate around the chromosome to the terminus without control or interruption during normal growth. DNA replication occurs through the entire division cycle in rapidly growing cells, but during only part of the cycle in slower growing cells as described in Section II. E. coli in glucose-salts medium have a division time of about 40 min, and under these conditions all the cells are making DNA at any moment [66]. Initiation occurs at the point of the division cycle at which the rate of DNA synthesis doubles. The rate per replication point is constant during the cycle [67].

DNA can replicate more slowly by two- to three-fold in thymineless mutants supplied with suboptimal concentrations of thymine [68]. This is because the intracellular pool of dTTP is low, particularly with low external thymine concentrations. Although DNA synthesis is slower, the division rate is not reduced, because of compensatory dichotomous replications (Section II). Thymine deprivation seems to offer a useful technique for studying coordination of DNA and cell replications.

Not all organisms seem to have the same pattern of DNA replication as do E. coli and B. subtilis. In an early study, Lark [69] reported that temperature-synchronized Alkaligenes faecalis make DNA periodically, and that additions of deoxynucleotides cause DNA synthesis to become exponential. Possibly the course of DNA synthesis is normally linear, but the temperature treatments modified the pattern by upsetting the precursor pools.

In contrast, DNA replication in Myxococcus xanthus, a bacterium that has a life cycle involving morphogenesis, appears to be very similar to that of E. coli growing on a poor carbon source [70]. In a defined medium, chromosome replication (4.9×10^9 daltons) begins immediately after division and proceeds at a constant rate for 80% of the division cycle. The D period lasts 40 min (10% of generation time), while cell separation required an additional 40 min.

Caulobacter crescentus provides a most remarkable example of coordination of DNA replication with a complex division pattern [71]. This stalked bacterium can attach to a surface. When it divides, the daughter cells are different; one is the nonmotile stalked cell and the other is the motile swarmer cell, which has a polar flagellum. The latter has to develop a stalk before it can again replicate DNA and divide; its division cycle in minimal medium is longer (180 min) than that of the attached cell (120 min) [71]. In the swimming cell there is a period of about 60 min before DNA synthesis begins. Thereafter the two cells have similar patterns consisting of about 85 min DNA synthesis followed by a D period of about 35 min [71]. The pattern of division in this organism has also been studied in rich medium [72].

Lark [8] and Kuempel [73] have written comprehensive reviews on initiation and control of DNA synthesis in E. coli. The thoughtful and comprehensive article by Pritchard et al. [74] is also well worth reading. Much work was presented in the Cold Spring Harbor Symposium of 1968.

A particularly detailed review emphasizing the genetic aspects of DNA replication is offered by Gross [58]. Information gained on DNA synthesis and the cell cycles of prokaryotes and eukaryotes has been reviewed by Mitchison [11].

IV. THE LINK BETWEEN DNA COMPLETION AND SEPTUM FORMATION

According to the Helmstetter-Cooper model, a period D is required for septum formation following completion of a round of DNA synthesis. D is experimentally found to be about 21 min for rapidly growing cells (generation times less than 60 min), and about one-third of the cycle for slower growing cultures [1].

Evidence that chromosome completion is required for septation in *E. coli* B/r was obtained by Helmstetter and Pierucci [77] and by Clark [75, 76]. They showed that when DNA synthesis was arrested in an exponentially growing culture, some of the cells continued to divide for 20 min. Experiments with synchronously growing cultures showed that the cells that divided were those that had completed DNA synthesis. Those that were still replicating their chromosomes at the time DNA synthesis was blocked, did not divide. Thus, DNA completion rather than initiation or some intermediate step, appears to be a necessary condition for cell division in *E. coli* B/r. Pierucci [78] claimed that the sole requirement for the timing of cell division in B/r is DNA completion. She observed a period of terminal division during DNA inhibition, followed after reversal in turn by periods of no division, a burst of division, and finally, a return to normal. But other data suggest further requirements for division.

Different relationships between DNA synthesis and septation are observed in other organisms. In *E. coli* K12, no division was seen following cessation of DNA synthesis [79]. Conversely, all cells of *B. subtilis* divided after DNA synthesis was inhibited [80].

A. Perturbed Linkage of DNA Completion with Septation

Septum formation is perhaps the cellular process most sensitive to inhibition. Numerous chemical and physical agents prevent septation, and as a result the cells grow into long filaments [81]. One large category includes agents that also affect DNA synthesis, including ultraviolet light [82]. Another group includes agents that modify the cell wall (see Section V), such as penicillin and cycloserine. Other agents such as low temperature [83], have not added much to our ideas concerning the mechanism of septation, being less well defined in their mode of action in filament production.

Numerous mutants of E. coli with modified cell division properties have now been isolated. A mutant has been found that produces large numbers of unusually small, anucleate cells (minicells) in the course of normal growth [84]. The normal minicells (Fig. 5a) contain no DNA, but do have RNA and protein. Enzymatic activity per mg protein comparable to the normal cells is found. In minicell-producing strains that carry episomes, the episomes will segregate into and replicate in the minicells [85]. Minicells do not divide.

The lon mutants [86], including E. coli B, which occurs in nature [87], form filaments when their DNA metabolism is briefly perturbed. A number of conditions are effective, of which ultraviolet irradiation is the most commonly used [88, 89]. The target appears to be DNA rather than the septum itself, according to photochemical experiments in which the DNA was made more sensitive to irradiation by means of incorporated bromodeoxyuridine [90]. The ability to synthesize DNA is recovered after the same time interval as is required by the wild type, but the lag before division that follows is about three times as long in the lon^- strain. Even individual cells that have completed their DNA cycle do not divide readily when DNA synthesis is blocked [91]. These results suggest that aberrant DNA metabolism results in building up an inhibitor of septation in the cell. Fisher et al. [92] have reported that cytoplasmic factors isolated from normal cells can stimulate septation in inhibited lon^- cells. Addition of a second mutation mon^- (formation of irregularly shaped cells) [93] to a lon^- strain results, after ultraviolet irradiation, in the production of large, amorphous giant cells with 500-1000 times the normal cell volume. These giant cells (Fig. 5B) contain an extensive network of intracellular membranes forming vacuoles, vesicles, and cisternae [94].

Temperature-sensitive mutants are particularly useful in the elucidation of a complex cellular activity such as septation. Caution is usually indicated, however, since at best mutations act like inhibitors with unique primary specificities; at worst, mutants can be considered pathological. Their failure to follow the normal pathways can lead to incorrect conclusions regarding what happens in the normal cell.

One large group of mutants in E. coli is blocked in DNA synthesis, either for initiation (dnaA) or for chain elongation (dnaB). As expected, these mutants tend to grow into filaments for a time after DNA synthesis stops. Addition of a second mutation (div) to either dnaA or B mutants results in the formation of normal-sized, DNA-less cells at the higher temperature [55]. Another E. coli mutant of the dnaB type forms septa at random and produces a variety of cell sizes with or without DNA at the higher temperature [79]. The effect of mutation can often be reversed by high NaCl concentrations [95, 96].

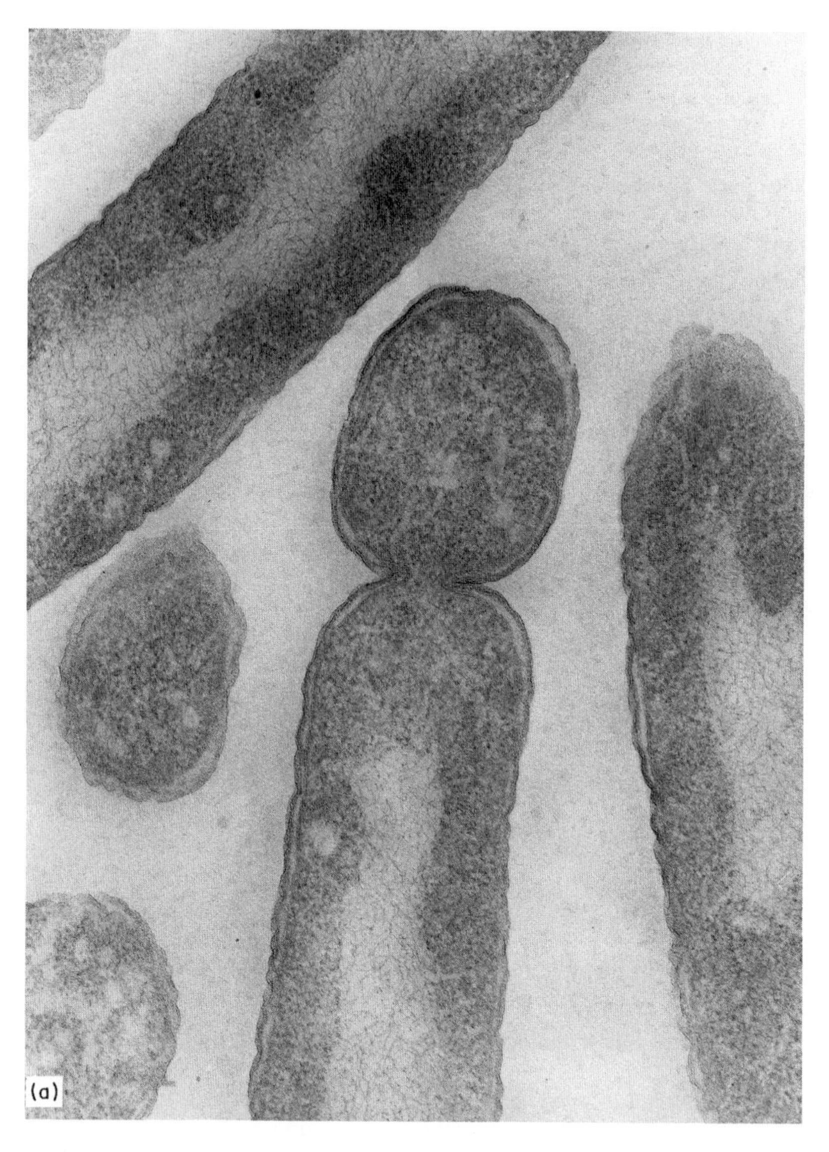

FIG. 5. (A) Thin-section electron micrograph showing a cell producing a minicell. Magnification: X66,000. Electron micrograph was prepared by D. Allison and A. Jacobson. Reprinted from reference 84 by courtesy of the National Academy of Sciences.

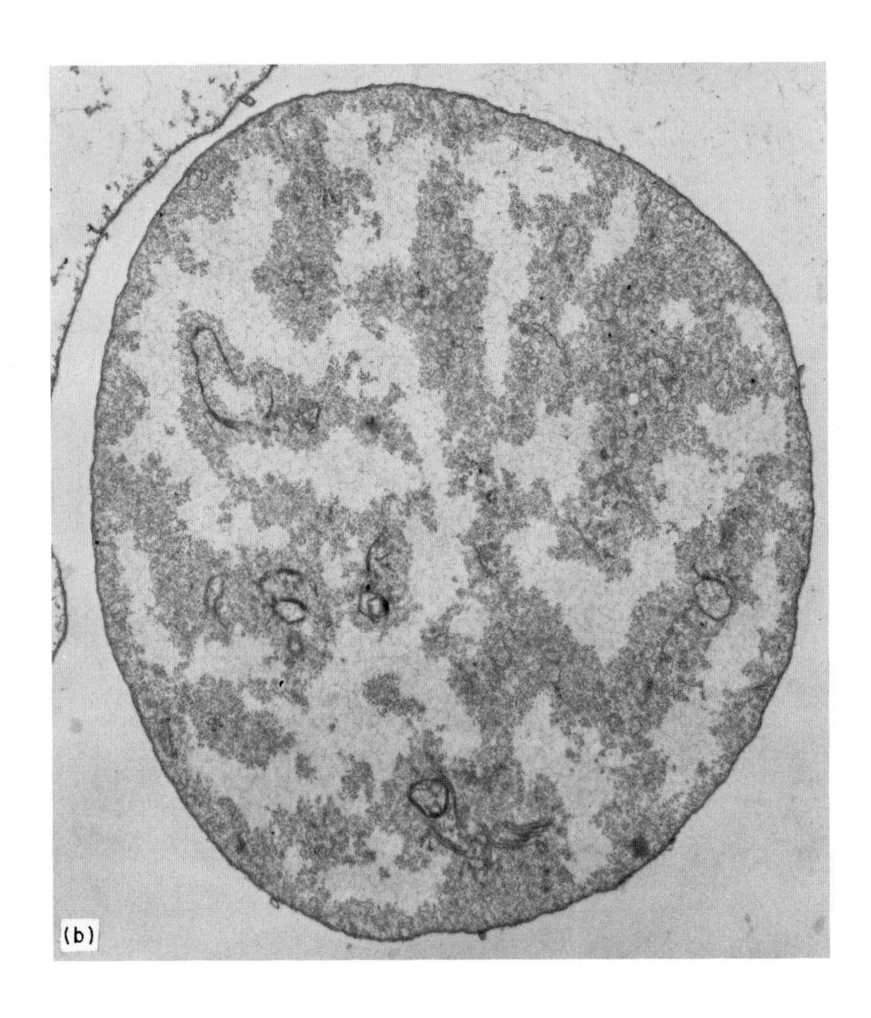

(B) Giant cell after irradiation. Note the extensive membrane involvement and the high ratio of nuclear to cytoplasmic material. Magnification: X10,000. Reprinted from reference 94 by courtesy of American Society for Microbiology.

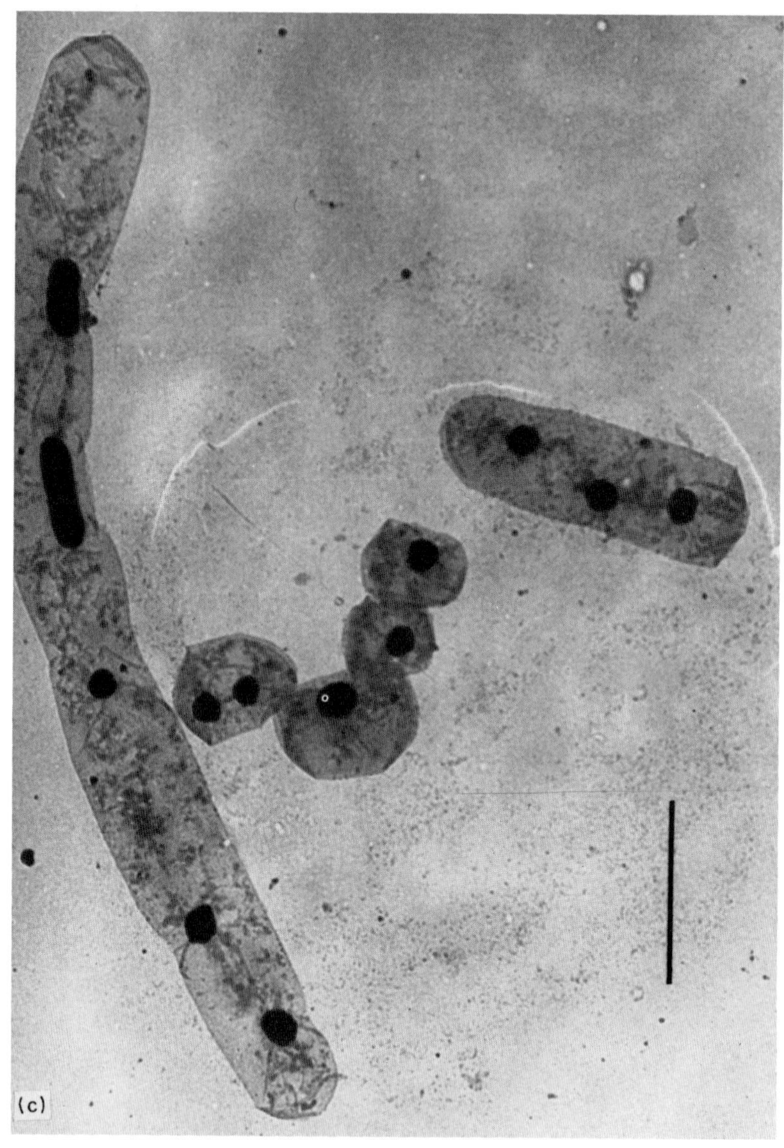

(C) Thermosensitive filament forming mutant ts52. Cells were grown at 41 °C in rich medium for 60 min, followed by chloramphenical (100 μg/ml) addition. The DNA was stained as described in reference 98. The bar represents 2μ.

Spratt and Rowbury [97] studied a mutant of Salmonella typhimurium, which appears to be temperature-sensitive for the initiation of DNA synthesis. This mutant, unlike dnaA mutants of E. coli, forms filaments that divide to yield normal-sized, anucleate cells after a period at the higher temperature. Residual DNA synthesis following the temperature shift is necessary for this delayed division. They suggest that termination of a round of replication is required for division.

Other mutants produce cells with altered structures, suggesting that the coupling of septation with DNA synthesis and other cellular events has been perturbed. Hirota [55] described several classes of mutants that show normal DNA synthesis at the higher temperature but divide irregularly (Fig. 6). The long filaments that result segregate nuclei normally in some mutants but not in others. Some of the phenotypes and map positions are shown in Fig. 6. One such mutant isolated by Zusman et al. [98] can be induced to divide by the addition of chloramphenicol even at the nonpermissive temperature (Fig. 5C). Filament formers presumably are defective in the synthesis of septum precursors or in the assembly of these precursors. Unfortunately, there is as yet no method for identifying septum components,

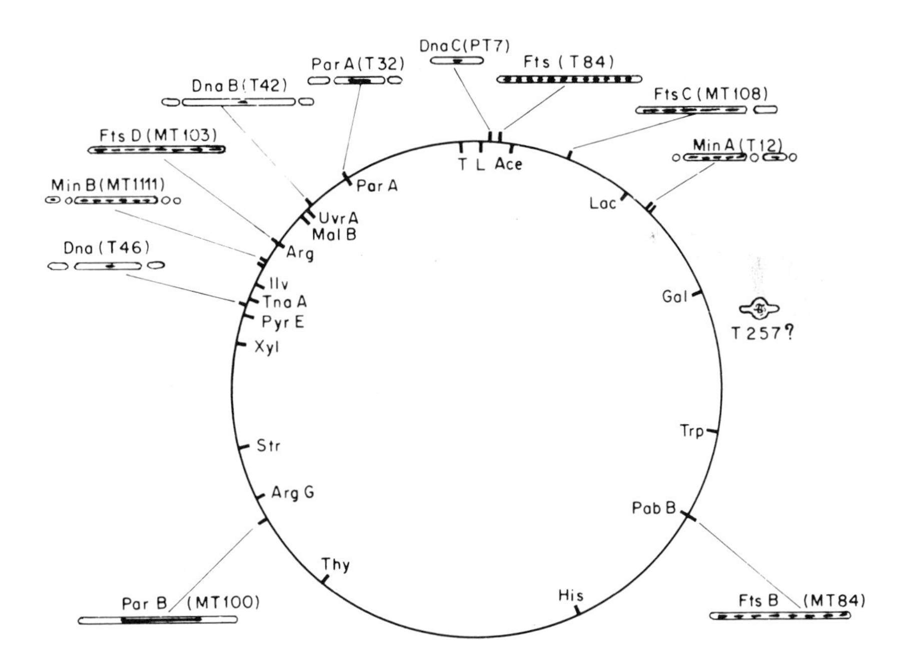

FIG. 6. A preliminary genetic map of the mutations altered at the different steps of cellular division. Reprinted from reference 182 by courtesy of Plenum Press.

let alone precursors, and so the distinction between these possibilities cannot be made directly. However, many filaments can septate after transfer to the permissive temperature in the presence of chloramphenicol [55, 99] or even chloramphenicol plus puromycin [98], and so at least the bulk of protein synthesis necessary for septation is accumulated at the restrictive temperature in these mutants.

Van Alstyne and Simon [100] have mapped and begun to characterize division mutants of B. subtilis. Four genetic and morphological classes of mutants were found; some produce incorrectly placed septa, while others have no septa at all (Fig. 7).

B. Other Requirements for Cell Division

Completion of a round of DNA synthesis might be a necessary condition for septation, and it might even be the crucial event in septum initiation,

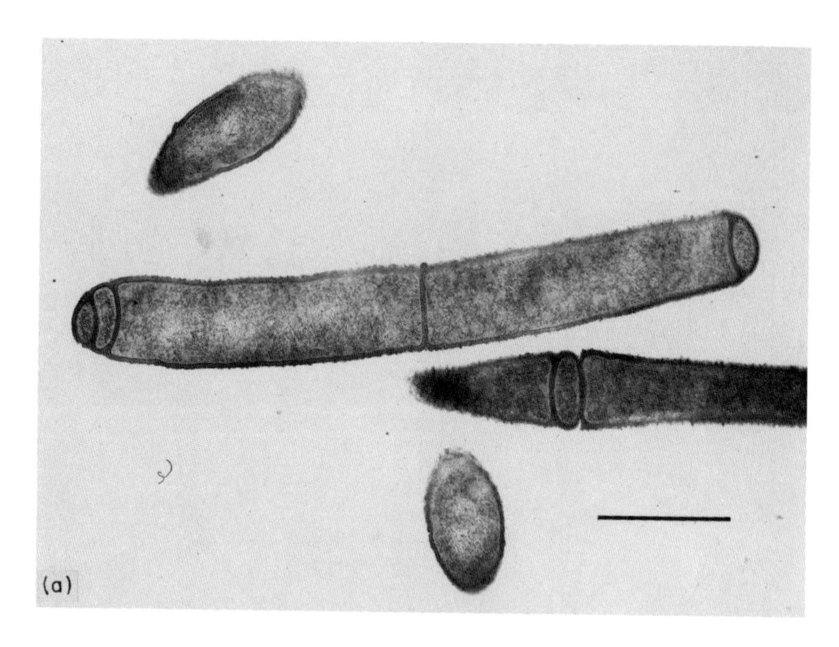

FIG. 7. Thin-section electron micrographs of mutant B. subtilis. (A) Div A mutant forming septa incorrectly; (B) div B mutant forming abnormal septa; (C) div D mutant lacking septa. The bars represent 1μ. Reprinted from reference 100 by courtesy of American Society for Microbiology.

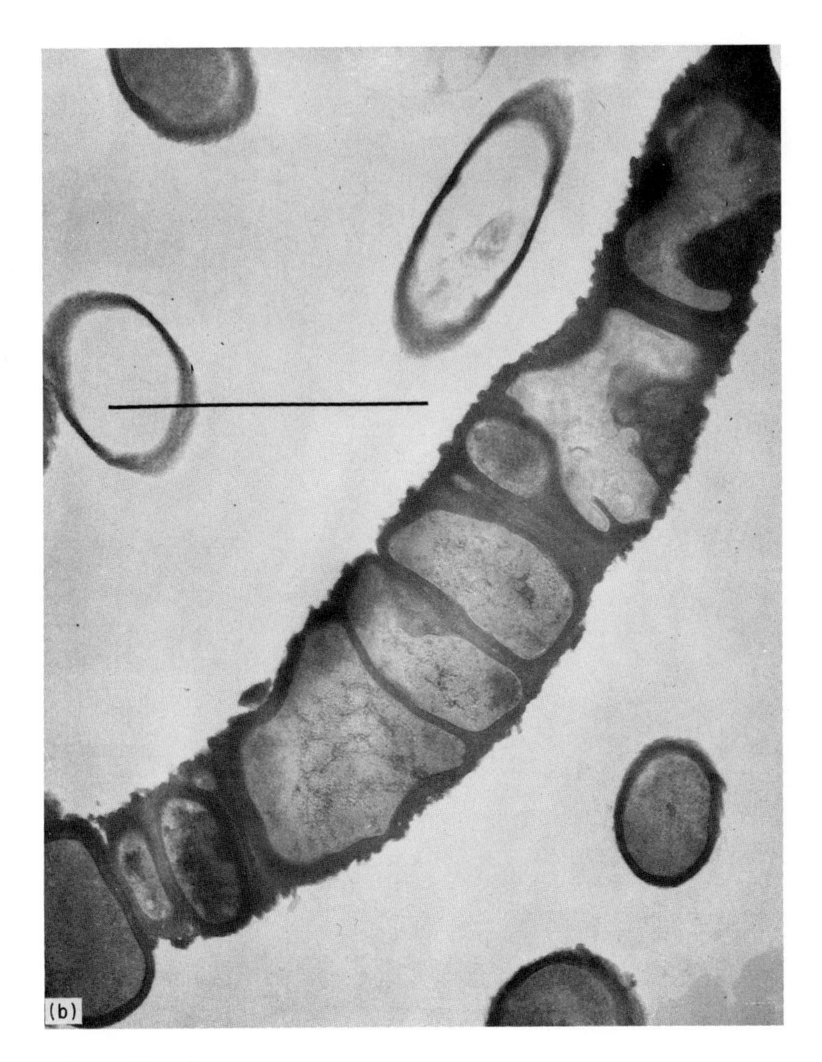

FIG. 7 (continued)

other events also appear to be required. The septum is a chemical structure; synthesis of its components at some time in the cycle is evidently necessary for its formation. Also, special cytoplasmic conditions for its assembly might have to be achieved. Thus the link between DNA completion and septation should be perturbed by modification of metabolic activities other than those related directly to DNA synthesis.

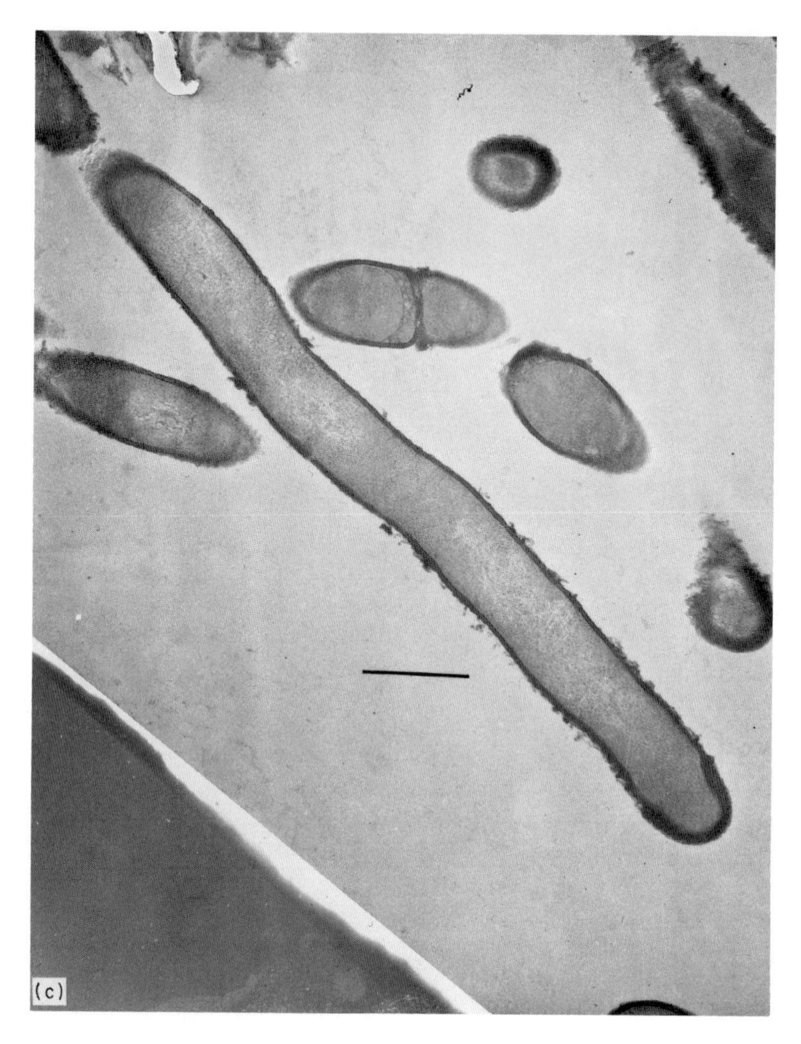

FIG. 7 (continued)

Septation depends on protein synthesis, as would be expected from membrane composition. The timing of synthesis of the necessary proteins could give clues regarding their roles. Regulatory and structural proteins might be made in different parts of the cycle. One example of the non-structural class might be proteases that were proposed to be responsible for a 10% drop in the protein content of E. coli at the time of division [101].

A similar drop in the rate of net protein synthesis at the time of division was observed in M. xanthus [102].

Two different approaches suggest that protein must be synthesized during a major part of the cell cycle in order that division can occur. Pierucci and Helmstetter [103] deprived synchronously dividing E. coli B/r of a required amino acid (histidine) for an interval and examined the consequences for subsequent division. It will be recalled that DNA synthesis proceeds to completion, but it does not reinitiate; hence, the requirement for DNA completion was satisfied in these cells. Restoration of histidine did not, however, permit cell division after a short time interval. Rather, there was a delay that was approximately proportional to the time that would have been required for chromosome completion at the time histidine starvation was initiated. Pierucci and Helmstetter conclude that protein synthesis must proceed during the entire C period of DNA synthesis in order that division can occur.

Smith and Pardee [104] also have suggested that a protein necessary for division must be synthesized throughout most of the cycle, up to the last few minutes. This is based on experiments with E. coli B/r and $15T^-$ in which a heat shock ($45°$ C for 16 min) was shown to produce synchronous division. Synchronized cells close to division were set back about 30 min by the heat shock; young cells were hardly affected. Other experiments showed that sensitivity was enhanced by incorporation during any part of the cycle of the unnatural, incorporatable amino acid p-fluorophenylalanine; this suggests that a protein that accumulates throughout the cycle is involved.

Little else has been reported regarding metabolic requirements for division. Inouye and Pardee [105] found that polyamines (putrescine and spermidine) in the cells, or added to the medium, profoundly affected the division pattern following amino acid starvation. Synchronous or exponential division could be achieved by varying the concentrations of the polyamines. It has long been known that a low Mg^{2+} concentration results in failure to form septa [106]. Thus, the major multiply charged cations of the cell appear important for division.

V. ROLE OF THE ENVELOPE IN SEPTATION

A. Properties of the Envelope

The bacterial envelope, with its wall and membrane layers, has two roles in cell division. The first role is an evident one: the formation of a cross-wall or septum by which the daughter cells are separated. The second role of the membrane is in the control and timing of events that precede division. The membrane also is the permeability barrier and site of transport systems, the site of synthesis of wall material, the site of energy-producing reactions, and the site on which DNA probably replicates and is partitioned. The peptidoglycan determines bacterial shape through its rigidity and the relative amounts of longitudinal and transverse growth (see Chapters 9 and 2). It physically holds the bacterium together by countering an internal osmotic pressure of many atmospheres. What biochemical properties determine the inward growth of wall and membrane at the point of septation? Is the septum site chemically different from the rest of the envelope? What determines the relative amounts of longitudinal and septal growth? How is the site of septation determined? Is this position marked by a special structure in the wall? Or is it a membrane or even a cytoplasmic component that measures off cell lengths? Answers to these questions are not available now, although they are actively being sought.

In general, our information is largely on the overall wall structure and composition (see Chapters 2 and 3). An exception is the work of Braun and Rehn [107]. They noted that incubation of a cell wall suspension of E. coli with trypsin results in a very rapid drop in optical density. They are able to demonstrate that a peptide bond especially sensitive to trypsin is present that joins a lipoprotein to the murein. The lipoprotein has been purified and its sequence, in part, determined.

Here we will assume that the wall is distinct from the underlying membrane. This is indicated in gram-negative cells by partial physical separations, as well as by electron micrographs in which the wall and membrane layers are seen to be separated [108], and in gram-positives by the removal of the wall to yield completely membrane bounded protoplasts. Other chapters of this book deal with this subject in detail, and there are numerous extensive reviews that can be referred to for further details. These include articles on membrane structure by Salton [109, 110], wall morphology by Glauert and Thornley [111], membrane biochemistry by Singer and Nicolson [112], and the role of the envelope in division by Rogers [113] and Higgins and Shockman [10]. Osborn [114] has reviewed the structure and biosynthesis of bacterial walls.

It ultimately will be essential to separate the various envelope layers. The difficult problem of obtaining clean membrane components of gram-negative organisms is being approached with E. coli by Schnaitman [115-117] among others. After removal of lipids, the membrane and outer wall components can be differentially solubilized with Triton-X100. Electrophoresis on acrylamide gels in a dimethylformamide system revealed about 30 major protein bands. These bands were very different in the wall and membrane fractions. In particular, Schnaitman found one major protein of molecular weight about 44,000 in the wall fraction.

Analysis shows the clean membrane to be composed of about 70% protein; most of the remainder is lipid. The proteins and lipids appear to be held together by relatively weak forces, all being solubilized by detergents or agents which do not break covalent bonds. Current ideas of how the membrane is put together favor a sort of solid solution in which molecules are held together by specific weak forces, and in which there is considerable fluidity [112]. This is in opposition to the older view of a membrane as a layered structure composed of a phospholipid bilayer with proteins on each side.

The search for chemical differences between the longitudinal membrane (hereafter referred to as "the membrane"), septum poles - which apparently are derived from septa, and mesosomes has not been fruitful. This is partly because no satisfactory way of separating septa from membranes has yet been devised. However, analysis of minicells, which should be richer in ends than are rods, has not revealed any distinctive biochemical differences [118, 119].

A few experiments have connected phospholipid changes to septation. Daniels [120] reported a transient increase in the rate of glycerol incorporation into lipids at about the time of cell division. This followed 3 hr of starvation for methionine in order to induce synchrony. Starká and Moravora [121] reported that penicillin or uv-induced filaments and also resting E. coli contain more cardiolipin and less phosphatidylglycerol than do growing cells. Conversely, Peypoux and Michel [122] found less cardiolipin and more (2x) phosphatidylglycerol in filaments formed by a temperature sensitive mutant at 41° C. Ballesta and Schaecter [122a] found that an increase in phosphatidylethanolamine synthesis occurs under conditions of increased membrane synthesis such as the breakdown of filaments of E. coli lon^- mutant or the recovery from a shift down in growth medium. Clearly more work is needed to unravel these differences, and to learn whether they are significant.

B. Sites of Envelope Growth

Data obtained in direct attempts to measure sites of growth are conflicting. Two general approaches have been used. One is briefly to incorporate a radioactive membrane precursor and then to follow its distribution with time along the membrane of one cell or in the membranes of a chain of cells formed by growth on a solid support. Thus, Morrison and Morowitz [123] took advantage of the large size of B. megaterium to find the location of briefly incorporated radioactive palmitate. Radioactivity was conserved in a small number of progeny cells, and was always located within about 0.2 μm of the ends of these cells. This result is consistent with a definite zone of membrane growth. In contrast, Lin et al. [124]

found in E. coli that oleate and glycerol were equidistributed among the daughter cells in a random way, a result suggesting growth of the membrane by intercalation. Diaminopimelic acid, a wall precursor, was similarly distributed. However, even thymidine in DNA continued to segregate until the cell had produced 16 progeny, a result that is inconsistent with earlier data [125] or the number of conserved DNA strands in these cells. Mindich is reported by Morrison and Morowitz [123] to have found random distribution of label from glycerol in a mutant of B. subtilis that cannot degrade glycerol, again suggesting intercalation.

The growth of the cytoplasmic membrane of B. subtilis was studied by using the flagella as membrane markers [126], since the flagella remain fixed to the membrane during cellular elongation. A temperature sensitive mutant, able to form flagella only at low temperatures, was shifted to the nonpermissive condition. The chains that resulted showed flagella only at the polar and central parts of the chains. These results suggest that membrane growth is localized. To test the linkage to DNA synthesis, thymine starvation was used to produce filaments. The distribution of flagella was found to be nearly random. However, pretreatment with chloramphenicol to allow completion of rounds of DNA replication and segregation of chromosomes, followed by thymine starvation, resulted in segregation of flagella to the ends of cells.

In another study, crystals of potassium tellurite were deposited on the membrane of B. subtilis, and their distribution following growth was observed [127]. Growth was localized at the center of the cell, but the experiments are open to the objection that the crystals could damage the membrane and prevent its growth at the ends where they are attached. Different organisms and methodologies again make results on the entire problem uncertain.

Contrary to the localized growth of membrane suggested (but not proven) by some of these observations, immunofluorescence data show that antigenic sites of rod shaped bacteria are uniformly distributed over the surface and are randomly dispersed as the cells grow [128]. The objection has been raised that these antigenic sites are not on the wall but on external layers that are laid down in a way that does not reflect growth of the wall itself [113, 10]. Wall growth in some cocci is localized, according to the same method.

A study on the regrowth of B. subtilis and B. megaterium after inhibition of protein synthesis (chloramphenicol or amino acid starvation) showed that the thicker peptidoglycan wall that was formed during the treatment was slowly sloughed off in fragments from the sides of the cell [129]. The ends, however, remained thickened. A correlation to growth was made and multiple growing points of the wall, except on the ends, were suggested.

The relation of wall growth to the rate of overall cell growth has not been intensively studied. Sud and Schaecter [130] found that when the growth rate of B. megaterium was changed by shifting to a different medium there was a 25 min lag before the peptidoglycan precursor incorporation rate changed. Grula and Smith [131] found that peptidoglycan invagination of Erwinia could occur before nuclear segregation. Timing and dependence of wall growth, both longitudinal and inward, in relation to other events in the cell division cycle deserve more study.

C. Site of Septation

Early experiments with E. coli by Lederberg [132] indicated that the "division septum" is especially sensitive to penicillin. Schwarz et al. [133] showed that at low concentrations of penicillin, bulges occur in the murein sacculus at sites where the new cell wall is normally formed. At high concentrations of penicillin, the sacculus split. These results were explained by assuming that the formation of a new septum requires the strictly localized action of murein hydrolases that are activated in the presence of penicillin. Interestingly, both penicillin and chloramphenicol have their strongest killing effects just at the time of division; presumably the septum site is particularly susceptible as the cells separate [134].

Donachie and Begg [9] have carried out original experiments with E. coli B/r and $15T^-$ to investigate the mode of elongation and the site of septation (Fig. 8). First they placed the cells on an agar surface and photographed them as they grew. In minimal medium they elongated from one end alone, until they reached a length of 3.4 μm. They then divided at their center, and growth continued only from each of the newly formed ends. In other experiments, application of a high concentration of penicillin caused a small bulge (due to the splitting of the sacculus) to appear in each cell, at a site that was shown to be where the septum eventually would form. This site was always 1.7 μm from the farther end of the growing cell.

These results are consistent with the concept of a "unit cell" of 1.7 μm length. This model predicts that the septum site is predetermined from the start of the cell cycle. Growth only takes place unidirectionally from this site according to the simplest model; otherwise the growth of membrane between septum site and old end would shift the site. These results have been combined with the Helmstetter-Cooper model and other data by Donachie and Begg [9] and Higgins and Shockman [10], and are also summarized in the last section of this chapter.

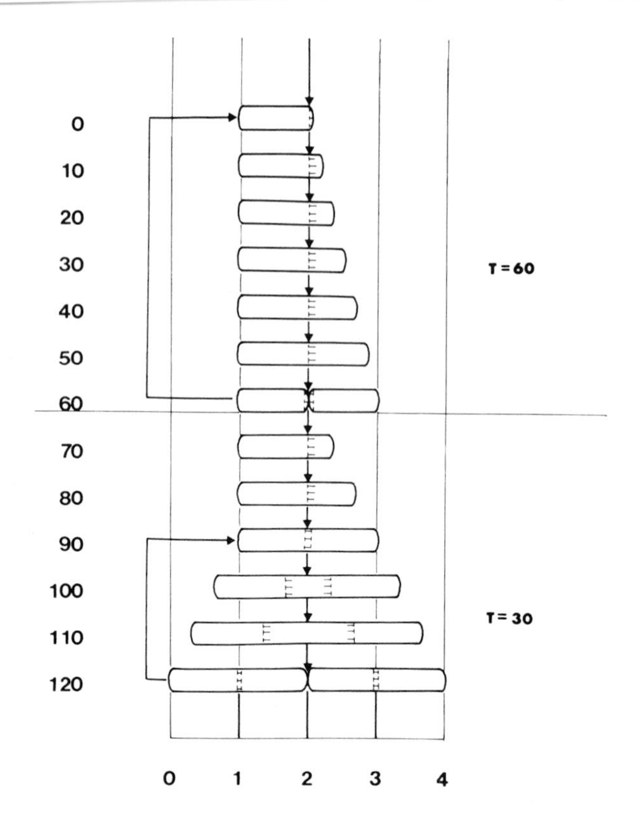

FIG. 8. Model of cell growth. In the imaginary experiment depicted, a unit cell is grown for one cell cycle in a medium where the mass doubling time is 60 min (T=60). At 60 min the daughter cell on the left is transferred to a richer medium where the mass doubling time is 30 min (T=30). The normal cell cycle for cells growing continuously in these two media is shown by the groups of cells connected by the heavy arrow to the left of them. During a shift-up of the kind shown here, however, there will be a 60-min interval between transfer to the new medium and the next cell division. Consequently, the length of cells at this division will be twice the length of dividing cells in the old medium. (For simplicity the increase in cell volume after the shift-up is assumed to result from increase in length without increase in diameter.) The growth sites are shown as dashed vertical lines across the cells, and the direction of growth at each site is shown by the horizontal lines attached to them. Each growth site gives rise to two new sites of opposite polarities when the cell reaches a length of 2 unit cells. The central vertical line (line 2) corresponds to the spatial location of cell divisions. The triangles show the penicillin-sensitive sites. Reprinted from reference 9 by courtesy of Nature.

D. The Wall and Septation

A definite role of the wall in division is gained from three sorts of experiments — with protoplasts, metabolites and inhibitors, and with mutants. The normal cell division pattern is greatly altered when the wall is completely removed (to form a protoplast) or partially removed (to form a spheroplast) from a growing cell in an osmotically protecting medium that prevents lysis. Only mycoplasma divide naturally without having a wall. Wall-less *B. subtilis* protoplasts grow and divide in the shape of rounded blobs, with almost random membranes dividing them. They can be made to revert to a normal shape and division pattern if they are grown in 25% gelatin [135], which is thought to retain wall precursor material next to the membrane and thus give new wall synthesis a start.

Nutritional changes can modify wall synthesis and also division. Thus, Weinbaum et al. [136] obtained filaments when *E. coli* B (which is *lon*$^-$) grew in a very rich medium. Glucosamine incorporation was diminished and three glycoproteins were produced in diminished amounts. Several D-amino acids, particularly D-serine and glycine, interfered with wall synthesis and caused septumless filament formation in *Erwinia* [137].

Escherichia coli mutants that are defective in wall growth have been isolated on the basis of their osmotic fragility [138, 139], or their altered resistance to antibiotics [140, 141]. An ampicillin-sensitive mutant, *envA* was isolated that grew as chains of cells during exponential phase but as single cells in stationary phase [140]. Transfer of these chains from rich to minimal medium resulted in fragmentation of the chains. Drug sensitivity and chain formation were both cured by 0.05% phenethylalcohol [141].

A combination of antibiotics was used to try to characterize the defect of an *E. coli* mutant, *BUG-6*, as being in a moderately late step of division [99]. At 42° C division stopped immediately. When the cells were returned to 30° C they septated and divided in proportion to cell mass. Recovery was not blocked by inhibitors of macromolecular synthesis or by cycloserine, which blocks an early step in cell wall synthesis. However, vancomycin, which blocks a later step, was partly effective, and penicillin, which blocks a terminal step, was completely effective in inhibiting division. High concentrations of NaCl (11 g/liter) partially reversed the phenotype. The results are interpreted as being caused by lability at 42° C of some component involved in cross wall synthesis. However, a double mutant of *BUG* 6 and *min*$^-$ was still able to divide and form minicells at the higher temperature.

While these experiments all show a role of the wall, it is not clear whether this role is in guiding and determining the site and direction of growth that leads to septation and separation, or whether the membrane has these roles and the wall is only a passive follower that lends rigidity and prevents osmotic forces from undoing the action of the membrane. Indeed, it is possible that both proposals are correct in different organisms. In a blue-green alga, Agmenellum quadruplicatum, the normal type divides by inward growth of the wall and membrane as does E. coli, but a mutant first puts a membrane septum across the cell and the wall formation follows [142].

A little evidence for independent inward growth of the membrane has been obtained from electron microscope pictures of plasmolyzed and sucrose-treated cells of Erwinia [131] and of E. coli [108]. Here the membrane is clearly seen to be separate from the wall and positioned by itself toward the center of the cell. However, one can ask if the preparative treatments did not distort the normal position of the wall and membrane in order to give the observed results.

Perhaps the first mutant isolated with impaired wall synthesis was a B. subtilis mutant defective in D-glutamate synthesis [143]. Other B. subtilis mutants with disturbed morphology and septation form defective walls [144]. Structural abnormalities of another B. subtilis mutant, rod^-, are repaired by 0.8 M NaCl or by glutamine and other amino acids found in rich medium [145, 146, 113]. Some B. subtilis mutants grow as L-forms in liquid medium containing 1.2 M NaCl [147]. Similarly, stable and unstable L-forms of Proteus vulgaris are known, and have differences in wall compositions [148]. These and other observations all point to dramatic effects on the cell division process when normal wall formation is disturbed.

E. Mechanism of Wall Action in Division

Higgins and Shockman [10] have carried out a beautifully detailed investigation of bacterial division with special emphasis on the role of the wall (see Chapter 2). Division of a coccal form, Streptococcus faecalis was considered first, since cocci are considered to be simpler in their mode of division than are rods. The latter exhibit longitudinal growth of wall and membrane as well as septation. In S. faecalis the sites of wall extension can be seen (Fig. 9) by noting the position of ridges on the surface, places at which the following round of wall invagination will occur. The electron microscopic studies of Higgins and Shockman [10] show tightly coupled inward growth of both wall and membrane at the equator of the cell as the primary event in wall growth. Their model is summarized in Fig. 9. In cross section, the wall at the septum site is shaped like a Y, with the stem representing the inward growing portion and the arms the recently formed walls. New material is constantly added at the base of the Y, so that inward growth occurs. But part of this inward growth is diverted at the fork into surface extension. The wall membrane thickens and simultaneously splits at the outside surface to extend the surface. The two processes can be experimentally separated. The inward extension does not continue when DNA synthesis is blocked, but rather synthesis is preferentially converted into external wall extension and thus filaments are formed. Conversely starvation for valine, an amino acid required for mass increase but not for wall growth, allows inward growth but not wall extension. Presumably the pressure produced by normal mass increase facilitates the splitting of the septum at the fork into walls. The site of new wall synthesis in S. faecalis is highly localized, being at the site of septum formation. This was demonstrated earlier in S. pyogenes using immunofluorescence [128].

F. Envelope Changes Related to Division

Envelope protein changes in E. coli related to division and DNA synthesis have been observed and studied principally by two groups. Inouye and Guthrie [149] first demonstrated a change in the membrane protein profile on SDS-acrylamide gels of a mutant defective in DNA synthesis. Further work with this mutant and with inhibitors showed that a protein X of molecular weight 49,000 is found in a greater amount in the membrane of cells whose septum formation is defective. Production of a second protein, Y, of molecular weight 44,000 decreased in cells the DNA synthesis of which had been blocked by any of a large variety of methods [150, 151]. That this change is more likely to be a consequence than a cause of blocked DNA synthesis was indicated by kinetic data.

Hirota et al. [152] have done very similar experiments and also have found changes in two membrane proteins. Their results are different in detail from those of Inouye, although at this early stage, differences are hardly surprising when different mutants and, more important, different methods of solubilizing the membranes before electrophoresis are used. Shapiro et al. [153] reported that temperature-conditional dnaA mutants (that stop DNA synthesis gradually at 41° C) contain at 41° C less of a protein with a molecular weight of 60,000 and more of a protein with a molecular weight of 30,000 than do normal cells. These cells ceased division at the higher temperature. The same result was obtained with dnaB mutants. The membranes were dissociated with SDS at 40° C in these experiments, in contrast to 70° C as used by Inouye and Guthrie [149]. When the temperature was increased from 40° to 100° C, the protein differences of dnaB mutants were not seen [154] and the entire pattern was greatly modified. It was concluded that the differences were based on an association phenomenon rather than a molecular weight change. However, these small differences might still be present but are obscured by other proteins on the gel. When dnaA mutants were used, the same two protein changes were observed, both with 40° and 100° C dissociation.

It is important to note that the gel patterns depend very much on the method of extraction and the conditions used [110]. Thus, most workers find about 30 protein bands in the envelope fraction, but the sizes and molecular weights of the peaks differ considerably. The use of internal fluorescent protein standards of known molecular weights [155] should greatly aid in identifying and calibrating the molecular weights of the bands. A further problem is to distinguish adsorbed proteins from those that are built into the membrane. This distinction might only be quantitative. Functional roles of the proteins have not, in general, been identified.

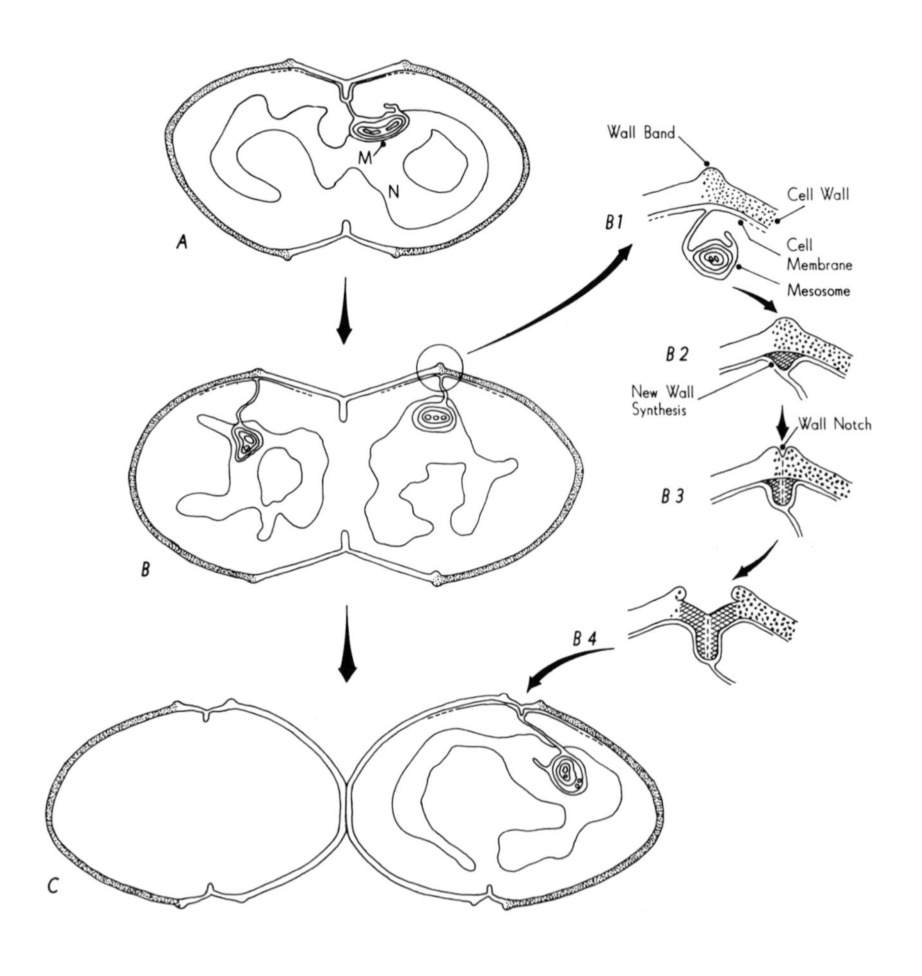

A particular interesting extension of this work was a study of membrane changes when the dnaB cells were grown at 41° C in the presence of 2% NaCl. Salt eliminated the phenotypic response; the cells grew normally. Nevertheless, Siccardi et al. [154] reported that the membrane changes were still found, and that therefore the membrane changes were a primary result of the mutation, rather than a consequence of metabolism in the non-DNA synthesizing, nondividing cell. The changes they observed though, were much smaller in the presence of NaCl, and so further data would be desirable to support this conclusion. In contrast, Inouye [96] found that salt reversed both phenotype and membrane changes in a dnaB mutant, ts27.

FIG. 9. Diagrammatic representation of the cell division cycle for Streptococcus faecalis. The model proposes that linear wall elongation is a unitary process, which results from wall synthetic activity at the leading edges of the nascent cross-wall. The diplococcus in A is in the process of growing new wall at its cross-wall and segregating its nuclear material to the two nascent daughter cocci. In rapidly growing exponential phase cultures before completion of the central cross wall, new sites of wall elongation are established at the equators of each of the daughter cells at the junction of old polar wall (stippled) and new equatorial wall beneath a band of wall material that encircles the equator (B). Beneath each band, a mesosome is formed while the nucleoids separate and the mesosome at the central site is lost. The mesosome appears to be attached to the plasma membrane by a thin membranous stalk (B1). Invagination of the septal membrane appears to be accompanied by centripetal cross-wall penetration (B2). A notch is then formed at the base of the nascent cross wall which creates two new wall bands (B3). Wall elongation at the base of the cross-wall pushes newly made wall outward. At the base of the cross-wall, the new wall peels apart into peripheral wall, pushing the wall bands apart (B4). When sufficient new wall is made so that the wall bands are pushed to a subequatorial position, e.g., from C to A to B, a new cross-wall cycle is initiated. Meanwhile the initial cross-wall centripetally penetrates into the cell, dividing it into two daughter cocci. At all times the body of the mesosome appears to be associated with the nucleoid. Doubling of the number of mesosomes seems to precede completion of the cross-wall by a significant interval. Nucleoid shapes and the position of mesosomes are based on projections of reconstructions of serially sectioned cells. Reprinted from reference 10 by courtesy of Chemical Rubber Company.

Other lines of evidence obtained by Hirota et al. [152] also demonstrated that the membranes of cells that are defective in DNA synthesis are modified. Inouye [151] found that a mutation in the DNA repair system of E. coli (recA) permitted cell division to continue when DNA synthesis was blocked by thymine deprivation. These results together with the membrane protein changes related to DNA synthesis noted by Inouye et al., and by Hirota et al., provide evidence that might eventually permit a biochemical description of the DNA-cell division link.

VI. CELL SEPARATION

A. Physiology of Cell Separation

The final event in the division cycle is separation of the daughter cells from one another. This occurs after septation is completed. The interval between the two depends on the organism. Thus, attached cells can frequently be seen with the microscope.

For E. coli about 7 min are required between the completion of septation and cell separation [76]. Completion of septation was taken as the time at which the daughter cells were physiologically independent, as measured by their independent killing by bacteriophage infection or sonic oscillation. Thus, the period D between completion of DNA synthesis and cell separation can be subdivided into intervals of about 13 min in rich media for septation and about 7 min for separation.

The picture is quite different in B. subtilis. Paulton [3] reports that about 138 min must elapse between septation and cell separation, at all growth rates studied (at 30°C). The consequence of this long period is that chains of unseparated cells are formed at rapid growth rates, but single cells are found at slow growth rates. Very slowly growing cells separate before the next round of septation occurs and hence have one septa or none. Very rapidly growing cells undergo a sort of dichotomous septation, analogous to dichotomous DNA replication, in which a second, and even a third wave of septa are completed before the first separation has occurred. This results in chains of cells with up to seven septa. Further studies by Paulton [156] demonstrated that the time between chromosome completion and cell separation was 150 min; thus, the time between DNA completion and septation was about 12 min. In this connection, one recalls that inhibition of DNA synthesis does not block new septum formation or separation of B. subtilis [80].

Cell separation has not been extensively studied using inhibitors or mutants. Tomasz [157, 158] showed that incorporation of ethanolamine in place of choline into the walls of Diplococcus pneumoniae resulted in chain formation. Mutants of E. coli that form chains and hence are apparently defective in the separation step have been found [99].

B. Autolytic Enzymes and Cell Separation

Cell separation probably requires both further wall thickening and action of enzymes to separate the two transverse walls (see Chapter 2). The role of autolytic enzymes in separation, in addition to their function in wall extension, has often been suggested. Weidel and Pelzer [159] discussed these ideas in relation to their concept of a murein sacculus, a bag-shaped macromolecule, that continually has to be opened up and reformed locally in order to permit cell growth. Autolytic enzymes undoubtedly exist in E. coli; little work has been done on them so far. Murein hydrolytic enzymes have been invoked to explain the localized penicillin sensitivity of cells, seen by Schwarz et al. [133] and Donachie and Begg [9].

In S. faecalis and B. subtilis, among others, the autolytic enzymes are quite firmly bound to the wall fraction. They can be released by 3 M LiCl [160, 160a]. Autolytic enzyme from S. faecalis largely exists in an inactive form which can be activated by proteolytic enzymes [161, 10]. Direct tests with isolated walls show that the active form of the enzyme is mostly located at the site of and preferentially hydrolyzes the most recent wall [161]. Interestingly, studies of cell wall turnover during normal growth of B. subtilis showed that the newly synthesized wall was not degraded until about a half-hour after it was deposited [162]. It is suggested that the site of new synthesis is resistant for this interval, but the basis for resistance is not known.

There are preliminary reports of mutants with altered autolysins or autolysin-resistant walls [10, 113]. An S. faecalis mutant with an autolysin-resistant wall formed chains of cells. Diplococcus pneumoniae with a suppressed autolytic system was found to be resistant to several antibiotics [163]. This result suggests a method for isolating other autolysin-negative mutants.

A B. subtilis mutant that makes a temperature-sensitive autolysin has been isolated by Fan [164]. It grows as long chains at 48°C. When the isolated autolysin (solubilized with LiCl) was added, the chains separated into individual bacteria. This work appears to demonstrate clearly a role of autolysin in cell separation.

See Ghuysen [165] and in Chapter 2 for extensive reviews on the known autolytic enzymes and other enzymes involved in wall synthesis, including discussion of their physiological roles and coordination. Higgins and

Shockman [10] and Rogers [113] also give prominence to these enzymes in their reviews.

VII. QUESTIONS AND MODELS

A. Overall View

The problem of bacterial division can be presented as a series of subproblems, largely owing to the Helmstetter-Cooper scheme. For E. coli at least, we visualize a four-step sequence of DNA initiator formation, DNA synthesis, septation, and separation. This sequence continually repeats in the overlapping rather than sequential manner demanded by the rule that each initiation commences immediately on the termination of the previous initiation process. For each part of the sequence, we want explanations on the chemical, genetic, and morphological levels. We want to be able to describe the low molecular weight precursors and their conversion into macromolecules, and hopefully we will be able to do this in the language of intermediary metabolism. But then we need to describe the conversion of the macromolecules into the supramolecular quaternary structures of the envelope, in order to provide a basis for the morphological processes that we see, e.g., in the inward growth of envelope at the point of septation. This sort of investigation, the linking of macromolecular structure with morphology is in its infancy; we hardly possess words or concepts for its description. For the moment we should look to other investigations of self-assembly — multisubunit enzymes, ribosomes, phages, flagella, organelles, etc., to gain clues as to how assembly might work for membrane and wall formation. Much of the process should depend on initial recognition of the structural macromolecules by each other, often followed by covalent bond formation (particularly for the membrane). In other cases (as for wall synthesis) enzyme-catalyzed bond formations that add subunits to growing template surfaces might be used much more frequently.

The second kind of problem is to understand the control of each stage of the process. Presumably, the precursor-macromolecule steps are controlled by the now-familiar metabolic regulatory systems of feedback inhibition and activation, induction and repression. Initiation of DNA synthesis could be based on one of these kinds of control, but it could also equally well be dependent on new sorts of controls, speculated on below. Controls of assembly processes and likewise of septation have hardly been explored, let alone explained. Help from other systems is possible though perhaps remote. We can recall the elaborate control systems that are involved in initiation of vegetative growth of phages such as lambda [166], development of phage ϕX174 [42], and the early and late gene-expression controls in T-even phage development [167]. More complex systems of interest here could include sporulation of bacteria in which analogies to bacterial division have been noted [168, 169] and on which considerable biochemistry has been done [170]. Physarum polycephalum provides some fascinating insights into control of DNA and nuclear replication at a higher organizational level, owing to the synchronous division of nuclei in this creature's plasmodium [171]. A new system of great interest for control of division is Caulobacter crescentis, with its bipartite division cycle and easy synchronization [72, 71].

How complex are the processes of DNA initiation, septation, and the overall cell cycle? Only a half-dozen or fewer mutations in E. coli have so far been identified that modify major portions of the DNA replication-cell division sequence [152]. The number of these genetic sites has been taken as evidence that the process must indeed be complex. However, one could take quite the other view, that the central process is quite simple, and depends primarily on only one or a few genes. The other mutations could modify the production or action of the key gene products either by altering one of several steps in the pathways of their formation or by peripheral effects on the state of the cell — its redox potential, osmotic pressure, fluidity of membrane, etc. Perhaps it is not too optimistic to hope that an explanation for cell division might not be harder to achieve than for complex processes such as oxidative phosphorylation phage assembly, etc.

B. Review of Principal Observations

Here some of the central observations presented above will be brought together with the aim of providing a working model for the division cycle of E. coli. These observations are (1) the Helmstetter-Cooper sequence, (2) the membrane attachment of DNA proposed by Jacob et al., (3) observations on the unit cell length and the site of penicillin sensitivity made by Donachie and Begg, and (4) morphological observations reported by Higgins and Shockman regarding inward growth at the septum area and splitting of the envelope. The model presented here is similar to that presented by Higgins and Shockman for growth of rod-shaped organisms [10].

The model assumes that DNA is attached to membrane at the longitudinal membrane growth sites (see Fig. 10). The DNA is circular, and is attached at its origin. It also passes through DNA replication sites that are attached to the membrane elsewhere. The part of the membrane at which longitudinal wall growth and septation occur, as well as the site of DNA attachment origin-terminus should have considerable flexibility in order to permit segregation in rapidly growing cells. It could play a role somewhat like the mitotic apparatus of eukaryotic cells in that it would serve to separate the daughter DNAs. The mesosome could well serve this role because various attachment sites could move apart as the mesosome unfolds. This is a variation of the original concept of Jacob et al. [2] in which the DNA attachment sites were assumed simply to be moved apart as the membrane grew. (A prediction from this model is that mesosomes would be more prominent in rapidly growing cells.)

We will first consider the simple case, seen under conditions of slow growth (Fig. 10A). Assume a division time of 60 min; the Helmstetter-Cooper timing predicts approximately 40 min of DNA replication starting at the time of cell division, followed by 20 min of septum formation and separation during which DNA is not made. It will be recalled that the division time of 60 min is also the interval required for formation of the initiator of DNA synthesis, I, and it is also the interval during which the membrane grows to twice its unit size. We start with a newly divided unit cell (1.7 μ) with its new end to the right of the figure. One double strand of DNA is attached at the membrane growth site, and its replication is just commencing at this same place. Membrane is assumed to be made continually at the growth site during the next 20 min; it pushes the newer cell end and the just previously formed membrane to the right. Meanwhile, the DNA is bidirectionally half-replicated, the origin remaining at its original attachment point. By 40 min, the membrane has grown to about 1.6 times its original length and DNA replication is complete. The replication points reach the DNA terminus. A new origin is created and septation commences. At 60 min, septation and separation are complete; two daughter cells have been produced. Each is in the same condition as the original cell. Note that the penicillin-sensitive potential septum site remains 1.7 μ from the older end. Meanwhile, in 60 min a round of I synthesis is completed, and the cycle commences again.

One conclusion seen more clearly from these combined results is that there need not be any length determining "ruler" to measure cell lengths. Rather, control of the elongation-division process is time-dependent. The septum site is preordained, as shown by Donachie's demonstration that penicillin-sensitive sites are determined from the beginning of the cycle. During the division time I, the mass and membrane length must double under conditions of balanced growth (by definition), and thus old and new cells are the same lengths. To explain this doubling, it is tempting to assign to the membrane itself the role of the "I" substance, which must double to allow DNA initiation to occur. This is of course far from proven, and it could simply be that the controlling substance and membrane elongation are always made in constant proportions. It is also tempting to surmise that upon completion of chromosome replication, the event that triggers septation is a change in membrane structure. A changed relation

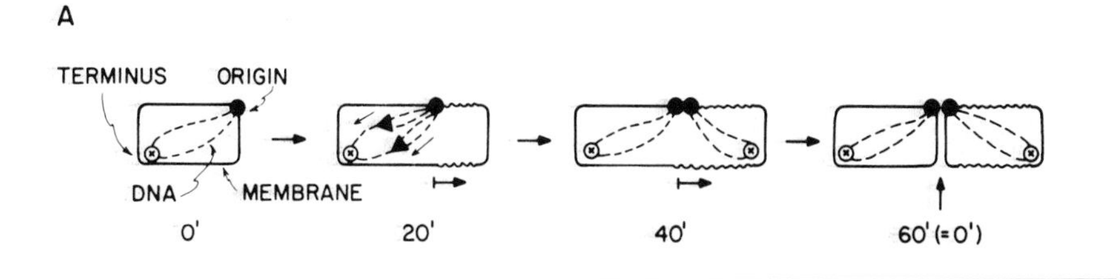

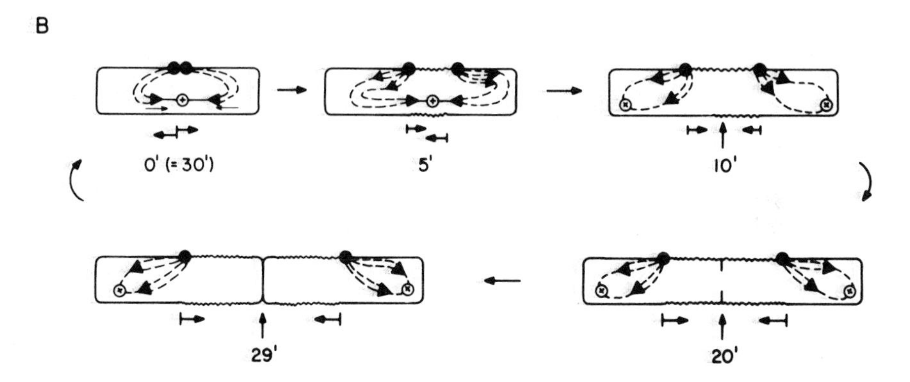

FIG. 10. Schematic representation of chromosome and membrane growth of E. coli B/r. (A) Division time of 60 min. (∿∿∿) new membrane growth, (____) old membrane growth, (----) DNA, (●) chromosome origin, (o) chromosome terminus and (Δ) chromosome growing point. (B) Division time of 30 min. These cells show dichotomous replication.

of DNA replication points on the membrane could modify the geometry of the cell membrane as the two daughter DNA strands are separately attached (see Fig. 10). Marvin [172] earlier suggested a "zone of influence" of DNA on the membrane, though his concept of how the DNA controls events is obscure.

A more complex case is seen when the division time is shorter. The sequence for a 30-min division time is shown in Fig. 10B. According to the Helmstetter-Cooper model, the DNA is two-thirds replicated at the time of division. Dichotomous replication occurs between zero and 10 min, with six growing points per cell. Then for the rest of the cycle there are two chromosomes per cell. According to Donachie, two unit cells are essentially growing back to back in a dichotomous division, and hence a rapidly growing cell increases from 3.4 μ to twice this length. At zero time, the cell has a prospective septum-forming site at its center.

We picture the two (double) origins of DNA as clustered on the membrane at this site and the replication points are elsewhere. There are unsolved problems of membrane and DNA site topology here [173]. Presumably the membrane would have to be somewhat folded to allow all of these sites to be fixed on the membrane (perhaps on a mesosome). DNA synthesis occurs at six points; membrane growth might occur at each of the two origins for 10 min as shown. Then at 10 min the older DNA round is completed with appearance of two termini and septation commences at the oldest septum site at the cell center, which must somehow remain structurally distinct. The DNA separation might require scission of DNA strands as reported by Yoshikawa [174], and reattachment of DNA to the membrane. During the last 20 min, septation occurs at the center, and DNA synthesis and membrane elongation continue.

Bleeken [175] has presented a mathematical treatment of the Helmstetter-Cooper model in connection with steady state growth and shifts. He concludes that the constant mass/DNA ratio found by Donachie [54] to be a requisite for division falls out of the calculations; that is, it is a result of the nature of the system rather than being a cause of division, and again the system is time-controlled. Margolis and Cooper [176] were able to confirm the Helmstetter-Cooper model using a computer simulation of DNA synthesis and growth.

C. Control of DNA Initiation

A particularly mysterious part of the overall scheme is the nature of the I event that on completion permits initiation of DNA synthesis. The chemical nature of "I" remains obscure although the requirement of protein synthesis for DNA initiation is very suggestive. I could be membrane material or it could be cytoplasmic. One infers that I has to reach a critical amount. It might be used stoichiometrically in a structural way, or as Pritchard et al. [74] suppose, I could be an inhibitor that is produced once per chromosome cycle, and that is titrated away by the replicating DNA. Thus, DNA would initiate at a constant cell mass per origin in accord with Donachie [54]. Pritchard et al. also discuss other models of DNA initiation.

Some models of DNA initiation are proposed to depend on soluble, cytoplasmic components that achieve critical concentrations at the time of chromosome initiation. This is similar to the activation of replication of defective lambda phage by gene N product [177]. These models depend on metabolic oscillations within the cell cycle. Cutler and Evans [178] have demonstrated transcription of different parts of the genome of E. coli during the cycle. Numerous enzymes have been demonstrated to fluctuate during the cycle [179]; unstable enzyme activities have been found to peak at a specific time during the cycle, whereas stable enzyme activities show either an exponential or a stepwise increase at a given cell age. These fluctuations are dependent on both the doubling of gene dosage when replication occurs at a given site, and on oscillations of metabolic pools during the cycle. Enzyme oscillations might be directly responsible for initiation, e.g., a key nuclease could be produced transitorily. However, the oscillations of nonprotein effector molecules such as nucleic acid precursors are more often favored. Huzyk and Clark [180] suggest that changes in the relative concentrations of nucleotides might be correlated with the division cycle. This seems to be consistent with experiments of Hertman and Luria [181] and of Kirby et al. [45] on both prophage induction and cell division after uv (see also Witkin [182]).

The study of DNA initiation is moving from the methodologies of whole cell study to subcellular approaches. *In vitro* DNA synthesizing preparations together with mutants that lack ability to initiate should give us the means of identifying components that are involved. The requirement of RNA synthesis [51] is an important recent clue.

D. Control of Septation

We have no real clues regarding the chemistry of septation. Schwarz and Leutgeb [183] have not found any chemical differences in the envelopes of *B. subtilis* that are dividing and those that are not. Previc [184] discusses the chemistry of bacterial division. He suggests that since the murein of bacilli contains diaminopimelic acid and that of cocci contains lysine, there might be more lysine in the septation region of bacilli, since the fundamental septation processes of the two types of organisms could well be similar [10]. Problems of bacterial shape are discussed extensively in the final chapter of this book.

One basic question is whether the rigid part of the envelope plays an active part or a passive one in septation. Higgins and Shockman [10] favor the former view of inward wall growth for both cocci and bacilli, and so does Previc [184], who bases his theoretical model for division on volume-surface shifts. One can easily imagine that the murein layer being assembled at the growth zone has a slight inward curvature that will eventually divide the cell unless it is split apart into the two wall layers [10]. The splitting force could be a pressure from within, generated by cell mass increase. Valine-starved cells, which can make murein but not internal proteins, i.e., mass, have a high ratio of inward wall growth to splitting. Conversely specific partial inhibitors of wall growth show predominantly a splitting-favored balance, and filaments are formed. The effect of heat shock in transitorily preventing division of *E. coli* [104] now seems in good part to be explained by a swelling of the cells at the higher temperature

[185]. This volume increase could push the incipient septa into a new geometry parallel to the elongating wall, and then new inward septum growth would be required.

There is less evidence favoring a primary role of the membrane in determining the direction of septum growth that is followed by wall deposition. The wall at least seems to play a necessary role in septation, since wall-less cells do not show inward membrane growth even in hypertonic medium.

The timing of septation presents another set of questions. DNA completion in E. coli [12, 76] normally appears to be a necessary but not sufficient condition, but in B. subtilis this does not appear to be a requirement [80]. Mutants of E. coli are known in which this link is deranged.

By what means does DNA influence septation in E. coli? Is it a direct action through membrane association? Considerable changes in this association must occur when the attachment sites are changed at the end of a DNA round. We know virtually nothing about the detailed structure of this region, including even the function of the mesosome that often lies there. In view of the complexity of the mitotic apparatus in cells of higher organisms, it would not be surprising that special machinery is located at the bacterial surface to rearrange the daughter DNA molecules and to signal the initiation of their separation.

In the view of Donachie [54], the DNA/mass ratio is the critical property for cell division. Since it is hard to conceive of a way in which the cell can measure its mass, except perhaps in terms of a mass to volume ratio with resulting surface tensions, one might prefer to imagine that mass really reflects the quantity of some constitutively produced cell protein or other component. There is indeed a requirement of protein synthesis for septation [1], and this protein can be made prior to septation itself [98], in fact, during most of the cycle (Helmstetter, personal communication). The protein reported by Smith and Pardee [104] to be required for septation (and which is heat-labile) could very likely be associated with the membrane, in view of the data correlating heat-shock delay of division and cell swelling [185].

Cytoplasmic molecules also probably have a role in septation. That the signal for septation does not depend on a cytoplasmic component alone is clear, because cells in rich medium contain both older and younger septation sites, and these septa are not completed at the same time, but rather in the order of their age. The nature of these factors is unknown. An intriguing observation regarding protein and cell division is the ability of cell extracts to permit septation of uv-irradiated lon^- cells [92]. A proper balance of putrescine and spermidine was shown to be important for division of E. coli, but the role of these compounds is unknown [105].

Molecules related to DNA replication or completion have been proposed to control division, but the restriction to a *cis* site of action puts limitations on models giving them a primary role. They could activate or inhibit processes at sites on the membrane that are otherwise primed to septate. Zusman et al. [98] obtained evidence for a nonprotein effector of septation in an *E. coli* mutant, *ts52*. Filaments form in this mutant at 41° C only in rich medium. They can be caused to divide when chloramphenicol or rifampicin are added, even at the nonpermissive temperature, but not when puromycin or amino acid analogs are added. Further experiments suggest that chloramphenicol changes the concentration of an effector in this mutant. Clearly, much more will have to be learned before these suggestive preliminary experiments can be converted to hard facts.

One final general question is whether the "trigger" for septation is a positive one, that is, one whose appearance activates septation [2, 185], or whether the control is a negative one, dependent on removal of an inhibitory molecule. Witkin [182] proposes a negative model for control in which a repressor normally prevents expression of genes whose products are needed for cell division; degradation of this repressor is prevented in *lon*$^-$ cells following uv irradiation and protein synthesis. The question of positive versus negative control is probably largely an academic one since the regulation of septation is likely to contain elements of both.

E. Epilogue

Almost all of the bacterial division problem is at a descriptive stage where facts are being collected and conditions are being sought. For example, mutants might permit the many phenomena in the tightly coupled intact cell to be studied relatively independently of one another. There is still room for much ingenuity in devising experiments at the level of cell physiology. But soon we should be able to move to the domains and techniques of biochemistry and molecular genetics. The value of a biological system for this sort of study is in proportion to the degree to which the part can be isolated from the whole. Cell fractionation and eventual isolation of individual components will surely supersede the pathfinding discoveries of the cell physiologists and yield a final certainty.

Bacterial division presents problems that are of interest not only to bacterial physiologists. Its material is useful for the study of pressing questions that border on both macromolecular biochemistry and morphology, questions that are vital to our eventually understanding development and malignant growth.

ADDENDUM

Many of the findings since January 1972 are summarized in a thoughtful brief article by W. D. Donachie, N. C. Jones, and R. Teather, in the 1973 Symposium of the Society for General Microbiology.

ACKNOWLEDGMENTS

This work was aided by Grant CA-AI-11595 from the National Cancer Institute. One of us (P. C. W.) was supported by a training grant from the United States Public Health Service, GM 962; D. R. Z. was supported by an American Cancer Society postdoctoral fellowship, PF-724.

REFERENCES

1. C. E. Helmstetter, in The Cell Cycle (G. M. Padilla, G. L. Whitson, and I. L. Cameron, eds.), p. 15, Academic Press, New York, 1969.

1a. M. Masters and P. Broda, Nature, New Biol., 232, 137 (1971).

2. F. Jacob, S. Brenner, and F. Cuzin, Cold Spring Harbor Symp. Quant. Biol., 28, 329 (1963).

3. R. J. Paulton, J. Bacteriol., 104, 762 (1970).

4. N. Sueoka, Molecular Genetics (J. H. Taylor, ed.), p. 1, Academic Press, New York, 1967.

5. P. L. Kuempel and A. B. Pardee, J. Cell Comp. Physiol., 62, (Suppl. 1), 15 (1963).

6. O. Maaløe, and N. O. Kjeldgaard, Control of Macromolecular Synthesis, W. A. Benjamin, New York, 1966.

7. A. Ryter, Y. Hirota, and F. Jacob, Cold Spring Harbor Symp. Quant. Biol., 33, 669 (1968).

8. K. G. Lark, Ann. Rev. Biochem., 38, 569 (1969).

9. W. D. Donachie, and K. J. Begg, Nature (London), 227, 1220 (1970).

10. M. L. Higgins and G. D. Shockman, CRC Critical Reviews in Microbiology, 1, 29 (1971).

11. J. M. Mitchison, The Biology of the Cell Cycle, Cambridge University Press, New York, 1971.

12. C. E. Helmstetter, and S. Cooper, J. Mol. Biol., 31, 507 (1968).

13. H. E. Kubitschek and M. L. Freedman, J. Bacteriol., 107, 95 (1971).

14. C. B. Ward and D. A. Glaser, Proc. Nat. Acad. Sci. U. S., 63, 800 (1969).

15. W. B. Wood, R. S. Edgar, J. King, I. Lielausis, and M. Henninger, Fed. Proc. Fed. Amer. Soc. Exp. Biol., 27, 1160 (1968).

16. C. Lark, Biochim. Biophys. Acta, 119, 517 (1966).

17. J. C. Wang, J. Mol. Biol., 55, 523 (1971).

18. A. Goldstein and B. J. Brown, Biochim. Biophys. Acta., 53, 19 (1961).

19. F. Jacob, A. Ryter, and F. Cuzin, Proc. Royal Soc. (London), 164, 267 (1966).

20. M. R. J. Salton, Ann. Rev. Microbiol., 21, 417 (1967).

21. D. A. Reaveley and H. J. Rogers, Biochem. J., 113, 67 (1969).

22. A. Ryter, Bacteriol. Rev., 32, 39 (1968).

23. W. van Iterson and J. B. Groen, J. Cell Biol., 49, 553 (1971).

24. R. D. Pontefract, G. Bergeron, and F. S. Thatcher, J. Bacteriol., 97, 367 (1969).

25. J. F. Whitfield and R. G. E. Murray, Can. J. Microbiol., 2, 245 (1956).

26. R. G. E. Murray and J. F. Whitfield, J. Bacteriol., 65, 715 (1953).

27. J. S. Ellison, C. F. T. Mattern, and W. A. Daniel, J. Bacteriol., 108, 526 (1971).

28. E. Kellenberger, A. Ryter, and J. Séchaud, J. Biophys. Biochem. Cytol., 4, 671 (1958).

29. C. Morgan, H. S. Rosenkranz, H. S. Carr, and H. M. Rose, J. Bacteriol., 93, 1987 (1967).

30. U. Henning, G. Dennert, K. Rehn, and G. Deppe, J. Bacteriol., 98, 784 (1969).

31. A. T. Ganesan and J. Lederberg, Biochem. Biophys. Res. Commun., 18, 824 (1965).

32. D. W. Smith and P. C. Hanawalt, Biochim. Biophys. Acta., 149, 519 (1967).

33. W. G. Quinn and N. Sueoka, Cold Spring Harbor Symp., Quant. Biol., 33, 695 (1968).

34. R. W. Snyder and F. E. Young, Biochem. Biophys. Res. Commun., 35, 359 (1969).

35. R. D. Ivarie and J. J. Pène, J. Bacteriol., 104, 839 (1970).

36. C. F. Schachtle, D. L. Anderson, and P. Rogers, J. Mol. Biol., 49, 255 (1970).

37. K. Yamaguchi, S. Murakami, and H. Yoshikawa, Biochem. Biophys. Res. Commun., 44, 1559 (1971).

38. C. L. Woldring, Biochim. Biophys. Acta, 224, 288 (1970).

39. G. Y. Tremblay, M. J. Daniels, and M. Schaecter, J. Mol. Biol., 40, 65 (1969).

40. D. Korn, and M. Thomas, Proc. Nat. Acad. Sci. U. S., 68, 2047 (1971).

41. F. W. Shull, J. A. Fralick, L. P. Stratton, and W. D. Fisher, J. Bacteriol., 106, 626 (1971).

42. R. L. Sinsheimer, Harvey Lectures, 64, 69 (1970).

43. C. F. Earhart, Virology, 42, 429 (1970).

44. W. O. Salivar and J. Gardinier, Virology, 41, 38 (1970).

45. E. P. Kirby, F. Jacob, and D. A. Goldthwait, Proc. Nat. Acad. Sci. U. S., 58, 1903 (1967).

46. K. G. Lark, in Molecular Genetics (J. H. Taylor, ed.), Vol. I, p. 153, Academic Press, New York, 1963.

47. T. S. Matney and J. C. Suit, J. Bacteriol., 92, 960 (1966).

48. K. G. Lark and H. Renger, J. Mol. Biol., 42, 221 (1969).

49. V. K. Vambutas and M. R. J. Salton, Biochim. Biophys. Acta., 203, 83 (1970).

50. D. R. Zusman, Fed. Proc. Fed. Amer. Soc. Exp. Biol., 31 (926 Abs.) (1972).

51. D. Brutlag, R. Schekman, and A. Kornberg, Proc. Nat. Acad. Sci. U. S., 68, 2826 (1971).

52. C. B. Ward and D. A. Glaser, Proc. Nat. Acad. Sci. U. S., 64, 905 (1969).

52a. S. Cooper and G. Weusthoff, J. Bacteriol., 106, 709 (1971).

53. W. D. Donachie, Nature (London), 219, 1077 (1968).

54. W. D. Donachie, J. Bacteriol., 100, 260 (1969).

55. Y. Hirota, A. Ryter, and F. Jacob, Cold Spring Harbor Symp. Quant. Biol., 33, 677 (1968).

56. D. Karamata and J. D. Gross, Mol. Gen. Genetics, 108, 277 (1970).

57. H. Yoshikawa and M. Haas, Cold Spring Harbor Symp. Quant. Biol., 33, 843 (1968).

58. J. D. Gross, Curr. Topics Microbiol. Immunol., (1972).

59. F. Cuzin and F. Jacob, Ann. Inst. Pasteur, 112, 529 (1967).

60. B. Hohn and D. Korn, J. Mol. Biol., 45, 385 (1969).

61. Y. Nishimura, L. Caro, C. M. Berg, and Y. Hirota, J. Mol. Biol., 55, 441 (1971).

62. E. Dubnau and W. K. Maas, J. Bacteriol., 95, 531 (1968).

63. F. Bonhoeffer and H. Schaller, Biochem. Biophys. Res. Commun., 20, 93 (1965).

64. S. R. Palchoudhury and V. N. Iyer, J. Mol. Biol., 57, 319 (1971).

65. A. J. Levine and R. L. Sinsheimer, J. Mol. Biol., 39, 655 (1969).

66. P. F. Pachler, A. L. Koch and M. Schaechter, J. Mol. Biol., 11, 650 (1965).

67. D. J. Clark and O. Maaløe, J. Mol. Biol., 23, 99 (1967).

68. R. H. Pritchard and A. Zaritsky, Nature (London), 226, 126 (1970).

69. K. G. Lark, Biochim. Biophys. Acta., 45, 121 (1960).

70. D. Zusman and E. Rosenberg, J. Mol. Biol., 49, 609 (1970).

71. S. T. Degnen and A. Newton, J. Mol. Biol., 64, 341 (1972).

72. L. Shapiro and N. Agabian-Keshishian, and I. Bendis, Science, 173, 884 (1971).

73. P. L. Kuempel, Adv. Cell Biology, p. 1 (1971).

74. R. H. Pritchard, P. T. Barth, and J. Collins, Symp. Soc. Gen. Microbiol., 19, 263 (1969).

75. D. J. Clark, J. Bacteriol., 96, 1214 (1968).

76. D. J. Clark, Cold Spring Harbor Symp. Quant. Biol., 33, 823 (1968).

77. C. E. Helmstetter and O. Pierucci, J. Bacteriol., 95, 1627 (1968).

78. O. Pierucci, Biophys. J., 9, 90 (1969).

79. M. Inouye, J. Bacteriol., 99, 842 (1969).

80. W. D. Donachie, D. T. M. Martin, and K. J. Begg, Nature New Biol., 231, 274 (1971).

81. L. E. Loveless, E. Spoerl, and T. H. Weisman, J. Bacteriol., 68, 637 (1954).

82. R. A. Deering, J. Bacteriol., 76, 123 (1958).

83. M. K. Shaw, J. Bacteriol., 95, 221 (1968).

84. H. I. Adler, W. D. Fisher, A. Cohen, and A. A. Hardigree, Proc. Nat. Acad. Sci. U. S., 57, 321 (1967).

85. S. B. Levy, J. Bacteriol., 108, 300 (1971).

86. P. Howard-Flanders, E. Simson, and L. Theriot, Genetics, 49, (1964).

87. R. F. Hill and E. Simson, J. Gen. Microbiol., 24, 1 (1961).

88. J. R. Walker and J. A. Smith, Mol. Gen. Genetics, 108, 249 (1970).

89. P. M. Leighton and W. D. Donachie, J. Bacteriol., 102, (1970).

90. J. R. Walker and A. B. Pardee, J. Bacteriol., 93, 107 (1967).

91. M. H. L. Green, J. Donch, and J. Greenberg, Mut. Res., 8, 409 (1969).

92. W. D. Fisher, H. I. Adler, F. W. Shull Jr., and A. Cohen, J. Bacteriol., 97, 500 (1969).

93. H. I. Adler, C. E. Terry, and A. A. Hardigree, J. Bacteriol., 95, 139 (1968).

94. D. P. Allison, J. Bacteriol., 108, 1390 (1971).

95. M. Ricard and Y. Hirota, Compt. Rend., 268, 1335 (1969).

96. M. Inouye, J. Mol. Biol., 63, 597 (1972).

97. B. G. Spratt and R. J. Rowbury, J. Gen. Microbiol., 65, 305 (1971).

98. D. R. Zusman, M. Inouye, and A. B. Pardee, J. Mol. Biol., 69, 119 (1972).

99. J. N. Reeve, D. J. Groves, and D. J. Clark, J. Bacteriol., 104, 1052 (1970).

100. D. van Alstyne and M. I. Simon, J. Bacteriol., 108, 1366 (1971).

101. A. Nishi and T. Kogoma, J. Bacteriol., 90, 88 (1965).

102. D. Zusman, P. Gottlieb, and E. Rosenberg, J. Bacteriol., 105, 811 (1971).

103. O. Pierucci and C. E. Helmstetter, Fed. Proc. Fed. Amer. Soc. Exp. Biol., 28, 1755 (1969).

104. H. S. Smith and A. B. Pardee, J. Bacteriol., 101, 901 (1970).

105. M. Inouye and A. B. Pardee, J. Bacteriol., 101, 770 (1970).

106. M. Webb, Science, 118, 607 (1953).

107. V. Braun and K. Rehn, Eur. J. Biochem., 10, 426 (1969).

108. E. H. Cota-Robles, J. Bacteriol., 85, 499 (1963).

109. M. R. J. Salton, in Biomembranes I, The Bacterial Membrane (L. A. Manson, ed.), p. 1, Plenum Press, New York.

110. M. R. J. Salton, CRC Crit. Rev. Microbiol., 1, 161 (1971).

111. A. M. Glauert and M. J. Thornley, Ann. Rev. Microbiol., 23, 159 (1969).

112. S. L. Singer and G. L. Nicolson, Science, 175, 720 (1972).

113. H. J. Rogers, Bacteriol. Rev., 34, 194 (1970).

114. M. J. Osborn, Ann. Rev. Biochem., 38, 701 (1969).

115. C. A. Schnaitman, J. Bacteriol., 104, 882 (1970).

116. C. A. Schnaitman, J. Bacteriol., 108, 545 (1971).

117. C. A. Schnaitman, J. Bacteriol., 108, 553 (1971).

118. N. Tsukagashi, P. Fielding, and C. F. Fox, Biochem. Biophys. Res. Commun., 44, 497 (1971).

119. G. Wilson and C. F. Fox, Biochem. Biophys. Res. Commun., 44, 503 (1971).

120. M. J. Daniels, Biochem. J., 115, 697 (1969).

121. J. Starká and J. Moravová, J. Gen. Microbiol., 60, 251 (1970).

122. F. Peypoux and G. Michel, Biochim. Biophys. Acta., 218, 453 (1970).

122a. J. P. G. Ballesta and M. Schaechter, J. Bacteriol., 107, 251 (1971).

123. D. C. Morrison and H. J. Morowitz, J. Mol. Biol., 49, 441 (1970).

124. E. C. C. Lin, Y. Hirota, and F. Jacob, J. Bacteriol., 108, 375 (1971).

125. F. Forro Jr., and A. Wertheimer, Biochim. Biophys. Acta., 40, 9 (1960).

126. A. Ryter, Ann. Inst. Pasteur, 121, 271 (1971).

127. A. Ryter, Folia Microbiol. (Prague), 12, 283 (1967).

128. R. M. Cole, Bacteriol. Rev., 29, 326 (1965).

129. C. Frehel, A. M. Beaufils, and A. Ryter, Ann. Inst. Pasteur, 121, 139 (1971).

130. I. J. Sud, and M. Schaechter, J. Bacteriol., 88, 1612 (1964).

131. E. A. Grula and G. L. Smith, J. Bacteriol., 90, 1054 (1965).

132. J. Lederberg, J. Bacteriol., 23, 144 (1957).

133. U. Schwarz, A. Asmus, and H. Frank, J. Mol. Biol., 41, 419 (1969).

134. G. C. Mathison, Nature (London), 219, 405 (1968).

135. O. E. Landman, in Microbial Protoplasts, Spheroplasts and L-forms (L. B. Guze, ed.), pp. 110, 319, Williams & Wilkins, Baltimore, Maryland, 1968.

136. G. Weinbaum, D. A. Fischman, and S. Okuda, J. Cell Biol., 45, 493 (1970).

137. E. A. Grula, G. L. Smith, and M. M. Grula, Can. J. Microbiol., 14, 293 (1968).

138. G. Mangiarotti and D. Schlessinger, Nature (London), 211, 761 (1966).

139. G. Mangiarotti, D. Apirion, and D. Schlessinger, Science, 153, 892 (1966).

140. S. Normark, H. G. Boman, and E. Matsson, J. Bacteriol., 97, 1334 (1969).

141. S. Normark, J. Bacteriol., 108, 51 (1971).

142. L. O. Ingram and E. L. Thurston, Protoplasma, 71, 55 (1970).

143. H. Momose, J. Gen. Appl. Microbiol., 7, 359 (1961).

144. R. M. Cole, J. J. Popkin, R. J. Boylan, and N. H. Mendelson, J. Bacteriol., 103, 793 (1970).

145. H. J. Rogers, M. McConnell, and I. D. J. Burdett, J. Gen. Microbiol., 61, 155 (1970).

146. H. J. Rogers and M. McConnell, J. Gen. Microbiol., 61, 173 (1970).

147. F. E. Young, P. Haywood, and M. Pollock, J. Bacteriol., 102, 867 (1970).

148. H.H. Martin, J. Gen. Microbiol., 36, 441 (1964).

149. M. Inouye and J. P. Guthrie, Proc. Nat. Acad. Sci. U. S., 64, 957 (1969).

150. M. Inouye and A. B. Pardee, J. Biol. Chem., 245, 5813 (1970).

151. M. Inouye, J. Bacteriol., 106, 539 (1971).

152. Y. Hirota, M. Ricard, and B. Shapiro, Biomembranes, 2, 13 (1971).

153. B. M. Shapiro, A. G. Siccardi, Y. Hirota, and F. Jacob, J. Mol. Biol., 52, 75 (1970).

154. A. G. Siccardi, B. M. Shapiro, Y. Hirota, and F. Jacob, J. Mol. Biol., 56, 75 (1971).

155. M. Inouye, J. Biol. Chem., 246, 4834 (1971).

156. R. J. L. Paulton, Nature, New Biol., 231, 271 (1971).

157. A. Tomasz, Proc. Nat. Acad. Sci. U. S., 59, 86 (1968).

158. E. B. Briles and A. Tomasz, J. Cell Biol., 47, 786 (1970).

159. W. Weidel and H. Pelzer, Adv. Enzymol., 26, 193 (1964).

160. H. M. Pooley, J. M. Porres-Juan, and G. D. Shockman, Biochem. Biophys. Res. Commun., 38, 1134 (1970).

160a. D. P. Fan, J. Bacteriol., 103, 489 (1970).

161. G. D. Shockman, H. M. Pooley, and J. S. Thompson, J. Bacteriol., 94, 1525 (1967).

162. J. Mauck, L. Chan, and L. Glaser, J. Biol. Chem., 246, 1820 (1970).

163. A. Tomasz, A. Albino, and E. Zanati, Nature (London), 227, 138 (1970).

164. D. P. Fan, J. Bacteriol., 103, 494 (1970).

165. J. Ghuysen, Bacteriol. Rev., 32, 425 (1968).

166. A. D. Hershey, ed., The Bacteriophage Lambda, Cold Spring Harbor Laboratory, Cold Spring Harbor, New York (1971).

167. A. Bolle, R. H. Epstein, W. Salser, and E. P. Geiduschek, J. Mol. Biol., 31, 325 (1968).

168. A. D. Hitchins and R. A. Slepecky, J. Bacteriol., 97, 1513 (1969).

169. A. D. Hitchins and R. A. Slepecky, Nature (London), 223, 804 (1969).

170. A. Kornberg, J. A. Spudich, D. L. Nelson, and M. P. Deutscher, Ann. Rev. Biochem., 37, 51 (1968).

171. J. E. Cummins, in The Cell Cycle. Gene-Enzyme Interactions (G. M. Padilla, G. L. Whitson, and I. L. Cameron, eds.), p. 141, Academic Press, New York, 1969.

172. D. A. Marvin, Nature (London), 219, 485 (1968).

173. H. Kasamatsu, D. L. Robberson, and J. Vinograd, Proc. Nat. Acad. Sci. U. S., 68, 2252 (1971).

174. H. Yoshikawa, Proc. Nat. Acad. Sci. U. S., 58, 312 (1967).

175. S. Bleeken, J. Theoret. Biol., 32, 81 (1971).

176. S. G. Margolis and S. Cooper, Comput. Biomed., 4, 427 (1971).

177. R. Thomas, J. Mol. Biol., 22, 79 (1966).

178. R. G. Cutler and J. E. Evans, J. Mol. Biol., 26, 91 (1967).

179. J. M. Mitchison, Science, 165, 657 (1969).

180. L. Huzyk and D. J. Clark, J. Bacteriol., 108, 74 (1971).

181. I. Hertman and S. E. Luria, J. Mol. Biol., 23, 117 (1967).

182. E. M. Witkin, Proc. Nat. Acad. Sci. U. S., 57, 1275 (1967).

183. U. Schwarz and W. Leutgeb, J. Bacteriol., 106, 588 (1971).

184. E. P. Previc, J. Theoret. Biol., 27, 471 (1970).

185. P. C. Wu and A. B. Pardee, J. Bacteriol., 114, 603 (1972).

186. M. Ycas, M. Sugita, and A. Bensam, J. Theoret. Biol., 9, 444 (1965).

Chapter 9

DETERMINANTS OF CELL SHAPE

Ulf Henning and Uli Schwarz

Max-Planck-Institut für Biologie
Tübingen, Germany

Friedrich-Miescher-Laboratorium der Max-Planck-Gesellschaft
Tübingen, Germany

I. INTRODUCTION

To ask the cause of any biological shape is to ask the question: How is genetic information translated into morphology? Here we are dealing with cellular morphology. "Cellular morphology" or "cell shape" are multivalent terms and they may be examined from three aspects. The first aspect is static and refers to a geometric classification (cylinder, sphere, etc.) as well as to the laydown of the dimensions. The second is dynamic and refers to variations in the dimensions of a given geometry that occur as a result of growth and division. The third is also dynamic. Since, e.g., a rod-shaped organism can develop from a spheroplast, somehow cells can manifest _de novo_ a certain geometry.

We thus can subdivide the general problem mentioned into three more specific questions. How is specificity expressed for a certain geometry and its dimensions? How can it be varied? How can a complete change occur in the specificity for one to another type of geometry? Because answers have not yet been obtained we should look for precedence, i.e., examine the hierarchy of biological structures to learn what pertinent answers are available.

The two known mechanisms responsible for the formation of structures possessing a defined and often highly specific morphology are self-assembly and sequential assembly [1] of polypeptides. Classification of a process as self-assembly requires that no other _specific_ information than the primary structure of a polypeptide chain determine the structure of the final product. The resulting structures, such as enzymes or a number of viruses, usually but not always have finite dimensions, e.g., the _in vitro_ reconstitution of bacterial flagella has been shown to be rather similar to the growth of a crystal [2-4]. Sequential assembly [5] was originally defined as an assembly in which, in the presence of all subunits, association of subunits can proceed only in a certain (single or branched) pathway. It has since become clear that enzymatic modification of subunits is required at a certain stage of the assembly process [6, 7]. In other words, additional information is required at a certain period in this process. The time parameter is invariant in such cases because the very formation of a subunit (e.g., head polypeptide chains of phage T4 [6, 7]) makes it a substrate for the modifying enzyme.

As long as we do not consider multicellular organs or organisms these two assembly mechanisms clearly would allow one to build biological structures of almost any morphology, be it as bizarre as that of the T phages. Self- and sequential assemblies do not, however, have the second important property we are looking for: they cannot effect any variability in composition and structure of a final product. However, one should probably not be too apodictic about this point because it has been shown that even in self-assembly other factors (e.g., "epigenetic control") than amino acid sequence seem to be able to direct protein subunit assembly [8]. Nevertheless, the desired variability at least is very easily conceivable. For example, a modifying enzyme or a modifying subunit could be made or activated only at a certain time during an assembly process. A modifier could then lead to the introduction, for example, of a curvature into a growing flagellum at a certain time of its assembly. In summary, it may not to be necessary *a priori* to suspect or ask for other *principles* than self-assembly and sequential assembly (including modification of subunits) of proteins to operate in the translation of genetic information into cellular morphology.

Cellular morphology, at least for bacteria, means morphology of the cell envelope. Manifestation of morphology must therefore occur during assembly of this envelope. In the following we shall examine the components of cell envelopes and ask whether or not any other principle may be applicable and whether or not we may be forced to dismiss our *a priori* considerations for the problem under discussion.

II. CELL ENVELOPE AND SHAPE

The anatomy of bacterial cell envelopes is fairly well known [9, 10], and can be briefly summarized as follows.

The innermost layer in all cells is a cytoplasmic membrane. The next main constituent, with some notable exceptions described subsequently, is a polymer, which for reasons given long ago [11] we call murein (also called peptidoglycan, mucopeptide, glycosaminopeptide, or glycopeptide). On top of the murein an outer membrane can be present (gram-negative species), or accessories such as teichoic acids or polysaccharides can be linked to the murein (gram-positive species).

The component to which a "shape-determining" role has often been ascribed is murein. We therefore will first attempt to evaluate its role in the manifestation of morphology and then draw our attention to other cell envelope components.

A. Murein

Weidel and collaborators [11] showed some time ago that this layer could be isolated intact from *Escherichia coli* cell envelopes by combined mechanical and detergent treatment. The isolated layer, the "sacculus," retains the shape and dimensions of the cells from which it was made. The original concept of the sacculus as a "bag-shaped macromolecule" [11] has sometimes been questioned [12, 13] on the grounds that the polysaccharide chains are much shorter than depicted in Weidel and Pelzer's article. However, these authors quite clearly stated that the chains need not be as long as their model might suggest. Certainly the sacculus _is_ a bag-shaped macromolecule.

There is no question that murein has shape-_maintaining_ properties — in rod-shaped organisms inhibition of murein synthesis or its degradation by lysozyme leads to the formation of spheroplasts. It should be pointed out, however, that it is not clear whether or not murein constitutes the only device for maintaining shape. At least in _E. coli_, spheroplasts, however they be induced, are not simply cells lacking murein or carrying partially degraded murein. They are much more sensitive to detergents than normal cells, which indicates the loss of one or several components of the cell envelope in addition to murein. Since this loss or degradation of murein has pleiotropic consequences simple causal conclusions cannot be drawn. In fact, it appears rather likely that envelope components other than murein also play some role in shape maintenance. It has been shown [14] that when under certain conditions plasmolyzed _E. coli_ B rods are treated with lysozyme, they become osmotically sensitive but remain rod-shaped. They become spherical upon dilution of the plasmolyzing sucrose. It is possible to draw similar conclusions from an earlier observation on _E. coli_ and _Pseudomonas aeruginosa_ [15]. It should also be noted that it is not clear how much murein need be made to permit the formation of a cylindrical cell. For example, growth of _Erwinia_ in the presence of D-cycloserine resulted in the formation of filaments containing about 60% of the murein of normal rods [16]. Similarly, another report claimed that penicillin-induced filamentous growth of _E. coli_ B leads to filaments containing about 20% less radioactive diaminopimelate per milligram dry weight, than control cells [17]. It was also observed that certain _E. coli_ strains showed nutrient-dependent filamentous growth [18]. These filaments contained considerably less murein than the corresponding short rods, and in fact, some evidence was obtained that in this case the murein in the filaments was discontinuous. Filamentous growth was accompanied by a strong inhibition (> 80%) of murein synthesis [19]. Evidence for discontinuous murein has also been presented for the rod-shaped vegetative cells of _Myxococcus xanthus_ [20].

One drawback of some of the experiments mentioned and of other similar reports in the literature, is the use of strains that are not diaminopimelate auxotrophs. It was shown some time ago [21] that there need not be an equilibrium between endogenously synthesized and exogenously supplied diaminopimelate. If indeed filamentous, i.e., rod-shaped growth is possible with discontinuous murein the very surprising conclusion would have to be drawn that a specific murein structure in the gram-negative species mentioned is required not even for shape maintenance. It appears that the role of murein in shape maintenance is not yet completely solved.

We have pointed out at the outset that what we are looking for is the expression of specificity for a certain morphology. Such specificity cannot reside in the sacculus nor in its precursors. This covalently closed net is made by enzymes. The information required to produce murein of a certain shape could only involve either the enzymes in their spatial arrangement, specificity, or relative activities, or a matrix onto which murein is laid down. Although thus murein cannot possess the specificity in question it may nevertheless contribute to it. First, it might be comparable to a cofactor. For example, it is possible that a protein able to assemble into a certain structure required murein as cofactor. Then a defect in biosynthesis of murein could cause the protein to make a defective assembly, an abnormal structure. Second, there may be restrictions in three-dimensional arrangement of murein. It is possible that in a rod shaped organism a certain type of murein structure at the two ends of a cell is compatible only with a more spherical shape. Finally, it should be recalled that murein chemistry is far from being completely known and specifically that the chemical constitution of a whole series of lysozyme split products described several years ago [22, 23] has not yet been determined. These split products are minor components in lysozyme digests, and they could reflect local changes in murein structure that occurred at some time during or after completed murein synthesis. Such changes, in turn, could cause a change in an assembly process in the neighboring membranes and thus, for example, perhaps induce septum formation. It is unknown whether any of these or similar possibilities do occur and at the time being it would appear preferable not to ascribe to murein any "shape-determining" or "shape-specifying" role.

It may well be that murein is unnecessarily complicating the issue — some extreme halophiles can grow as perfect rods in the absence of murein. An early report [24] stated that low amounts of murein were present in Halobacterium salinarium. A separate report [25] at the same time found no murein in H. halobium and salinarium (both rods). In a recent reinvestigation [26] with H. salinarium, cell envelopes were found to be devoid of muramic acid and to contain no more than 1.2% of the amount of glucosamine present in corresponding E. coli preparations (in E. coli murein probably constitutes a monolayer). Evidence for the absence of diaminopimelate is meaningless since mureins lacking it are known [27]. Additional evidence for the absence of murein sacculi or a structural analog in these bacteria is presented below (p. 421). The Halobacteria thus nicely demonstrate that murein sacculi are not at all required for the expression of specificity for a certain morphology (coccoid halophiles are also known not to possess murein [28]).

We may add that murein, regardless of its role, has one potentially very useful property. At least in E. coli, once murein has been synthesized it does not show considerable turnover, and thus labeled murein can be used experimentally as a marker for events in cell-envelope synthesis. For example, strictly localized enzyme action (murein hydrolysis) has been demonstrated at the site of septum formation during cellular division [29]. Also, specific labeling of murein is simple and once incorporated the label is not lost from it [30].

The main other constituents of cell envelopes are lipids (mainly phospholipids), proteins, and "accessories," such as polysaccharides, lipopolysaccharides, or teichoic acids. What conclusions regarding the translation of genetic information into morphology can be drawn for any class of these substances?

B. Lipid

In principle, the same arguments apply to lipids as those we have put forward concerning murein and the expression of specificity for cellular morphology (p. 417). Lipid, or rather phospholipid, cannot possess this specificity but could very well contribute to it. There appears to be evidence favoring this possibility.

A requirement for a certain lipid composition in the formation of cylindrical shape has been demonstrated in Mycoplasma species. These do not possess a cell wall or murein. Yet they can grow as long, thin filaments [31]. The fatty acid composition of M. laidlawii phospholipids reflects, within wide ranges, that of the medium [32, 33]. Pronounced effects of fatty acid composition on Mycoplasma morphology have been observed [33-35]. It has been shown that unsaturated fatty acids must be present in adequate amounts for filamentous growth of this organism. In an unsupplemented tryptose medium, the organism grows as short coccal chains. In medium containing only saturated fatty acids, cells exist as spheres and as short swollen filaments. An interesting correlation was revealed between these morphological changes and the state of the lipid in M. laidlawii membrane in a study showing that thermal phase transitions exist in these membranes and depend on the fatty acid composition [36]. Cells grown in oleate-supplemented medium grew as filaments and showed a transition point of -20°C. Cells grown in unsupplemented medium grew as cocci with a transition point of +45°C, and cells grown in stearate-supplemented medium swelled and lysed and showed a transition point of +65°C. Filamentous growth of Mycoplasma thus appears to be only possible as long as the growth temperature is above the "melting point" of the crystalline state of membrane lipids.

In addition, the report just described states that the swelling seen in stearate-grown cells is reversed by 0.3 M sucrose, and also that cellular division ceases when cells become coccoid. Razin et al. [33] had previously shown that the presence of unsaturated fatty acids in the growth medium markedly increases the osmotic stability of M. laidlawii, even though the resulting cylindrical cells should be theoretically more sensitive to osmotic shock than spherical cells [37]. Thus it appears that the membrane lipid in these cells is involved in maintenance of shape. It would certainly be interesting to see if and how unsupplemented (coccoid) cells grow in the presence of sucrose. Lipid could play a comparable role in many other bacteria and suggestive evidence exists in Bacillus stearothermophilus. Lysozyme-induced protoplasts of this organism were obtained that often still retained a rod shape [38]. No biochemical evidence is available as to how much murein still remained on these protoplasts. It is nevertheless tempting to believe in a pertinent role of the lipid, especially since B. stearothermophilus grows at 58°-60°C, and the lysozyme protoplasts were made at 26°C and 4°C, i.e., possibly below a melting point such as that described above for Mycoplasma.

If now, however, we consider these facts in the light of present knowledge about membrane structure (cf. reference 39) we have to admit that rather trivial explanations are possible. Singer and Nicolson [39] summarize "Cell membranes are viewed as two-dimensional solutions of oriented globular proteins and lipids." "Lipids" means a phospholipid bilayer. At lower temperatures the frozen bilayer could maintain shape in B. stearothermophilus and at temperatures below the melting point Mycoplasma may simply be unable to assemble correctly its cytoplasmic membrane.

Phospholipid involvement in shape maintenance was suggested by a study [40], which showed that a Clostridium phospholipase preparation could induce sphere formation in E. coli B and Chromobacterium violaceum. It has since been shown [41], however, that such phospholipase preparations can contain a murein-degrading endo-N-acetylglucosaminidase.

In summary, there does not yet appear to be any clear-cut evidence for a contribution of lipids to the expression of specificity for cellular morphology.

C. Lipopolysaccharides and Teichoic Acids

In view of what we have argued about murein and lipid concerning their possible role in specifying cellular shape we need only ask whether evidence exists that the "accessories" such as teichoic acids or lipopolysaccharides may contribute to this specificity.

A B. subtilis mutant has been described, which at low temperature grows as a rod, and at high temperature, as an irregular sphere [42-44a]. Sphere formation is accompanied by a marked reduction of the teichoic acid content of the cell envelope. However, another B. subtilis mutant has been found [45], which completely lacks teichoic acid, and yet is a rod. Of the many mutants [46] known to lack more or less of the saccharide moiety of their lipopolysaccharide none seems to exhibit an altered cellular morphology. We do not know of mutants completely missing lipopolysaccharide and thus the role of the lipid part appears to have remained unknown.

D. Protein

We believe that it has become fairly clear that at least primarily the specificity we are searching for must reside in proteins, be it as enzymes, as structural proteins, or both. We may first ask if any type of protein assembly could yield a closed container (cylinder, sphere, etc.), which is capable of growth, division, and other variations in its dimensions. As one answer we should like to quote the following from Caspar and Klug's classic article [47].

> Large containers could be constructed from identical units quasi-equivalently bonded into a large number of 6-coordinated units (themselves probably hexamers) together with twelve 5-coordinated units (either pentamers or statistically bonded hexamers). Such a container, constructed of globular units would be thin, compared to its diameter, and would thus be intrinsically more flexible than a shell built of a smaller number of units. ...the structural distinction between a rigid shell and a flexible membrane is not sharp. The shape of a flexible membrane is more likely to be determined by its contents than by considerations of the lowest energy arrangement of the structural units.

Thus, although the protein container is possible it is not clear what may dictate its dimensions. Nevertheless, such protein containers have one attractive feature in that they can grow. They may be treated as surface crystals and Harris and Scriven [48] have shown how dislocations in surface crystals can serve as growth centers (Fig. 1). Let us take the model one step further and consider a cylindrical or spherical protein surface crystal. Dislocations in these crystals could be expected to be distributed randomly. Adding subunits would then result in their random incorporation, proceed via "nonconservative climbing" (Fig. 1), and lead to an elongation of the cylinder with constant diameter as well as an increase in the diameter of the sphere. It is of interest to note that such growth would be initiated at random sites and would proceed, microscopically, as a growth zone. Any biochemical investigation would naturally record this type of growth as random.

It may be highly relevant that in a large variety of bacterial species (for a compilation of the literature see reference 49) surface crystals of the type just discussed do exist. A study with rather remarkable results has been performed on such an outside layer (T-layer) of Bacillus brevis [49, 50]. The material the T-layer consists of has been isolated and shown to consist of identical polypeptide chains. It has been possible to reassociate this protein in vitro, and it has been found that under certain circumstances it will form hollow cylinders (open at the ends), which can have the same diameter as the cells the protein was obtained from. It is quite clear then that indeed the "large containers" Caspar and Klug have been speculating on in their 1962 article [47] can be self-assembly systems. This preliminary finding of Brinton and colleagues obviously may constitute part of the solution to the question on determination of cylindrical shape. If a pure self-assembly system can lead to the formation of a cylinder with the properties mentioned this finding could also be extended to explain how, for example, the cylinder-derivative, spirillar shape is made. A variable diameter could be the result of the incorporation of varying amounts, possibly even low amounts, of another different subunit.

It could very well be that a similar situation can be found with some of the extremely halophilic bacteria growing as rods. It has been reported [24] that cells of H. salinarium upon dilution (from about 4 M to 1-2 M NaCl) immediately become spherical. This transition occurs within seconds, has no metabolic requirements, and cannot be reversed by increasing the salt concentration again [51]. Thus, these bacteria cannot possess any structural analog of murein. Stoeckenius and Rowen [52] found, in contrast to earlier claims, that the envelope of H. halobium consists of an inner, cytoplasmic membrane and an outer cell wall. It clearly is this cell wall that exhibits the long known [53] regular surface pattern observed in shadowed preparations of whole cells or empty envelopes. Stoeckenius and Rowen also showed that upon stepwise reduction of the salt concentration, the cell wall is lost before the cytoplasmic membrane begins to break down. The wall material apparently is mainly protein containing some hexosamines but no significant amount of lipid [54]. Reassembly studies with cell-wall protein have not yet been reported and it is also not known whether loss of rod shape and loss of cell wall are causally related. We feel that it is rather likely that this causal relationship does exist. Halobacteria do not, of course, require very strong forces to maintain cylindrical shape. Almost all the cell wall may have to withstand could be the surface tension of the cytoplasmic membrane, and loss of the cell wall would lead to the formation of spheres only because of this surface tension.

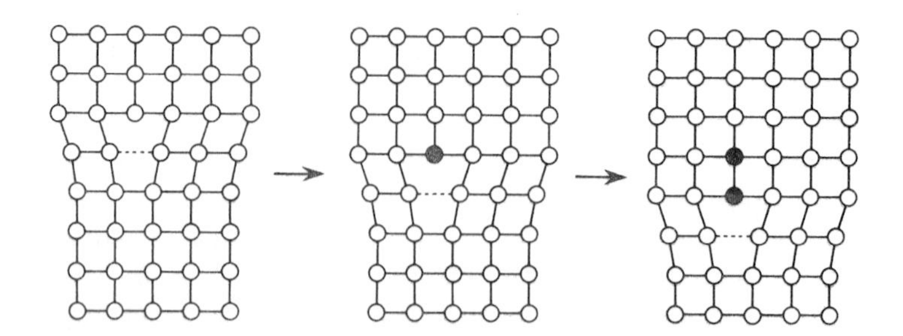

FIG. 1. Dislocation at growth center. After reference 48. New morphological subunits (filled dots) are added at a dislocation in a square lattice, and the dislocation climbs one lattice spacing to a new position as each new subunit is added. The climb is "nonconservative" because the amount of material in the crystal increases.

There is one drawback, which does not permit one to wholeheartedly accept the idea that protein containers — protein surface crystals — may always be the final product of the expression of specificity for cellular shape. One of the first examples discovered of such cell envelope layers is the outside layer of Spirillum serpens. It has been examined by Houwink and by Murray and his colleagues [55-57]. It exhibits primarily hexagonal subunit packing and consists mostly, if not only of protein. Murray found that this rod-shaped organism does not have this layer when it is grown in medium with low Ca^{2+} concentration and that variants exist which do not manifest it at all. It thus clearly is not involved in shape determination. Yet, one important point should not be neglected. The hexagonal packing of subunits found by Murray presents no problem for the side walls of the cylindrical cells, but the packing could be "expected to show discontinuities marked by six pentagons at each hemispherical end" [56]. His electron microscopic observations did indicate such discontinuities.

E. Conclusion

Examination of the composition of bacterial cell envelopes has shown that there is no reason to assume that anything else but proteins are the primary carriers of specificity for cellular morphology. There are very few hard facts, however. The striking exception consists in Brinton's finding that hollow cylinders of the diameter of the cell can be reconstituted from a cell-envelope protein layer. Although it is rather tempting to assume that such protein surface crystals do at least partially specify the shape of a cell there is no proof for this assumption. Murray's observations on such a protein layer actually tend to argue against the assumption.

One should probably keep in mind, however, that different organisms may not use the same pathway in translating genetic information into cellular morphology. It cannot be deduced whether, e.g., cylindrical shape has been invented only once in the evolution of bacterial species. If such shape were a case of convergence it may be specified via quite different means: a rather pessimistic outlook for an experimental attack.

Cells grow, divide, and occasionally can change shape completely. In the following sections we examine whether these dynamic cellular properties can give us new insights.

III. SINGLE CELL GROWTH AND DIVISION

A. Experimental Evidence

The mechanisms of these processes are not known, but since the early studies of Bisset [58] much effort has been invested in studying their topology [12, 13, 59, 60]. Is new material inserted randomly or at one or more defined growth zones when, e.g., a rod is elongated or a sphere increases its diameter? How is new material incorporated during cellular division? Answers to these questions could allow at least some meaningful speculation on the flow of specificity from the gene into cell shape. Unfortunately, however, the situation is fairly complicated, and actually some of the available evidence suggests that the processes mentioned are not achieved in the same way in all bacteria. Instead of reviewing the often conflicting results from the literature we shall try to explain what the experimental difficulties are (or may be), and we then will select a few examplex where unequivocal answers have been obtained.

There are two main problems for pertinent studies, which are difficult to overcome. First, not knowing what the sequence is of what events in the growth of a cell envelope one does not know what to look for. One cannot expect that all components of the complex envelope are inserted in the same way (zonal, randomly), and since one does not know the pacemaker, if there is a single one, information on the topology of incorporation of any one class of components can be meaningless or at least without significance for our problem. The second difficulty, which appears to be almost unsurmountable, is that it has become clear that lipids [61] as well as proteins [62] can be randomized in membranes by lateral diffusion. It actually appears that not only randomization but also that some sort of topological concentration of cell envelope proteins (enzymes in this case) can occur [63, 64].

As though these difficulties were not enough, there is another one when it comes to the interpretation of data on the topology of incorporation of cell envelope material. Let us assume that for a rod shaped organism the mode of incorporation is different for cell elongation and cell division, it could be random for elongation but zonal (equatorial) for division. The rate of such equatorial synthesis may be faster than the rate of a presumed additional random incorporation. Also, equatorial growth will lead to a higher concentration of newly incorporated material per unit surface than random incorporation. The latter therefore may easily remain undetected and a wrong conclusion could be reached, namely, that cell mass increase of the organism occurs only by equatorial growth.

Yet, some clear experimental results have been obtained. (We wish to apologize at this point for discussing only a very selected number of such experiments, i.e., for neglecting others. We saw that reviewing all available work would not bring us closer to a solution of the problem we are discussing, and we do not mean that the examples we are citing are more illuminating than others.)

One interesting conclusion was reached in a study with E. coli [65]. Cells fully labeled in their murein or lipid were allowed, in the absence of label, to grow in a highly viscous medium for five-six generations under conditions in which the daughter cells could not swim away. The cells line up like pearls on a string. Quantitative autoradiography revealed that all the parental label had been randomly distributed among the daughter cells. For reasons mentioned above, the results with lipid labeling concerning our problem may be meaningless. Unless the assumption is made that intracellular turnover of murein occurs overall murein synthesis cannot be confined to one or a few growth zones. A note of caution has to be added, however. High-resolution radioautography of sacculi from cells pulse labeled with radioactive diaminopimelate has recently shown [30] that the label is incorporated almost exclusively at a distinct equatorial zone. Surprisingly enough, preliminary evidence could be obtained that the zonally incorporated radioactivity is randomized over the whole sacculus well within half a generation time.

Different results have been obtained with several gram-positive organisms. Work with Diplococcus pneumoniae has used fluorescent antibodies [66] and specific radioactive labeling [67]. Tritiated choline was employed, which is exclusively incorporated into teichoic acid, an envelope component covalently linked to murein [68]. The outcome of these studies was that new envelope material is added at distinct equatorial zones of the organism. Unequivocal evidence for equatorial growth has also been shown for Streptococcus faecalis. Cell envelope growth has been studied by use of immunofluorescence [69] with labeled antibodies [70] and with a combination of biochemical and electron microscopical methods [13]. The work on the cellular growth of this organism has been reviewed ([13]; see also

Chapter 2 in this volume) and the essential outcome is that this growth is completely conservative. Cross-envelope formation during division and enlargement of the coccus are identical processes. One hemisphere of a newly made cell appears to consist practically only of parental and the other hemisphere practically only of new material.

It need not be pointed out that there is a wealth of other experimental data on cellular division of bacteria. We believe that for our problem one general feature is of interest.

The pioneering electron microscope analysis of ultrathin sections of B. cereus by Chapman and Hillier [71] initiated a series of detailed morphological studies of cross envelope formation. If one ignores species specific differences, the general conclusion can be drawn that cross envelope formation always starts with a change in direction of the assembly of the cytoplasmic membrane. (It should be clear that this is a morphological description which has no implications concerning the site of initiation of septum formation.) The formation of a cross envelope has not been observed to occur de novo, i.e., new septum membrane and preexisting membrane always appear to be continuous.

The same situation has been found for the formation of endospores the process of which constitutes a special case of cellular division [72-77]. Spore morphogenesis starts with the axial accumulation of DNA and subsequent separation of chromosomes in the mother cell. One nucleus, later found in the spore, becomes separated from vegetative cell cytoplasm by a septum. The formation of this septum, which is the first clearly visible step in sporogenesis, is initiated by an invagination of the cytoplasmic membrane close to one cell pole. As in "normal" cell division, the prespore septum is closed by centripetal ingrowth and membrane fusion. In many instances, more than one potential septum is formed. However, only one of them, the one which contains DNA, continues to grow [73] until the growing edges of the membrane meet and fuse. The germ cell wall and the cortex, both of which contain murein [78], are laid down followed by the outer spore coat. According to these morphological observations, the only component of the vegetative cell envelope physically participating in spore formation is the cytoplasmic membrane [79, 80]. The following evidence shows that the outer layers of the spore are laid down de novo onto the prespore membrane. Biochemical studies demonstrated that labeled murein from the vegetative cell is not incorporated into the spore coat [81], and the enzyme systems synthesizing spore and vegetative cell murein are different [82]. Additional support comes from the finding that protoplasts, practically devoid of vegetative cell murein, can produce complete spores [83, 84]. Also, the chemical composition of vegetative cell and spore murein may be different [85, 86].

In summary, it appears that in cellular division or sporulation morphological specificity is expressed during the assembly of the cytoplasmic membrane.

B. Conclusions

Phenomenologically we see two types of processes: Elongation of a cylinder and division of a cylinder or a sphere and cross envelope formation and enlargement of a spherical cell can be identical processes. The questions regarding our problem are as follows. How can a cylinder grow and maintain its diameter? How is the change for division specified in the direction of the assembly of the cytoplasmic membrane, and how can the site of this change be found? What are the biochemical processes directing cross-envelope formation in a coccus so that a new hemisphere is formed of the right size? There are almost no biochemical answers to these questions, and it does not yet appear very fruitful to present more or less detailed speculations on each of them. The following considerations, however, may be useful.

It seems paradoxical that most of these questions are posed to one or the other cellular membrane. If nothing is basically wrong with the present view of the structure of such membranes [39], they themselves appear to be quite unable to perform any of the required functions in directional specificity. We should like to quote the following from the article by Singer and Nicolson [39].

> ..... long-range order clearly exists in certain cases where differentiated structures (for example, synapses) are found within a membrane. We suggest, in such special cases, either that short-range specific interactions among integral proteins result in the formation of an unusually large two-dimensional aggregate or that some agent extrinsic to the membrane (either inside or outside the cell) interacts multiply with specific integral proteins to produce a clustering of those proteins in a limited area of the membrane surface. In other words, we suggest that long-range random arrangements in membranes are the norm; wherever nonrandom distributions are found, mechanisms must exist which are responsible for them.

We cannot add much if the process of a change of direction of membrane assembly (division, sporulation, divisional growth of a coccus) is considered; a phospholipid bilayer matrix possessing no long-range order obviously cannot do it. Long-range order must therefore be created at certain sites in the membrane or it must be existent or made outside it.

Nevertheless, we feel that, e.g., the detailed observations of Shockman and his colleagues on divisional growth of S. faecalis make it very difficult to propose pertinent mechanisms. The main problem we believe one is faced with is that more than one directional specificity appears to be expressed on the same substrate: there is an ingrowing septum, the cross-envelope base is peeled apart, and the newly made envelope is "pushed" out in the right direction to become a new hemisphere of the daughter coccus (see Fig. 34 of Chapter 2). It certainly could very well be that the mechanisms operating in this type of shape specification are entirely different from those under work in a cylinder.

If the pattern of murein synthesis in E. coli (p. 424) reflects that of the whole envelope than long-range order must exist over the whole cylinder. We have argued that long-range order of this type cannot (and in some organisms actually does not) come from murein, and it cannot come either from membranes proper if the fluid mosaic model of their structure is correct. One certainly is tempted to postulate that in the E. coli envelope a protein surface crystal exists such as those found in many bacterial species (cf. pp. 421-422) and that it is this self assembling container which provides the long range order required. We have pointed out in Section I that elongation of such a surface crystal can proceed with a topology of incorporation of new subunits which biochemically would be scored as random.

It is not clear whether or not in E. coli a protein container of the type required does exist. A hexagonal lattice structure in cell envelopes of E. coli B has been described [87]. It could be that this lattice has to be ascribed to the lipoprotein covalently bound to murein [88, 89]. If so, it does not appear to be the best candidate for the container we are asking for because it has been shown that it is not present in two rod-shaped Proteus species [89]. One could, of course, argue that the protein does exist in Proteus, but is not covalently bound to murein, and thus has escaped detection.

Since we can discuss facts almost only in a speculative way we may as well add one pure speculation concerning the question as to how the divisional equator could be found in a cylindrical cell, a very intriguing question if elongation of a rod-shaped cell is achieved by random incorporation of new material. One simple way to find the equator of a cylinder is to conceive of a vectorial process initiated at its two ends and traveling with the same speed towards the middle. If the opposingly directed vectors meet they will be annihilated and this event can define the equator. A cylindrical protein surface crystal could be suited to this purpose. In order to find an equator in such a lattice some sort of measuring device must exist. It does exist at the two ends of the cell where the type of subunit packing may change from, e.g., hexagonal to pentagonal. Recognition of an equator potential can thus be achieved at these two sites. Let us assume it is recognized by an effector, which transiently and simultaneously induces at these two sites a vectorial process proceeding along the hexagonally packed lattice. This process could be a vectorial allosteric transition. The process might also exist in any other sort of gradient compatible with the behavior of surface crystals. In any event, an "equatorial signal" would be generated.

In addition to the physiological processes discussed so far, others concerning changes of bacterial morphology exist. Also, "pathological" morphology has been extensively studied. In the last sections we shall examine such phenomena from the point of view of morphological specificity.

IV. MORPHOLOGICAL VARIATION

A. Abnormal Morphology

Studies have been performed on morphological mutants [42-44, 90-98], nutritionally induced variations [99-102], and variations induced by drugs (see, e.g., reference 103). (Results with penicillin are treated in Section IV, B.) To date, the results have not contributed much to the problem under discussion. However, except for certain aspects mentioned below, one usually can hardly expect such experiments to reveal causal relationships at the present time because (1) it very often is not possible to pinpoint the primary mutational defect or nutritional effect, and (2) even if the primary defect were known it would almost always remain obscure what other structural or functional consequences it had. The cell envelope is simply too complex, or alternatively, we have only poor understanding of the interdependence of its various components.

One fact concerning the chemistry of murein has emerged from all these studies: consistent correlations between murein chemistry and alterations in shape have not been found. Changes in murein composition are known which are not accompanied by shape alterations [104, 105]. Gross changes in shape are known where alterations in murein composition could not (not yet?) be detected [95, 105]. However, changes in shape are also known with concomitant alterations in murein chemistry [106]. The latter case particularly well illuminates the fact that at present the problem of causality mentioned is almost impossible to overcome.

Arthrobacter crystallopoietes can grow as a rod or as a sphere, depending on nutritional conditions [101]. It was found [106] that the murein of spheres had shorter polypeptide chains, and very likely contains longer peptide cross-links than that of rods (the spherical murein peptide cross-links contain glycine and alanine residues while the rod murein cross-links only contain alanine). Finally, it could be shown [106] that spheres possessed more N-acetyl-muramidase activity than rods and that this activity decreased sharply during the sphere to rod transition. The case appears rather clear-cut. Yet how can one know whether the differences in murein composition are causally related to the changes in shape? (The authors did not make that claim.) We know of no case where similar considerations do not apply.

Some phenomenological observations are of potential significance. They concern changes in shape and DNA synthesis. Escherichia coli C, which normally grows more spherically than rod-shaped, undergoes rather drastic morphological changes when treated with mitomycin C or when deprived of thymine [103]. The cells become filamentous with highly variable diameters, become branched, and contain grossly swollen areas. Comparable observations have been made in our laboratory with a conditional rod mutant of E. coli K12; the mutant at high temperature grows as a sphere, and at lower temperature as a rod [107]. A "normal" transition from sphere to rod, during a downward temperature shift, is inhibited by Mitomycin C or thymine deprivation. Remarkable monsters develop from which, as one might now expect, murein sacculi can be isolated exhibiting the shape of the monsters (Fig. 2). These findings indicate the possibility that DNA synthesis can be coupled with envelope assembly; what the coupling may be is, naturally, unknown (cf. Pardee's article in this volume).

B. Changes Induced by Penicillin

In Enterobacteria low doses of penicillin inhibit cellular division and filamentous growth results [17]. When penicillinase is added to such filaments, cellular division occurs after a certain lag period and the filaments are dissected into small cells of about equal length. If chloramphenicol is added together with penicillinase no division is observed [108]. Penicillin at the same low dosage has also been found to completely inhibit spherical growth in several different conditional rod mutants of E. coli. At higher temperatures these cells are spherical but if penicillin is added, spherical growth ceases and the cells begin to form rods [108]. The mode of the transition from sphere to rod differs from one mutant to another. The action of penicillin at low dosage is unknown, but it does not inhibit net murein synthesis at all in the strains we have studied. One explanation is that the antibiotic interferes with the synthesis of a certain type of murein (e.g., a certain tertiary structure), which is compatible only with a spherical geometry, be it that of the pole of the cell or that of a spherical cell. It would certainly be very informative to know what penicillin does under the conditions discussed.

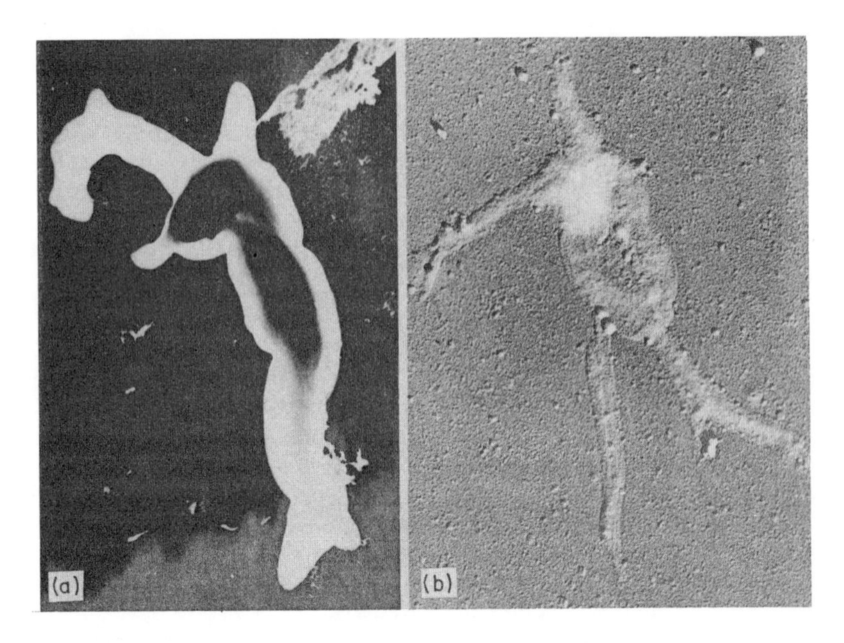

FIG. 2. Monster developed from a conditional E. coli rod mutant. The mutant cells (thymine-requiring) were pregrown at 42° C at which temperature they grow as perfect spheres [107]. The spheres were placed into a thymine-less medium and the temperature was lowered to 30° C. (A) Shadowed preparation of a cell 6 hr after the temperature shift; the still visible spherical part of the cell has grown out at four sites into filaments of rather irregular diameters. (B) Murein sacculus isolated from such monster cells. Final magnification of the electron micrographs: X 10,600.

C. De novo Manifestation of Cylindrical Shape

1. Spheroplasts

As early as 1956, Lederberg showed that penicillin-induced spheroplasts from E. coli and S. typhimurium are able to return to rod shaped cells in the absence of penicillin [109]. Such "reversion" resembles a trial and error process. First an amoeboid type of outgrowth occurs from the spherical cell, and eventually one or more filaments are formed from which rods are separated. The initial stages of this process have recently been investigated in more detail with E. coli spheroplasts obtained by diaminopimelate starvation of the appropriate auxotroph [105]. When diaminopimelate was added back to such spheroplasts, murein synthesis was resumed and the spheres became osmotically stable again. It was found that complete spherical sacculi had been reformed. To the limits of the analysis, the murein was indistinguishable chemically from the murein of rod shaped cells. Reformation of rods occurred only much later. The example shows that sacculus synthesis in E. coli is not necessarily coupled with cylindrical geometry. (It should be kept in mind that the absence of differences in this murein compared to murein from rod-shaped cells means no more than the absence of measurable differences (see p. 417).

2. L-Forms

L-forms, which have been known since 1935 [110], can arise from many bacteria [111]. Three types can be distinguished on the basis of ability to "revert" to the bacillary form: (1) stable L-forms, which have not been observed to revert, (2) stable L-forms, which can be made to revert, and (3) unstable L-forms, which require the continued presence of the inducing agent (penicillin, lysozyme) for growth as an L-form. Unfortunately, the situation is very complex and does not allow one to conclude much regarding the envelope components involved in reversion to the bacillary form. Too many permutations are known, in particular stable, nonreverting L-forms possessing both cell wall and murein, and stable, reverting L-forms without cell wall [111-113]. For our discussion, two findings are of interest. In B. subtilis, stable L-form growth can be induced by the action of lysozyme, and these L-form cells have been shown to lack cell wall (biochemical evidence for complete absence of murein apparently has not yet been published). They will, on soft agar, continue to grow as L-form indefinitely in the absence of lysozyme but will almost quantitatively revert to the bacillar form in media with certain physical properties [114-117]. What these media (hard agar, membrane filters, addition of killed cells of other organisms such as pseudomonads or yeast) have in common is mechanical support. This mechanical support, however, is primarily required for reformation of cell wall. As in the case of E. coli spheroplasts mentioned before, rod-shaped cells develop only *after* cell wall has been resynthesized [115].

The other finding comes from studies with penicillin induced unstable L-forms of Proteus mirabilis [113]. L-form growth requires the continued presence of penicillin. These cells possess murein whose chemical composition was not distinguishable from murein of normal cells. (It had been reported that L-form murein behaved differently toward lysozyme than normal Proteus murein [113], but it has been found that this difference is not causally related to the L-form state [118].)

D. Conclusions

Concerning murein one cannot help having the impression that it may not provide anything else but osmotic protection. We have discussed before that cylindrical shape can be maintained in E. coli without a murein sacculus when the osmotic pressure in the medium is high enough. We have seen in the last sections that murein sacculi of virtually any shape can be made (see Fig. 2). From all that is known to date murein may best be viewed as a bacterial bone; the chemistry of the substance may really have nothing to do with its gross shape. It remains to be seen, of course, if the situation truly is that extreme.

One other fact of interest emerges. When rod shape has been destroyed (spheroplasts, lysozyme induced L-form in B. subtilis) restoration of a cylinder is not an all-or-none process. The sphere itself cannot be reshaped into a cylinder nor does the sphere, once a complete envelope has been resynthesized, grow out as a perfect rod. Normal cylindrical geometry is only reached via formation of more or less bizarre structures such as have been described by Lederberg [109]. Such a phenomenology might not be inconsistent with the behavior of a protein surface crystal. If the cylindrical cell is forced into a sphere, the lattice of the crystal must be more or less distorted and may also be lost. This will not influence the synthetic capacities (concerning murein, cell wall) of the cytoplasmic membrane and thus the sphere becomes transiently "calcified" as a sphere. Reformation of a cylinder is difficult as the lattice has to be reconstructed de novo or from remaining "seed crystals."

V. CONCLUDING REMARKS

We have, throughout this contribution, put much emphasis on the concept of a protein surface crystal as one candidate for the expression of morphological specificity. It is almost needless to point out that one of the reasons for doing so was Brinton's demonstration that hollow protein cylinders can be self assembly systems. Yet if it is correct that a cylindrical cell grows by random incorporation of new envelope material it then appears fairly difficult to conceive of an entirely different system which would be able to do so and maintain the diameter of the cylinder. We do not wish, however, to leave the impression that the surface crystal we are looking for must be exactly one like Brinton's T-layer from

B. brevis. It is not difficult to visualize that, e.g., the presence of lipid in such a crystal could provide very important additional properties.

It clearly is, on the other hand, impossible that such a self assembly system could suffice for the manifestation of morphology, e.g., the *B. brevis* T-layer protein cannot, of course, spontaneously assemble to form *closed* cylinders of uniform length. It is very obvious that no single self-assembly system is able to achieve changes in the direction of envelope assembly and do so at a certain time of the cell cycle. We have discussed in Section III, A that for a membrane to undergo a specific change of direction of assembly some long-range order must be established at certain sites in or on that membrane. Although the biochemical processes specifying such changes are entirely unknown there is no question that any such specificity must primarily reside in proteins, be it as modifiers of an assembly as we have discussed it in the Introduction or be it via a whole sequence of reactions with "intermediates" not necessarily consisting of protein (such as murein, cf. Section II, A).

ACKNOWLEDGMENTS

The invaluable criticisms of Dr. L. Leive, Dr. J.-M. Ghuysen, and Dr. G. D. Shockman have been most appreciated. Dr. M. Achtman has provided decisive help in the preparation of the manuscript and we also would like to thank Dr. K. Rehn for the electron micrographs, and Dr. V. Braun, Dr. V. Nüsslein, and Dr. K. Rehn for critical comments. Our thanks go to Dr. J.-M. Ghuysen, Dr. Carol Henry, Dr. Y. Hirota, Dr. P. Overath, Dr. G. D. Shockman, and Dr. F. E. Young for preprints.

REFERENCES

1. D. J. Kushner, *Bacteriol. Rev.*, *33*, 302 (1969).
2. S. Asakura, C. Eguchi, and T. Iino, *J. Mol. Biol.*, *10*, 42 (1964).
3. S. Asakura, C. Eguchi, and T. Iino, *J. Mol. Biol.*, *16*, 302 (1966).
4. S. Asakura, C. Eguchi, and T. Iino, *J. Mol. Biol.*, *35*, 227 (1968).

5. W. B. Wood, R. S. Edgar, J. King, L. Lielausis, and M. Henninger, Fed. Proc. Fed. Amer. Soc. Exp. Biol., 27, 1160 (1968).

6. E. Kellenberger and C. Kellenberger-Van der Kamp, FEBS Letters, 8, 140 (1970).

7. U. K. Laemmli, Nature (London), 227, 680 (1970).

8. M. Rosenbaum, Nature New Biol., 230, 12 (1971).

9. M. J. Osborn, Ann. Rev. Biochem., 38, 501 (1969).

10. N. Nanninga, J. Bacteriol., 101, 297 (1970).

11. W. Weidel and H. Pelzer, Advan. Enzymol., 26, 193 (1964).

12. H. J. Rogers, Bacteriol. Rev., 34, 194 (1970).

13. M. L. Higgins and G. D. Shockman, in Critical Reviews in Microbiology (I. Laskin and H. Lechevaliev, eds.), Vol. 1, p. 29, 1971.

14. D. C. Birdsell and E. H. Cota-Robles, J. Bacteriol., 93, 427 (1967).

15. J. G. Voss, J. Gen. Microbiol., 35, 313 (1964).

16. M. M. Grula and E. A. Grula, Can. J. Microbiol., 11, 453 (1965).

17. J. Stárka and J. Moravová, Folia Microbiol., Acad. Sci. Bohemoslov., 12, 240 (1967).

18. G. Weinbaum, J. Gen. Microbiol., 42, 83 (1966).

19. G. Weinbaum and S. Okuda, J. Biol. Chem., 243, 4358 (1968).

20. D. White, M. Dworkin, and D. J. Tipper, J. Bacteriol., 95, 2186 (1968).

21. L. Leive and B. D. Davis, J. Biol. Chem., 240, 4370 (1965).

22. J. Primosigh, H. Pelzer, D. Maass, and W. Weidel, Biochim. Biophys. Acta, 63, 229 (1962).

23. W. Leutgeb, D. Maass, and W. Weidel, Z. Naturforsch., 18b, 1062 (1963).

24. V. Mohr and H. Larsen, J. Gen. Microbiol., 31, 267 (1963).

25. A. D. Brown, Bacteriol. Rev., 28, 296 (1964) and references cited therein.

26. U. Henning, V. Braun, and B. Hoehn, unpublished.

27. J.-M. Ghuysen, Bacteriol. Rev., 32, 425 (1968).

28. A. D. Brown and K. Y. Cho, J. Gen. Microbiol., 62, 267 (1970).

29. U. Schwarz, A. Asmus, and H. Frank, J. Mol. Biol., 41, 419 (1969).

30. A. Ryter, Y. Hirota and U. Schwarz, J. Mol. Biol., 77, 183 (1973).

31. S. Razin and B. J. Cosenza, J. Bacteriol., 91, 858 (1966).

32. S. Razin, B. J. Cosenza, and M. E. Tourtelotte, J. Gen. Microbiol., 42, 139 (1966).

33. S. Razin, M. E. Tourtelotte, R. N. McElhaney, and J. D. Pollack, J. Bacteriol., 91, 609 (1966).

34. A. W. Rodwell and A. J. Abbot, J. Gen. Microbiol., 25, 201 (1961).

35. R. N. McElhaney and M. E. Tourtelotte, Science, 164, 433 (1969).

36. J. M. Steim, M. E. Tourtelotte, J. C. Reinert, R. N. McElhaney, and R. L. Rader, Proc. Nat. Acad. Sci., U. S., 63, 104 (1969).

37. P. Mitchell and J. Moyle, cited by R. N. McElhanty and M. E. Tourtelotte in ref. 35.

38. D. Abram, J. Bacteriol., 89, 855 (1965).

39. S. J. Singer and G. L. Nicolson, Science, 175, 720 (1972).

40. G. Weinbaum, R. Rich, and D. A. Fischman, J. Bacteriol., 93, 1693 (1967).

41. H. H. Martin and S. Kemper, J. Bacteriol., 102, 347 (1970).

42. R. J. Boylan and N. H. Mendelson, J. Bacteriol., 100, 1316 (1969).

43. R. M. Cole, T. J. Popkin, R. J. Boylan, and N. H. Mendelson, J. Bacteriol., 103, 793 (1970).

44. R. J. Boylan, N. H. Mendelson, D. Brooks, and F. E. Young, J. Bacteriol., 110, 281 (1972).

44a. F. E. Young, personal communication.

45. J. van Heijenoort, D. Menjon, B. Flouret, J. Szulmajster, J. Laporte, and G. Batelier, Eur. J. Biochem., 20, 442 (1971).

46. O. Lüderitz, Angew. Chemie, 82, 708 (1970), and references cited therein.

47. D. L. D. Caspar and A. Klug, Cold Spring Harbor Symp. Quant. Biol., 27, 1 (1962).

48. W. F. Harris and L. E. Scriven, Nature (London), 228, 827 (1970).

49. J. E. McNary, C. C. Brinton, J. Carnahan, and C. M. Henry, J. Mol. Biol., 1973, in press.

50. C. C. Brinton, J. C. McNary, and J. Carnahan, Bacteriol. Proc., p. 48, 1969.

51. U. Henning and B. Hoehn, unpublished experiments.

52. W. Stoeckenius and R. Rown, J. Cell Biol., 34, 365 (1967).

53. A. L. Houwink, J. Gen. Microbiol., 15, 146 (1956).

54. W. Stoeckenius and W. H. Kunau, J. Cell Biol., 38, 337 (1968).

55. A. L. Houwink, Biochim. Biophys. Acta, 10, 360 (1953).

56. R. G. E. Murray, Can. J. Microbiol., 9, 381 (1963).

57. F. L. A. Buckmire and R. G. E. Murray, Can. J. Microbiol., 16, 1011 (1970).

58. K. A. Bisset, J. Gen. Microbiol., 5, 155 (1951).

59. A. Ryter, Bacteriol. Rev., 32, 39 (1968).

60. R. M. Cole, Bacteriol. Rev., 29, 326 (1965).

61. P. Overath, F. F. Hill, and I. Lamnek, Nature, New Biol., 234, 264 (1971).

62. C. D. Frye and M. Edidin, J. Cell Sci., 7, 313 (1970).

63. B. K. Wetzel, S. S. Spicer, H. F. Dvorak, and L. A. Heppel, J. Bacteriol., 104, 529 (1970).

64. H. F. Dvorak, B. K. Wetzel, and L. A. Heppel, J. Bacteriol., 104, 543 (1970).

65. E. C. C. Lin, Y. Hirota, and F. Jacob, J. Bacteriol., 108, 375 (1971).

66. K. L. Chung, R. Z. Hawirko, and P. K. Isaac, Can. J. Microbiol., 10, 43 (1964).

67. E. B. Briles and A. Tomasz, J. Cell Biol., 47, 786 (1970).

68. I. L. Mosser and A. Tomasz, J. Biol. Chem., 245, 287 (1970).

69. K. L. Chung, R. Z. Hawirko, and P. K. Isaac, Can. J. Microbiol., 10, 473 (1964).

70. I. Swanson, K. C. Hsu, and E. C. Gottschlich, J. Exp. Med., 130, 1063 (1969).

71. G. B. Chapman and I. Hillier, J. Bacteriol., 66, 362 (1953).

72. C. F. Robinow, in The Bacteria (I. C. Gunsalus and R. Y. Stanier, eds.), Vol. I, p. 207, Academic Press, New York, 1960.

73. A. Ryter, Ann. Inst. Pasteur, 108, 40 (1965).

74. H. O. Halvorson, I. C. Vary, and W. Steinberg, Ann. Rev. Microbiol., 20, 169 (1966).

75. A. Kornberg, I. A. Spudich, D. L. Nelson, and M. P. Deutscher, Ann. Rev. Biochem., 37, 51 (1968).

76. P. Schaeffer, Bacteriol. Rev., 33, 48 (1969).

77. I. Mandelstam, in Symp. Soc. Exp. Biol., 25, Cambridge, at the University Press, 1971, p. 1.

78. A. D. Warth, D. F. Ohye, and W. G. Murrell, J. Cell Biol., 16, 593 (1963).

79. P. C. Fitz-James, J. Biophys. Biochem. Cytol., 8, 505 (1960).

80. D. F. Ohye and W. G. Murrell, J. Cell Biol., 14, 111 (1962).

81. V. Vinter, Fol. Microbiol., Acad. Sci. Bohemoslov., 8, 147 (1963).

82. V. Vinter, Experientia, 19, 307 (1963).

83. M. R. J. Salton, J. Gen. Microbiol., 13, IV (1955).

84. J. Stárka and J. Trpisovská, Int. Congr. Microbiol., Montreal, 1962. Cited by J. W. Newton in ref. 97.

85. A. D. Warth and J. L. Strominger, Proc. Nat. Acad. Sci., U. S., 64, 528 (1969).

86. K. D. Hungerer and D. J. Tipper, Biochemistry, 8, 3577 (1969).

87. D. A. Fischman and G. Weinbaum, Science, 155, 472 (1967).

88. V. Braun and K. Rehn, Eur. J. Biochem., 10, 426 (1969).

89. V. Braun, K. Rehn, and H. Wolff, Biochemistry, 9, 5041 (1970).

90. H. I. Adler, C. E. Terry, and A. A. Hardigree, J. Bacteriol., 95, 139 (1968).

91. S. Normark, J. Bacteriol., 98, 1274 (1969).

92. H. J. Rogers, M. McConnell, and I. D. J. Burdett, Nature (London), 219, 285 (1968).

93. H. J. Rogers, M. McConnell, and I. D. J. Burdett, J. Gen. Microbiol., 61, 155 (1970).

94. H. J. Rogers and M. McConnell, J. Gen. Microbiol., 61, 173 (1970).

95. H. J. Rogers, M. McConnell, and R. C. Hughes, J. Gen. Microbiol., 66, 297 (1971).

96. J. W. Newton, Biochim. Biophys. Acta, 141, 633 (1967).

97. J. W. Newton, Biochim. Biophys. Acta, 165, 534 (1968).

98. J. W. Newton, Nature (London), 228, 1100 (1970).

99. J. L. Stevenson, Can. J. Microbiol., 8, 655 (1962).

100. D. C. Gillespie, Can. J. Microbiol., 9, 509, 515 (1963).

101. J. C. Ensign and R. S. Wolfe, J. Bacteriol., 87, 925 (1964).

102. D. White, M. Dworkin, and D. J. Tipper, J. Bacteriol., 95, 2186 (1968).

103. J. C. Suit, T. Barbee, and S. Jetton, J. Gen. Microbiol., 49, 165 (1967).

104. K. H. Schleifer, L. Huss, and O. Kandler, Arch. Mikrobiol., 68, 387 (1969).

105. U. Schwarz and W. Leutgeb, J. Bacteriol., 106, 588 (1971).

106. T. A. Krulwich, J. C. Ensign, D. J. Tipper, and J. L. Strominger, J. Bacteriol., 94, 734, 741 (1967).

107. U. Henning, K. Rehn, V. Braun, B. Hoehn, and U. Schwarz, Eur. J. Biochem., 26, 570 (1972).

108. C. Schmitges and U. Henning, unpublished observations.

109. J. Lederberg, Proc. Nat. Acad. Sci., U. S., 42, 574 (1956).

110. E. Klieneberger, J. Pathol. Bacteriol., 11, 93 (1935).

111. For pertinent information see L. B. Guze, ed., Microbial Protoplasts, Spheroplasts, and L-Forms, Williams & Wilkins, Baltimore, Maryland, 1968.

112. P. H. Hofschneider and H. H. Martin, J. Gen. Microbiol., 51, 23 (1968).

113. W. Katz and H. H. Martin, Biochem. Biophys. Res. Commun., 39, 744 (1970).

114. O. E. Landman and S. Halle, J. Mol. Biol., 7, 721 (1963).

115. O. E. Landman, A. Ryter, and C. Frehel, J. Bacteriol., 96, 2154 (1968).

116. O. E. Landman and A. Forman, J. Bacteriol., 99, 576 (1969).

117. D. Clive and O. E. Landman, J. Gen. Microbiol., 61, 233 (1970).

118. H. H. Martin, personal communication.

NOTES ADDED IN PROOF

Chapter 1

The prediction made in Section V that acetyl CoA carboxylase activity is regulated by ppGpp has been borne out by the studies of Polakis, S. E., Guchait, R. B., and Lane, M. D., "Stringent Control of Fatty Acid Synthesis in *Escherichia coli*: Possible regulation of Acetyl CoA Carboxylase by ppGpp," *J. Biol. Chem.* (in press). These investigators have shown that physiological concentrations of ppGpp inhibit the *in vitro* activity to the same extent as found *in vivo* during amino acid deprivation.

Chapter 3

Recent observations include the following.

(1) A cell envelope protein containing covalently linked glucosamine, KDO, and phosphate has been isolated from *E. coli* (M. C. Wu and E. C. Heath, *Fed. Proc. Fed. Amer. Soc. Exp. Biol.*, *32*, 481 Abs, 1973). This finding suggests that covalent linkage between LPS and this protein exists at least for some LPS molecules or during certain stages of LPS biosynthesis.

(2) A *S. typhimurium* mutant defective in KDO synthesis has been isolated. The bacterial population cannot grow unless the lesion is phenotypically repaired by the addition of D-arabinose-5-phosphate (P. D. Rick and M. J. Osborn, *Proc. Nat. Acad. Sci. U.S.A.*, *69*, 3756, 1972).

(3) A class of mutants partially defective in heptose biosynthesis incorporate some D-*glycero*-D-*manno*heptose into LPS. It is inferred that the "normal" heptose, L-*glycero*-D-*manno*heptose, is synthesized by the epimerization of the D-*glycero* isomer and that the mutants are partially defective in this process (V. Lehmann, G. Hämmerling, M. Nurminen, I. Minner, E. Ruschmann, O. Lüderitz, T.-T. Kuo, and B.A.D. Stocker, *Eur. J. Biochem.*, *32*, 268, 1973).

(4) The "major" protein of the outer membrane was actually shown to be comprised of several different proteins with a similar molecular weight (C. Moldow, J. Robertson, and L. Rothfield, *J. Membrane Biol.*, *10*, 137, 1972; G. F. Ames, *J. Biol. Chem.*, submitted for publication).

(5) "Deep rough" mutants of S. typhimurium were found to have greatly reduced amounts of some outer membrane proteins (G. F. Ames, E. N. Spudich, and H. Nikaido, J. Bacteriol., submitted for publication). This suggests a close interaction between the proteins and the inner core region of LPS. The absence of proteins may also be related to the sensitivity of these mutants toward various dyes and antibiotics (see Section V, B).

(6) Vitamin B_{12} was shown to bind to a specific site on the outer membrane of E. coli (D. R. Di Masi, J. C. White, and C. Bradbeer, Fed. Proc. Fed. Amer. Soc. Exp. Biol., 32, 600 Abs, 1973).

Chapter 6

(1) It is not excluded that the single-strand breaks in DNA may be produced by colicin E2 itself. Early tests for nuclease activity of colicin E2 in vitro were not sensitive or exhaustive enough to detect a few localized one-strand scissions. Using as substrate the supercoiled circles of phage DNA, Saxe [56a] has observed in the presence of purified colicin E2 a reproducible shift of the DNA molecules from the supercoiled to the relaxed circle form, a shift that shows the opening of one or more phosphodiester bonds and is readily detectable by a change in sedimentation in sucrose gradients. This finding can only be considered as preliminary in view of the well-known difficulties in excluding the presence of unknown nucleases. Because of the similarity and serological cross-reactivity between colicin E2 and colicin E3 [67] and the fact that the latter is presumed to have a specialized endonucleolytic activity on certain RNA molecules [45, 46], the hypothesis that colicin E2 may itself be a nuclease remains most attractive.

(2) Saxe [56a] reports that even at later stages the addition of trypsin, with rescuing cell viability, arrests the process of conversion of DNA molecules from circles to relaxed circles to linear molecules.

(3) It appears therefore that colicin molecules adsorbed to a bacterium shift to the effective, cell-damaging state by random events whose probability is correlated with the "fluidity" of the bacterial envelope. Whether this shift, which is completely prevented by inhibitors and uncouplers of energy metabolism, is a physical transfer from one molecular situation to another — for example, from combination with a receptor to the cytoplasmic membrane — remains to be decided by experiment. There is at present no explanation for the discrepancy between the earlier findings of trypsin rescuability of cells already damaged by colicin K [4] and the more recent demonstration of concomitant cell damage and loss of trypsin rescue [92b].

AUTHOR INDEX

Numbers in brackets are reference numbers and indicate that an author's work is referred to although his name is not cited in the text. Underlined numbers give the page on which the complete reference is listed.

A

D

E

G

H

M

N

O

P

Q

R

S

T

U

V

W

Y

SUBJECT INDEX

A

C

F

H

M

N

O

P

U

V

X